BIOTIC INTERACTIONS IN PLANT–PATHOGEN ASSOCIATIONS

Biotic Interactions in Plant–Pathogen Associations

Edited for the British Society for Plant Pathology by

M.J. Jeger
T.H. Huxley School
Imperial College at Wye
Wye, Ashford, Kent, UK

and

N.J. Spence
Department of Plant Pathology and Microbiology
Horticulture Research International
Wellesbourne
Warwick, UK

CABI *Publishing*

CABI is a trading name of CAB International

CABI Head Office
Nosworthy Way
Wallingford
Oxfordshire OX10 8DE
UK

Tel: +44 (0)1491 832111
Fax: +44 (0)1491 833508
Email: cabi@cabi.org
Web site: www.cabi.org

CABI North American Office
875 Massachusetts Avenue
7th Floor
Cambridge, MA 02139
USA

Tel: +1 617 395 4056
Fax: +1 617 354 6875
Email: cabi-nao@cabi.org

A catalogue record for this book is available from the British Library, London, UK.

Library of Congress Cataloging-in-Publication Data
Biotic interactions in plant-pathogen associations / edited for the British Society for Plant Pathology by M.J. Jeger and N.J. Spence.
p. cm.
Includes bibliographical references.
ISBN 0-85199-512-8 (alk. paper)
1. Plant-pathogen relationships. 2. Biotic communities. I. Jeger, Michael J. II. Spence, N.J. (Nicola J.) III. British Society for Plant Pathology.

SB732.7 .B62 2001
632'.3—dc21 00-064174

ISBN-13: 978-0-85199-512-0
ISBN-10: 0-85199-512-8

First published 2001
Transferred to print on demand 2007

Printed and bound in the UK by CPI Antony Rowe, Eastbourne.

Contents

Contributors

C. Alabouvette, CMSE-INRA, U.M.R. Biochimie, Biologie Cellulaire et Ecologie des Interactions Plantes Microorganismes, F 21034 Dijon Cedex, France

M.P. Ayres, Dartmouth College, Hanover, NH 03755, USA

J. Colvin, Natural Resources Institute, Central Avenue, Chatham Maritime, Chatham, Kent ME4 4TB, UK

R. Cook, Institute of Grassland and Environmental Research, Aberystwyth, Ceredigion SY23 3EB, UK

W. de Boer, Department of Plant–Microorganism Interactions, Netherlands Institute of Ecology, Centre for Terrestrial Ecology, PO Box 40, 6666 ZG Heteren, The Netherlands

V. Edel, CMSE-INRA, U.M.R. Biochimie, Biologie Cellulaire et Ecologie des Interactions Plantes Microorganismes, F 21034 Dijon Cedex, France

S.M. Garnsey, USDA-ARS (retired), 2313 Sherbrooke Road, Winter Park, FL 32792, USA

T.R. Gottwald, USDA-ARS, US Horticultural Research Laboratory, 2001 South Rock Road, Fort Pierce, FL 34945, USA

J. Hallmann, Institut for Plant Diseases, University of Bonn, Nußallee 9, 53115 Bonn, Germany

P.E. Hatcher, Department of Agricultural Botany, School of Plant Sciences, The University of Reading, 2 Earley Gate, Whiteknights, Reading RG6 6AU, UK

R.J. Hillocks, NRI-University of Greenwich, Central Avenue, Chatham Maritime, Chatham, Kent ME4 4TB, UK

R.F. Hoekstra, Laboratory of Genetics, Department of Plant Sciences, Wageningen University, Dreijenlaan 2, NL-6703 HA, Wageningen, The Netherlands
R.W. Hofstetter, Dartmouth College, Hanover, NH 03755, USA
J. Holt, Natural Resources Institute, Central Avenue, Chatham Maritime, Chatham, Kent ME4 4TB, UK
G. Hughes, Institute of Ecology and Resource Management, University of Edinburgh, Edinburgh EH9 3JG, UK
M. Hyakumachi, Faculty of Agriculture, Gifu University, Gifu 501-1193, Japan
M.J. Jeger, T.H. Huxley School, Imperial College at Wye, Wye, Ashford, Kent TN25 5AH, UK
L. Kiss, Plant Protection Institute, Hungarian Academy of Sciences, H-1525 Budapest, PO Box 102, Hungary
K.D. Klepzig, USDA Forest Service, Pineville, LA 71360, USA
H. Lecoq, INRA, Station de Pathologie Végétable, Domaine Saint Maurice, BP 94, 84143 Montfavet Cédex, France
P. Lemanceau, CMSE-INRA, U.M.R. Biochimie, Biologie Cellulaire et Ecologie des Interactions Plantes Microorganismes, F 21034 Dijon Cedex, France
G.C. Lewis, Institute of Grassland and Environmental Research, North Wyke, Okehampton, Devon EX20 2SB, UK
M.J. Lombardero, Dartmouth College, Hanover, NH 03755, USA
T.P. McGonigle, Faculty of Agriculture, Gifu University, Gifu 501-1193, Japan and Department of Biological Sciences, Idaho State University, Campus Box 8044, ID 83209, USA
J.C. Moser, USDA Forest Service, Pineville, LA 71360, USA
C. Olivain, CMSE-INRA, U.M.R. Biochimie, Biologie Cellulaire et Ecologie des Interactions Plantes Microorganismes, F 21034 Dijon Cedex, France
N.D. Paul, Division of Biology, Institute of Environmental and Natural Sciences, Lancaster University, Lancaster LA1 4YQ, UK
B. Raccah, ARO, Department of Virology, The Volcani Center, PO Box 6, 50250 Bet Dagan, Israel
G. Recorbet, CMSE-INRA, U.M.R. Biochimie, Biologie Cellulaire et Ecologie des Interactions Plantes Microorganismes, F 21034 Dijon Cedex, France
G.S. Saddler, CABI Bioscience UK Centre (Egham), Bakeham Lane, Egham, Surrey TW20 9TY, UK
J.J. Smith, CABI Bioscience UK Centre (Egham), Bakeham Lane, Egham, Surrey TW20 9TY, UK
N.J. Spence, Department of Plant Pathology and Microbiology, Horticulture Research International, Wellesbourne, Warwick CV35 9EF, UK
C. Steinberg, CMSE-INRA, U.M.R. Biochimie, Biologie Cellulaire et Ecologie des Interactions Plantes Microorganismes, F 21034 Dijon Cedex, France
W.H. Van der Putten, Multitrophic Interactions Department, Netherlands Institute of Ecology, NIOO-CTO, PO Box 40, 6666 ZG Heteren, The Netherlands

J.A. van Veen, Netherlands Institute of Ecology, Centre for Terrestrial Ecology, Department of Plant–Microorganism Interactions, PO Box 40, 6666 ZG Heteren, The Netherlands

C.J. Walkinshaw, USDA Forest Service, Pineville, LA 71360, USA

Preface

In December 1999 a joint meeting was organized by the British Society of Plant Pathology and the Association of Applied Biologist's Virology Group on the topic of 'Biotic Interactions in Plant Pathogen Associations'. Speakers made presentations in a series of sessions on: (i) within-taxon interactions; (ii) interactions with fungi; (iii) interactions with procaryotes; (iv) virus–vector associations, Homoptera; (v) virus–vector associations, other vectors; (vi) biological control, within-taxon; (vii) biological control, across-taxa; (viii) complex diseases and diseases of complex aetiology; and (ix) methodology and modelling. This book represents a selection of the key papers given in sessions (i)–(iii) and sessions (vi–ix). Papers on topics addressed by speakers in sessions (iv) and (v), which deal with virus–vector interactions largely at the molecular level, are published elsewhere (Plumb *et al.*, 2001). Speakers were originally charged to address the genetical, physiological and ecological interactions influencing plant–pathogen associations according to the specific standpoint of each of these topics. Additionally the editors have written two introductory chapters to the book: one addressing the importance of biotic interactions generally; the second addressing the topic of plant virus transmission specifically from an epidemiological perspective.

M.J. Jeger, Wye
N. Spence, Wellesbourne

Plumb, R. (ed.) (2001) Interactions between plant viruses and their vectors. In: *Advances in Botanical Research.* Academic Press Publications, London.

Biotic Interactions and Plant–Pathogen Associations

1

M.J. Jeger

T.H. Huxley School, Imperial College at Wye, Wye, Ashford, Kent TN25 5AH, UK

Introduction

The history of plant pathology has been dominated by the search for single agents of disease, how these can be identified and how they can be shown to cause a distinctive set of symptoms in a particular crop. The reasons for this strong focus are understandable given the many examples of devastating plant disease epidemics (Horsfall and Cowling, 1978), and the imperative to find control measures that will alleviate the economic, social and human consequences. It can be argued however that it is misleading to emphasize these examples as either typical or even the *raison d'être* for the study of plant diseases. More typical, then, are the complexes of diseases that occur within agricultural, horticultural and forest tree crops, and the interactions between the causal pathogens, other biotic components associated with crops and the physical (abiotic) environment. It can be argued that to promote long-term plant health and crop sustainability it is these interactions, whether occurring at the genetical, physiological or ecological levels of integration, that should set the framework for future research in plant pathology. This is not an argument against the single agent–single disease approach but rather for a structured integrative approach that does not lose contact with the complexity of plant disease in the field. Such an approach is highly relevant for managed but non-agricultural landscapes that are likely to increase in scale in Europe as a consequence of land use policies and agricultural reform.

Interactions can, of course, have negative and positive effects with respect to crops. Biotic interactions result sometimes in complex diseases and sometimes in the suppression of particular pathogens (Darpoux, 1960). Insects and other animals, including humans, may play an essential role in the dispersal of pathogens (Broadbent, 1960). The questions posed by Cowling and Horsfall (1979), 'If I were a pathogen how would I attack a plant? Alone? Simultaneously with other pathogens? With assistance from a mobile vector?', although anthropomorphic are still relevant in forcing attention to the life-history strategies that pathogens have (and continue to) evolved. Interactions may occur externally to the plants and result in situations in which the activation of pathogens is enhanced by the biotic component of the environment (Dickinson, 1979); or within plants, to such an extent that diseases of complex aetiology may be more common than those of specific aetiology (Powell, 1979). Indeed it has been claimed by Powell (1979) that 'until synergistic complexes receive appropriate attention, the discipline of plant pathology will remain fragmented, incomplete and irrelevant to nature'. The problem is that synergy has proved an elusive term to define in operational terms.

Biotic interactions may also suppress the activity of particular pathogens, either with respect to specific biological control mechanisms involving mycoparasitism, antibiosis and hypovirulence, or to more general ecological mechanisms involving competition. Indeed the intelligent use of biotic interactions and manipulations of the environment have been seen as essential to a future integrated agriculture with the use of fewer pesticides and fertilizers (Beemster *et al.*, 1991).

Both positive and negative interactions as described by Darpoux (1960) will be reviewed in this introductory chapter. Interactions within populations of the same taxonomic grouping, often involving novel means of horizontal genetic exchange between individuals, are first examined. Examples of interactions between plant-associated fungi and bacteria and other organisms, including insects, are then considered, including mutualistic associations and endophytes as well as pathogens. Many of these interactions occur in natural plant populations as frequently as in crops. The virus–vector association is one of the most important biotic interactions in terms of plant disease and is reviewed in detail elsewhere (Plumb, 2001). Spence (Chapter 2) considers specifically how knowledge of virus transmission contributes to an understanding of disease epidemiology. Biological control may occur as a consequence of within-taxon interactions, often mediated through the host, or as direct antagonistic or competitive interactions within and across taxa, and examples of both kinds of interactions are reviewed. Finally, disease complexes and diseases of complex aetiology are considered, together with some of the methodologies required in the face of such complexity, including modelling.

Within-taxon Population Interactions

Within the main plant pathogen taxa of fungi, prokaryotes and viruses, interactions occur at the population level that can significantly influence the occurrence of plant disease. With the true fungi, basidiomycetes and ascomycetes, the occurrence of vegetative incompatibility as a form of (non-) self-recognition is virtually universal (Leslie, 1993). The significance of incompatibility is that it prevents the somatic exchange of genetic material between conspecifics. Genetic evidence suggests that vegetative compatibility group is an indicator of evolutionary origin at least in the vascular wilt pathogen *Fusarium oxysporum* f. sp. *lycopersici* (Elias *et al.*, 1993). Population genetic explanations for the high levels of polymorphism for vegetative incompatibility found in many species have been postulated (Nauta and Hoekstra, 1994, 1996) but the significance for plant-pathogenic fungi may have less to do with the constraints on genetic exchange than with the potential for biological control through transmission of deleterious cytoplasmic elements (Hoekstra, Chapter 3). Vegetative incompatibility can prevent or retard the transmission of dsRNAs associated with hypovirulence in the chestnut blight fungus, *Cryphonectria parasitica*. The spatial patterns and dispersion of vegetatively compatible groups are of fundamental importance (Milgroom *et al.*, 1990) in the spread of hypovirulent strains. Also of significance in the context of biological control is the phenomenon of non-self-anastomosing isolates found in *Rhizoctonia solani* during sugarbeet monoculture (Hyakumachi and Ui, 1987). Isolates of AG-2 were found which neither anastomosed with each other nor self-anastomosed. These isolates, although abundant in the field, have lost the ability to form sclerotia and have poor parasitic fitness: their increase in the field may be closely related to the decline in sugarbeet root rot during monoculture.

For plant pathogenic bacteria, evolutionary change can be rapid and an understanding of the genetic structure of bacterial populations provides a framework within which epidemics can be monitored and tracked, and biological control strategies can be rationally designed especially where frequent large-scale recombination occurs (Haubold and Rainey, 1996). However, the role and value of population genetics in the study of pathogenic and saprobic bacterial populations – and their interrelationships – in the environment is often neglected. Equally the opportunities for and scale of natural recombination among plant viruses (MacFarlane, 1997) needs a better understanding, as do the implications for utilization of virus transgenic resistance (Foster and Taylor, 1998).

Interactions with Fungi

Interactions of plant pathogens with other organisms, including insects, take many forms and have been investigated for both natural and crop populations, although it is for the former that most ecological investigations have been made. The rust *Puccinia monoica* infects wild mustards, notably *Arabis* species, leading to a systemic infection that radically affects the host's growth and morphology, including infected rosettes, or pseudoflowers, which are highly attractive to flower-visiting insects (Roy, 1993). In so doing, insect visitation promotes rust fertilization by bringing together spermatia of opposite mating types. This is an extreme example of floral mimicry involving gross changes in host morphology that goes far beyond the known effects on insects of the exudates produced by rust fungi, the secretions by *Monilinia* spp. on blueberry and the smut spores dispersed by insect pollinators in a range of crops. Hybrids between two sedge *Carex* spp. were found to strongly influence the incidence of the floral smut fungus *Anthracoidea fischieri* (Ericson *et al.*, 1993). High occurrence of the disease on hybrid plants appears to reinforce an incomplete fertility barrier between the parent species. The situation is complicated further by an interaction with a beetle, *Phalacrus substriatus*, which feeds on the smut teliospores.

Many grasses form associations with fungal species of the plant-pathogenic genus *Epichloë* and the related asexual endophytes classified in *Acremonium* (Clay, 1993). *Acremonium* endophytes occur worldwide and cause a range of debilitating symptoms and toxicoses on animals grazing on these grasses, but cause no disease in the grass host and may improve tolerance to biotic and abiotic stress (Siegel, 1993). The continuum between mutualism and antagonism in relation to the host plants and between these fungi is further complicated by potential hybridization of the two fungi, as found in perennial ryegrass (Schardl *et al.*, 1994). During the sexual cycle in *Epichloë*, fertilization only occurs with spermatia of the opposite mating type. This is facilitated by specialized flies of the genus *Botanophila* which feed on the fungal stomata, ingest spermatia and pass them through their gut. Thus, a disease in grass (choke disease) interacts with the endophytic asexual stage causing toxicoses in animals, mediated in part by interactions with a symbiotic fly.

Fungal endophytes may affect grass seed predation as found in the interaction between fescue seeds, infected or not infected by *Acremonium*, and seed harvesting ants (Knoch *et al.*, 1993). Seed infestation enhanced the probability of germination in favourable sites, i.e. (ant) refuse piles.

Endophytic fungi (*Neotyphodium* spp.) also have an impact on invertebrate as well as livestock herbivory. There is some evidence, for example, that endophyte-infected grasses express very effective resistance to root parasitic nematodes (Cook and Lewis, Chapter 4). In terms of negative (from the plant's perspective) interactions with nematodes, the synergism between nematodes and *Fusarium* species in causing wilt has long been documented (Hillocks, 1986; Abawi and Chen, 1998; Hillocks, Chapter 14), although rarely described as a dynamic process (Starr *et al.*, 1989).

Many interactions between mycorrhizal fungi and other soil organisms have been noted (Fitter and Garbaye, 1994). Bacteria may promote mycorrhizal formation but also soil invertebrates may graze external mycelium. Mycorrhizal fungi modify the interaction of plants with other soil biota, including fungal pathogens and plant parasitic nematodes (Pinochet *et al.*, 1993). Foliar herbivores such as the stem and cone-boring moth, *Dioryctria albouitella*, attacking *Pinus edulis* negatively affect the ectomycorrhizal mutualism in susceptible trees (Gehring and Whitham, 1991). Invertebrate grazing on fungi has long been documented for saprobic and mycorrhizal systems and the effects on ecosystem function and fungal community structure have been described (McGonigle, 1997). Similar studies have now been conducted for pathogenic fungi and their co-occurrence with saprotrophic and mycorrhizal fungi (McGonigle and Hyakumachi, Chapter 5).

Interactions with Prokaryotes

Bacteria and non-culturable mollicutes, or phytoplasmas, are pathogens of numerous wild and cultivated species in virtually all plant families. Bacteria are involved in many positive and negative interactions with other biota on plant surfaces and in the soil as discussed later. They also interact with insects, being often transmitted by insect vectors. The plant-pathogenic bacteria *Erwinia carotovora* var. *atroseptica* (Eca) and var. *carotovora* (Ecc) are both transmitted by fruit flies but the success of transmission for each depends on temperature (Kloeper *et al.*, 1981), with the proportion of Ecc transmitted being greater at 27°C whereas at 15°C they were approximately equal. Equally some prokaryotes are endosymbionts or pathogens of insects and in these cases the plant may be considered to be the vector from insect to insect. Such associations may represent either a stage in the evolution of an intracellular symbiosis with an insect host or alternatively parasitization of plant and insect hosts via insect transmission (Purcell *et al.*, 1994). Endosymbiotic bacteria of aphids and whiteflies have also been shown to play a critical role in the stabilization and retention of luteoviruses and begomoviruses in insect vectors.

Endophytic bacteria are ubiquitous in most plant species, residing in living plant tissues without doing substantive harm to the plant. Endophytic bacteria have historically been considered as weakly virulent plant pathogens but have recently been shown to have several beneficial effects on their host plants (Hallman *et al.*, 1997). Beneficial effects can include: (i) direct antagonism or niche exclusion of pathogens; (ii) induction of systemic resistance; and (iii) increasing tolerance to biotic stresses (Hallman, Chapter 6). Equally, however, endophytic bacteria may form apparently neutral associations, or remain latent until active in later stages of plant development. Endophytic bacteria originate from epiphytic rhizosphere or phylloplane communities or from infected seed or planting material. At the moment it remains speculation as to whether they are saprophytes evolving towards pathogens, or have evolved further than pathogens by conserving shelter and nutrient supply without causing disease.

Virus–Vector Associations

Many pathogens are dependent upon spread by a vector for an epidemic to occur in a plant population. This is certainly the case for plant-infecting viruses where the majority depend on arthropod vectors for transmission between hosts, including homoptera, thrips, chrysomellid beetles and eriophyid mites (Spence, Chapter 2). According to Nault (1997), an understanding of virus transmission is the key to interpreting virus epidemiology and the control of virus disease. Knight and Webb (1993) claim that an understanding of the origin and evolution of vector–virus relationships and the 'predictability of potential vectors' is largely dependent upon an understanding of vector phylogeny. Thus, an understanding of virus–vector associations is itself predicated upon an interaction between plant virologists and entomologists. The value of understanding virus–vector transmission characteristics can be seen in the recent global expansion of whitefly-transmitted viruses to new host plants with the emergence of the 'B' biotype of *Bemisia tabaci* (Markham *et al.*, 1994).

Virus transmission also occurs through soil-inhabiting organisms (Jones, 1993), mostly ectoparasitic nematodes (Ploeg *et al.*, 1992; Brown *et al.*, 1995) and chitrid or plasmodiophorid fungi (Adams, 1991; Spence, Chapter 2). Although much research has identified genomic components associated with virus transmission by these organisms, as with the chrysomellid beetles (Bakker, 1971; Wang *et al.*, 1992), it is the case that the epidemiological implications are far less appreciated than for homopteran vectors. In some cases, such as with helper-dependent virus complexes (Pirone and Blanc, 1996), interactions at the genetic and physiological level can be scaled up and the epidemic dynamics modelled (Zhang *et al.*, 2000).

Over a period of 20 years, plant virus epidemiology has developed from a largely descriptive account of observed epidemics (Thresh, 1980) to one in which the basic information on vector–virus associations can be interpreted in terms of epidemiology and potential control measures (Jeger *et al.*, 1998; Madden *et al.*, 2000).

Biological Control

From the plant's perspective, the most important positive biotic interactions are those which contribute to biological control of pathogens. The distinction is made again between those interactions which occur within-taxon and are often manifest through effects on the plant, and those which involve direct interactions involving the same or different taxonomic groupings. The phenomenon of induced resistance is not considered more generally across taxonomic groupings, e.g. the effects of fungal infection on arthropod feeding, except to note that this may be due to pathogen-related reductions in host plant quality rather than an induced defence response (Jongebloed *et al.*, 1992).

The fungal pathogen species *Fusarium oxysporum*, causing fusarium wilt, is present worldwide affecting a wide range of plant species in every type of soil. Pathogenic strains are characterized as special forms, so-called *formae speciales*, according to the plant species they are able to infect. There are also apparently non-pathogenic strains of *F. oxysporum* whose presence is associated with the suppressiveness of several soils to fusarium wilt, although the designation of a non-pathogenic strain in relation to fusarium wilt on a given crop does perhaps illustrate the illogicality of the special form terminology. Interactions among strains of *F. oxysporum* in soil and the rhizosphere undoubtedly occur, as does competition at the root surface and indirect interaction through the plant (Alabouvette and Couteaudier, 1992; Alabouvette *et al.*, Chapter 8). The challenge for biological control is to develop procedures for screening and determining modes of action for efficient strains, and for the mass production, formulation and delivery of effective inoculum.

Successful biological control of crown gall caused by *Agrobacterium* using an avirulent strain of *Agrobacterium* has long been in commercial practice (Kerr, 1980, 1991). The avirulent strain produces a bacteriocin, which in combination with good colonization ability, leads to highly effective control of many pathogenic *Agrobacterium* strains. In a similar way avirulent mutants of the bacterial pathogen *Ralstonia solanacearum* have been produced and studies have developed and evaluated biological control effectiveness against bacterial wilt disease in tomato and potato (Trigalet and Trigalet-Demery, 1990; Smith and Saddler, Chapter 9).

For viruses, the potential for mild strain protection or cross-protection has been well documented. One of the best known studies has been with zucchini yellow mosaic potyvirus, one of the most damaging viruses in cucurbits but which was first reported as recently as 1981 and has since spread worldwide to become a major constraint wherever cucurbits are grown. A mild strain of ZYMV was found to be stable and effective in controlling the disease (Lecoq *et al.*, 1991; Spence *et al.*, 1996) and in recent years the technology developed for treating crops has increasingly been accepted by growers (Lecoq and Raccah, Chapter 10).

Direct mycoparasitic interactions between pathogenic fungi and their hyperparasites also represent within-taxon biological control and there are many examples for both foliar pathogens (Jeffries, 1995) and root or soil-inhabiting pathogens (Adams, 1990). Increasingly the importance of recognizing the multitrophic nature of the interactions, at the very least the host–parasite–hyperparasite relationship, has been appreciated (Kiss, 1998; Kiss, Chapter 12). It may be possible to introduce the vectoring of foliar biological control organisms to manage foliage and fruit disease (Sutton and Peng, 1993), an example of a positive interaction.

Bacteria have long been studied for their antifungal properties, whether through competition for nutrients, antibiotic production or mycolytic properties. Chitin is an important constituent of the cell walls of fungi and it is possible that chitinolytic soil bacteria may be effective antagonists through chitinase activity. Alternatively, chitinolytic soil bacteria may compete for chitin successfully in soil (de Boer *et al.*, 1998). In fact it turns out that chitinolytic bacteria have specific rather than general anti-fungal activity and that differences between and within species were most likely attributable to the action of antibiotics in combination with or succeeded by lyctic enzymes (de Boer and van Veen, Chapter 7).

If parasitic weeds are considered to be plant pathogens then the accumulated experience of plant pathologists could be directed towards research in biological control of these major constraints to crop production. In fact there has been little research done on this topic despite the undoubted importance of the parasitic weeds *Striga* and *Orobanche* in the tropics and sub-tropics. In a study of the fungi occurring on *Orobanche* spp. and their potential for *Orobanche* control, a *Ulocladium atrum* isolate was found to be promising under appropriate temperature and humidity conditions (Linke *et al.*, 1992). *U. atrum* is indeed a relatively weak pathogen on some crop plants but is actively being developed as an antagonist against *Botrytis* diseases (Kessel *et al.*, 1999). This is one example I am aware of where a potential biological control agent has potential against two pathogen taxa. *Fusarium* spp. are often associated with roots of plants infected by parasitic weeds. Perhaps there may prove to be wilt-causing isolates of *Fusarium oxysporum* which are pathogenic on the parasitic weed *Striga* but are non-pathogenic and confer biological control against fusarium wilt of a *Striga* host.

Biological control of (non-parasitic) weeds is also a topic of much research effort at present with increased attention being given to interactions between potential biocontrol fungi and the outcome in terms of additive or synergistic effects (Morin *et al.*, 1993a). Similarly, interactions in weed biocontrol may be present between fungi and insect herbivores (Hatcher *et al.*, 1994) although the interaction may not even be additive with herbivory reducing the incidence of disease (Hatcher and Paul, 2000; Chapter 11). There may also be interactions between mycoparasites and weeds, at least with the absence of weeds due to herbicide usage (Teo *et al.*, 1992).

Complex Interactions, Disease Complexes and Complex Aetiology

There is a fine distinction between the co-occurrence of a range of plant diseases found on crops in the field, where sometimes a particular combination of pathogenic agents is associated with a characteristic symptomatology, and diseases which are only expressed if one or more pathogens act in concert. Synergy in terms of effects on the host plant is sometimes claimed within the same taxonomic grouping, e.g. for nematodes (Nyczepir *et al.*, 1993) and fungal pathogens (Morin *et al.*, 1993b). Sometimes an organism considered beneficial can be mildly pathogenic in the presence of other pathogens. For example, *Trichoderma harzianum*, a beneficial soil saprophyte, can cause severe necrosis on roots in the presence of root-knot nematodes. In other situations *T. harzianum* can suppress reproduction of the nematode (Windham *et al.*, 1993). Some have argued that new and unique control strategies must develop for biological system management of soil-borne diseases, including nematodes (Sikora, 1997). In particular, it is claimed, the antagonist potential of agricultural ecosystems must be managed by means other than inundative release (Sikora, 1992).

The complexities of ecological interactions, when unravelled, for example in the southern pine beetle system that kills healthy trees through mass colonization and involves the bark beetle, three main fungi and phoretic mites which have symbiotic relationships with the fungi, may provide novel opportunities for control of the pest complex (Klepzig, 1998; Klepzig *et al.*, Chapter 13). In perhaps a simpler and more direct interaction the larch sawfly and the needlecast fungus (*Mycosphaercella laricinia*) both defoliate European larch, but in fact fungal infection may limit sawfly populations where both species co-exist, although at a considerable cost to larch productivity (Krause and Raffa, 1992). Forest ecosystem interactions between pathogen and non-pathogenic agents (notably air pollutants) must, it has been argued, be evaluated in terms of whole ecosystem health – a new perspective for plant pathology (Smith, 1984). In recent years much ecological insight has been obtained through studies of disease complexes in natural vegetation. In particular, studies on sand dune vegetation in The Netherlands have shown how plant parasitic nematodes and soil-borne fungi interact in relation to successional processes (Van der Putten *et al.*, 1993; Van der Putten and Van der Stoel, 1998; Van der Putten, Chapter 15). If sand dune sites, indeed any non-agricultural site, are to be managed for scientific interest, aesthetic or 'heritage' reasons then clearly the role of plant pathogens, either negatively or positively, must be appreciated in securing these landscape objectives.

Methodology and Modelling

The presence of biotic interactions at the same or different trophic levels presents major methodological problems and challenges for the modelling of disease epidemics. For vector-borne diseases the methodologies used simply to monitor and/or sample for disease incidence may need to change depending on the vector involved, for example with citrus tristeza (Hughes and Gottwald, 1998; Hughes *et al.*, Chapter 16). The incorporation of probabilistic or stochastic elements introduces additional complexities into methodology and modelling (Rouse, 1991; Shaw, 1994). The explicit consideration of vector population dynamics into models of virus epidemics has a major impact on the level of mathematical sophistication required in developing and analysing the models (Szymanski and Caraco, 1994; Holt *et al.*, 1997; Jeger *et al.*, 1998; Madden *et al.*, 2000). It is only recently that mathematical models of biological control and the various interactions involved have been devised for both soil-borne (Gubbins and Gilligan, 1996) and foliar diseases, including those with a vector involvement such as Dutch elm disease, where interactions between *Ophiostoma* species or virus-like d-factors can be explored using the model framework developed (Swinton and Gilligan, 1996).

Conclusion

The broad range of examples of biotic interactions and their complexities included in this chapter and expanded on in this book, it is hoped, will help to consolidate and stimulate further, more detailed studies in this important area. In many cases it will be necessary for theory and experimentation to be developed simultaneously (Jeger, 2000), and indeed interact, for plant pathologists to appreciate fully the significance of biotic interactions in understanding and manipulating plant–pathogen associations in agriculture and other managed landscapes.

References

Abawi, G.S. and Chen, J. (1998) Concomitant pathogen and pest interactions. In: Pederson, G.A. and Windham, G.L. (eds) *Plant and Nematode Interactions. Agronomy Monograph* 36. American Society of Agronomy, Madison, Wisconsin.

Adams, M.J. (1991) Transmission of plant-viruses by fungi. *Annals of Applied Biology* 118, 479–492.

Adams, P.B. (1990) The potential of mycoparasites for biological control of plant diseases. *Annual Reviews of Phytopathology* 28, 59–72.

Alabouvette, C. and Couteaudier, Y. (1992) Biological control of fusarium wilts with non pathogenic *Fusaria*. In: Tjamos, E.C., Cook, R.H. and Papavizas, G.C. (eds) *Biological Control of Plant Diseases*. Plenum Press, New York, pp. 415–426.

Bakker, W. (1971) Three new beetle vectors of rice yellow mottle virus in Kenya. *Netherlands Journal of Plant Pathology* 77, 201–206.

Beemster, A.B.R., Bollen, G.J., Gerlagh, M., Ruissen, M.A., Schippers, B. and Tempel, A. (eds) (1991) Biotic interactions and soil-borne diseases. *Proceedings of the First Conference of the European Foundation for Plant Pathology*. Elsevier, Amsterdam.

Broadbent, L. (1960) Dispersal of inoculum by insects and other animals, including man. In: Horsfall, J.G. and Dimond, A.E. (eds) *Plant Pathology: an Advanced Treatise*, Vol. III, *The Diseased Population, Epidemics and Control*. Academic Press, New York, pp. 97–135.

Brown, D.J.F., Robertson, W.M. and Trudgill, D.L. (1995) Transmission of viruses by plant nematodes. *Annual Review of Phytopathology* 33, 223–249.

Clay, K. (1993) The ecology and evolution of endophytes. *Agriculture, Ecosystems and Environment* 44, 39–64

Cowling, E.B. and Horsfall, J.G. (1979) Prologue: how pathogens induce disease. In: Horsfall, J.G. and Cowling, E.B. (eds) *Plant Pathology: an Advanced Treatise*, Vol. IV, *How Pathogens Induce Disease*. Academic Press, New York, pp. 1–21.

Darpoux, H. (1960) Biological interference with epidemics. In: Horsfall, J.G. and Dimond, A.E. (eds) *Plant Pathology: an Advanced Treatise*, Vol. III, *The Diseased Population, Epidemics and Control*. Academic Press, New York, pp. 521–565.

de Boer, W., Klein Gunnewiek, P.J.A., Lafeber, P., Janse, J.D., Spit, B.E. and Woldendorp, J.W. (1998) Anti-fungal properties of chitinolytic dune soil bacteria. *Soil Biology and Biochemistry* 30, 193–203.

Dickinson, C.H. (1979) External synergisms among organisms inducing disease. In: Horsfall, J.G. and Cowling, E.B. (eds) *Plant Pathology: an Advanced Treatise*, Vol. IV, *How Pathogens Induce Disease*. Academic Press, New York, pp. 97–111.

Elias, K.S., Zamir, D., Lichtman-Pleban, T. and Katan, T. (1993) Population structure of *Fusarium oxysporum* f. sp. *lycopersici*: restriction fragment length polymorphisms provide genetic evidence that vegetative compatibility group is an indicator of evolutionary origin. *Molecular Plant–Microbe Interactions* 6, 565–572.

Ericson, L., Burdon, J.J. and Wennström, A. (1993) Inter-specific host hybrids and phalacrid beetles implicated in the local survival of smut pathogens. *Oikos* 68, 393–400.

Fitter, A.H. and Garbaye, J. (1994) Interactions between mycorrhizal fungi and other soil organisms. *Plant and Soil* 159, 123–132.

Foster, G.C. and Taylor, S.C. (eds) (1998) *Plant Virology Protocols: From Virus Isolation to Transgenic Resistance*. Humana Press, Totowa, New Jersey.

Gehring, C.A. and Whitham, T.G. (1991) Herbivore-driven mycorrhizal mutualism in insect-susceptible pinyon pine. *Nature* 353, 556–557.

Gubbins, S. and Gilligan, C.A. (1996) Population dynamics of a parasite and hyper-parasite in a closed system: model analysis and parameter estimation. *Proceedings of the Royal Society London* 263, 1071–1078.

Hallmann, J., Quadt-Hallmann, A., Mahaffee, W.F. and Kloepper, J.W. (1997) Bacterial endophytes in agricultural crops. *Canadian Journal of Microbiology* 43, 895–914.

Hatcher, P.E. and Paul, N.D. (2000) Beetle grazing reduces natural infection of *Rumex obtusifolius* by fungal pathogens. *New Phytologist* 146, 325–333.

Hatcher, P.E., Paul, N.D., Ayres, P.G. and Whittaker, J.B. (1994) Interactions between *Rumex* spp., herbivores and a rust fungus: *Gastrophysa viridula* grazing reduces subsequent infection by *Uromyces rumicis*. *Functional Ecology* 8, 265–272.

Haubold, B. and Rainey, P.B. (1996) Genetic and ecotypic structure of a fluorescent *Pseudomonas* population. *Molecular Ecology* 5, 747–761.

Hillocks, R.J. (1986) Localised and systemic effects of root-rust nematode on incidence and severity of fusarium wilt in cotton. *Nematologica* 32, 202–208.

Holt, J., Jeger, M.J., Thresh, J.M. and Otim-Nape, G.W. (1997) An epidemiological model incorporating vector population dynamics applied to African cassava mosaic virus disease. *Journal of Applied Ecology* 34, 793–806.

Horsfall, J.G. and Cowling, E.B. (1978) Some epidemics man has known. In: Horsfall, J.G. and Cowling, E.B. (eds) *Plant Disease: an Advanced Treatise*, Vol. II, *How Disease Develops in Populations*. Academic Press, New York, pp. 17–32.

Hughes, G. and Gottwald, T.R. (1998) Survey methods for assessment of citrus tristeza virus incidence. *Phytopathology* 88, 715–725.

Hyakumachi, M. and Ui, T. (1987) Non-self-anastomosing isolates of *Rhizoctonia solani* obtained from fields of sugarbeet monoculture. *Transactions of the British Mycological Society* 89, 155–159.

Jeffries, P (1995) Biology and ecology of mycoparasitism. *Canadian Journal of Botany* 73 (Suppl. 1), S1284–S1290.

Jeger, M.J. (2000) Theory and plant epidemiology. *Plant Pathology* 49, 2–11.

Jeger, M.J., Van den Bosch, F., Madden, L.V. and Holt, J. (1998) A model for analysing plant virus transmission characteristics and epidemic development. *IMA Journal of Mathematics Applied in Medicine and Biology* 14, 1–18.

Jones, A.T. (1993) Virus transmission through soil and by soil-inhabiting organisms in diagnosis. In: Matthews, R.E.F. (ed.) *Diagnosis of Plant Virus Diseases*. CRC Press, Boca Raton, Florida, pp. 73–99.

Jongebloed, P.H.J., Elgersma, D.M. and Sabelis, M.W. (1992) Does a vascular fungus of tomato induce a defence response or a change in host plant quality that also affects the oviposition of spider mites? *Experimental and Applied Acarology* 16, 227–236.

Kerr, A. (1980) Biological control of crown gall through production of Agrocin 84. *Plant Disease* 64, 28–30.

Kerr, A. (1991) The genus *Agrobacterium*. In: Balows, A., Trüper, H.G., Dworkin, N., Harder, W. and Schleifer, K.-H. (eds) *The Procaryotes: a Handbook on the Biology of Bacteria*, 2nd edn. Springer-Verlag, Berlin, pp. 2214–2235.

Kessel, G.J.T., de Haas, B.H., Lombaers-Van der Plas, C.H., Meijer, E.M.J., Dewey, F.M., Goudriaan, J., Van der Werf, W. and Kohl, J. (1999) Quantification of mycelium of *Botrytis* spp. and the antagonist *Ulocladium atrum* in necrotic leaf tissue of cyclamen and lily by fluorescence microscopy and image analysis. *Phytopathology* 89, 868–876.

Kiss, L. (1998) Natural occurrence of *Ampelomyces* intracellular mycoparasites in mycelia of powdery mildew fungi. *New Phytologist* 140, 709–714.

Klepzig, K.D. (1998) Competition between a biological control fungus, *Ophiostoma piliferum*, and symbionts of the southern pine beetle. *Mycologia* 90, 69–75.

Kloepper, J.W., Harrison, M.D. and Brewer, J.W. (1981) Effect of temperature on differential persistence and insect transmission of *Erwinia carotovora* var. *atroseptica* and *Ewinia carotovora* var. *carotovora*. *American Potato Journal* 58, 585–592.

Knight, W.J. and Webb, M.D. (1993) The phylogenetic relationships between virus vector and other genera of macrosteline leafhoppers, including descriptions of new taxa (Homoptera: Cicadellidae: Deltocephalinae). *Systematic Entomology* 18, 11–55.

Knoch, T.R., Faeth, S.H. and Arnott, D.L. (1993) Endophytic fungi alter foraging and dispersal by desert seed-harvesting ants. *Oecologia* 95, 470–473.

Krause, S.C. and Raffa, K.F. (1992) Comparison of insect, fungal, and mechanically induced defoliation of larch: effects on plant productivity and subsequent host susceptibility. *Oecologia* 90, 411–416.

Lecoq, H., Lemaire, J.M. and Wipf-Scheibel, C. (1991) Control of zucchini yellow mosaic virus in squash by cross protection. *Plant Disease* 75, 208–211.

Leslie, J.F. (1993) Fungal vegetative compatibility. *Annual Review of Phytopathology* 31, 127–150.

Linke, K.-H., Scheibel, C., Saxena, M.C. and Sauerborn, J. (1992) Fungi occurring on *Orobanche* spp. and their preliminary evaluation for *Orobanche* control. *Tropical Pest Management* 38(2), 127–130.

MacFarlane, S.A. (1997) Natural recombination among plant virus genomes: evidence from tobraviruses. *Seminars in Virology* 8, 25–31.

Madden, L.V., Jeger, M.J. and van den Bosch, F. (2000) A theoretical assessment of the effects of vector–virus transmission mechanism on plant virus disease epidemics. *Phytopathology* 90, 576–594.

Markham, P.G., Bedford, I.D., Liu, S.J. and Pinner, M.S. (1994) The transmission of geminiviruses by *Bemisia-tabaci*. *Pesticide Science* 42(2), 123–128.

McGonigle, T.P. (1997) Fungivores. In: *The Mycota*, Vol. IV, *Environmental and Microbial Relationships*. Springer-Verlag, Berlin, pp. 237–248.

Milgroom, M.G., MacDonald, W.L. and Double, M.L. (1990) Spatial pattern analysis of vegetative compatibility groups in the chestnut blight fungus, *Cryphonectria parasitica*. *Canadian Journal of Botany* 69, 1407–1413.

Morin, L., Auld, B.A. and Brown, J.F. (1993a) Interaction between *Puccinia xanthii* and facultative parasitic fungi on *Xanthium occidentale*. *Biological Control* 3, 296–310.

Morin, L., Auld, B.A. and Brown, J.F. (1993b) Synergy between *Puccinia xanthii* and facultative parasitic fungi on *Xanthium occidentale*. *Biological Control* 3, 288–295.

Nault, L.R. (1997) Arthropod transmission of plant viruses: a new synthesis. *Annals of the Entomological Society of America* 90, 521–541.

Nauta, M.J. and Hoekstra, R.F. (1994) Evolution of vegetative incompatibility in filamentous ascomycetes. I. Deterministic models. *Evolution* 48, 979–995.

Nauta, M.J. and Hoekstra, R.F. (1996) Vegetative incompatibility in ascomycetes: highly polymorphic but selectively neutral? *Journal of Theoretical Biology* 183, 67–76.

Nyczepir, A.P., Riley, M.B. and Sharpe, R.R. (1993) Dynamics of concomitant populations of *Meloidogyne incognita* and *Criconemella xenoplax* on peach. *Journal of Nematology* 25, 659–665.

Pinochet, J., Camprubi, A. and Calvet, C. (1993) Effects of the root-lesion nematode *Pratylenchus vulnus* and the mycorrhizal fungus *Glomus mosseae* on the growth of EMLA-26 apple rootstock. *Mycorrhiza* 4, 79–83.

Pirone, T.P. and Blanc, S. (1996) Helper-dependent vector transmission of plant viruses. *Annual Review of Phytopathology* 34, 227–247.

Ploeg, A.T., Brown, D.J.F. and Robinson, D.J. (1992) Acquisition and subsequent transmission of tobacco rattle virus isolates by *Paratrichodorus* and *Trichodorus* nematode species. *Netherlands Journal of Plant Pathology* 98, 291–300.

Plumb, R. (ed.) (2001) *Interactions between Plant Viruses and their Vectors*. Advances in Botanical Research series, Academic Press, London.

Powell, N.T. (1979) Internal synergisms among organisms inducing disease. In: Horsfall, J.G. and Cowling, E.B. (eds) *Plant Pathology: an Advanced Treatise*, Vol. IV, *How Pathogens Induce Disease*. Academic Press, New York, pp. 113–133.

Purcell, A.H., Suslow, K.G. and Klein, M. (1994) Transmission via plants of an insect pathogenic bacterium that does not multiply or move in plants. *Microbial Ecology* 27, 19–26.

Rouse, D.I. (1991) Stochastic modeling of plant disease epidemic processes. In: Arora, D.K., Rai, B., Mukerji, K.G. and Knudsen, G.R. (eds) *Handbook of Applied Mycology*, Vol. 1, *Soil and Plants*. Marcel Dekker, New York, pp. 647–665.

Roy, B.A. (1993) Floral mimicry by a plant pathogen. *Nature* 362, 56–58.

Schardl, C.L., Leuchtmann, A., Tsai, H.-F., Collett, M.A., Watt, D.M. and Scott, D.B. (1994) Origin of a fungal symbiont of perennial ryegrass by interspecific hybridization of a mutualist with the ryegrass choke pathogen, *Epichloë typhina*. *Genetics* 136, 1207–1317.

Shaw, M.W. (1994) Modeling stochastic processes in plant pathology. *Plant Review of Phytopathology* 32, 523–544.

Siegel, M.R. (1993) *Acremonium* endophytes: our current state of knowledge and future directions for research. *Agriculture, Ecosystems and Environment* 44, 301–321.

Sikora, R.A. (1992) Management of the antagonistic potential in agricultural ecosystems for the biological control of plant parasitic nematodes. *Annual Review of Phytopathology* 30, 245–270.

Sikora, R.A. (1997) Biological system management in the rhizosphere an *Inside-out/ Outside-in* perspective. Mededelingen Faculteit Landbouw wetenschappen, Rijks Universiteit, Gent, 62/2a.

Smith, W.H. (1984) Ecosystem pathology: a new perspective for phytopathology. *Forest Ecology and Management* 9, 193–219.

Spence, N.J., Mead, A., Miller, A., Shaw, E.D. and Walkey, D.G.A. (1996) The effect on yield in courgette and marrow of the mild strain of yellow zuccini yellow mosaic virus used for crop protection. *Annals of Applied Biology* 129, 247–259.

Starr, J.L., Jeger, M.J., Martyn, R.D. and Schilling, K. (1989) Effects of *Meloidogyne incognita* and *Fusarium oxysporum* f. sp. *vasinfectum* on plant mortality and yield of cotton. *Phytopathology* 79, 640–646.

Sutton, J.C. and Peng, G. (1993) Manipulation and vectoring of biocontrol organisms to manage foliage and fruit diseases in cropping systems. *Annual Review of Phytopathology* 31, 473–493.

Swinton, J. and Gilligan, C.A. (1996) Dutch elm disease and the future of elm in the UK: a quantitative analysis. *Philosophical Transactions of the Royal Society of London, Series B. Biological Sciences* 351, 605–615.

Szymanski, B. and Caraco, T. (1994) Spatial analysis of vector-borne disease: a four-species model. *Evolutionary Ecology* 8, 299–314.

Teo, B.K., Verma, P.R. and Morrall, R.A.A. (1992) The effects of herbicides and mycoparasites at different moisture levels on carpogenic germination in *Sclerotinia sclerotiorum*. *Plant and Soil* 139, 99–107.

Thresh, J.M. (1980) The origins and epidemiology of some important plant virus diseases. *Applied Biology* 5, 1–65.

Trigalet, A. and Trigalet-Demery, D. (1990) Use of avirulent mutants of *Pseudomonas solanacearum* for the biological control of bacterial wilt of tomato plants. *Physiological and Molecular Plant Pathology* 36, 27–38.

Van der Putten, W.H. and Van der Stoel, C.D. (1998) Plant parasitic nematodes and spatio-temporal variation in natural vegetation. *Applied Soil Ecology* 10, 253–262.

Van der Putten, W.H., Van Dijk, C. and Peters, B.A.M. (1993) Plant-specific soil-borne diseases contribute to succession in foredune vegetation. *Nature* 362, 53–56.

Wang, R.Y., Gergerich, R.C. and Kim, K.S. (1992) Noncirculative transmission of plant-viruses by leaf-feeding beetles. *Phytopathology* 82, 946–950.

Windham, G.L., Windham, M.T. and Pederson, G.A. (1993) Interaction of *Trichoderma harzianum*, *Meloidogyne incognita*, and *Meloidogyne arenaria* on *Trifolium repens*. *Nematropica* 23, 99–103.

Zhang, X.S., Holt, J. and Colvin, J. (2000) Mathematical models of host plant infection by helper-dependent virus complexes: why are helper viruses always avirulent? *Phytopathology* 90, 85–93.

Virus–Vector Interactions in Plant Virus Disease Transmission and Epidemiology

2

N.J. Spence

Department of Plant Pathology and Microbiology, Horticulture Research International, Wellesbourne, Warwick CV35 9EF, UK

Introduction

The virus–vector association is one of the most important biotic interactions in the epidemiology of plant virus diseases. Vector transmission provides the main method of disease spread in the field that causes severe economic losses, although disease incidence depends on many factors including number and behaviour of vectors, host resistance to virus and vector, and the transmission process. There is considerable biological interest in the relationships that exist between vector, virus and host. In many cases the virus multiplies in the vector and even with those that do not, the relationship is usually complex, involving the virus, vector, host and the environment. Advances in the understanding of the relationship between virus and vector in the transmission process have led to epidemiological studies and the development of sophisticated mathematical models which could lead to improved selection of management strategies for virus diseases in the future. There have been several recent studies to determine the influence of virus–vector interaction on disease epidemics that are described in this chapter.

Arthropod Transmission of Plant Viruses

A majority of the plant-infecting viruses are dependent on arthropod vectors for transmission between hosts and/or alternative hosts. The viruses have evolved specific associations with their vectors, and the underlying mechanisms that regulate the virus transmission process are beginning to be understood. A majority of plant viruses are carried on the cuticle lining of a vector's mouthparts or foregut. This initially appeared to be simple mechanical contamination, but it is now known to be a biologically complex interaction between specific virus proteins and as yet unidentified vector cuticle-associated compounds. Numerous other plant viruses and the majority of animal viruses are carried within the body of the vector. These viruses have evolved specific mechanisms to enable them to be transported through multiple tissues and to evade vector defences (Gray and Banerjee, 1999). In response, vector species have evolved so that not all individuals within a species are susceptible to virus infection or can serve as a competent vector. Not only are the virus components of the transmission process now being identified, but also the genetic and physiological components of the vectors that determine their ability to transmit viruses successfully are being elucidated. The mechanisms of arthropod–virus associations are many and complex, but common themes are beginning to emerge which may allow the development of novel strategies to ultimately control epidemics caused by arthropod-borne viruses.

The most important arthropod vectors of plant viruses are four families of homopterans (aphids, whiteflies, leafhoppers and delphacid planthoppers), thrips, chrysomellid beetles and, among the acarines, the eriophyid mites. More than 380 viruses from 27 plant virus genera are transmitted by the *Homoptera*. Nault (1997) described two systems to group homopteran-borne plant virus diseases distinguished by the transmission characteristics and the nature of the interaction of the virus with the vector. One is based on persistence of transmissible virus in the vector and the other on the mechanism of transmission or route of virus transport in the vector. The two systems are combined to create the following four plant virus disease transmission groups. For non-persistently transmitted viruses (NP), the virus is usually restricted to the stylet of the insect. For persistently transmitted viruses, the virus is ingested, passes through the gut wall into the haemolymph, and then moves to the salivary glands where it can potentially be transmitted to other plants. Persistently transmitted viruses, have two subclasses termed circulative (CP) if there is no multiplication in the insect vector and propagative (PP) if there is. A fourth class, semi-persistent (SP), which is intermediate between non-persistent and persistent, is generally recognized; in this class the virus moves to the foregut of the insect. Homopteran-transmitted viruses are also characterized by the following traits: (i) no virus is transmitted by vectors from more than one homopteran family; (ii) most plant virus genera have vectors from one family of homopterans, although some are characterized by vectors from more than one family, usually from the same homopteran suborder; (iii) the persistence and mode of transmission for all viruses within a virus genus is almost always the same and appears to be a stable evolutionary trait. Certain aspects of the feeding behaviour and the morphology of the mouthparts and digestive systems of the *Homoptera* and other vectors have relevance to vector transmission. The thrips-borne viruses are transmitted much in the same way as the persistently transmitted, propagative viruses are by the *Homoptera*. At least 42 plant viruses are transmitted by the *Coleoptera*, their mode of transmission differs markedly from that described for the homopterans and thrips. So, too, does the transmission of plant viruses by eriophyid mites, which are not discussed in this chapter.

The Effects of Vector–Virus Transmission Mechanism on Plant Virus Disease Epidemics

A model has been developed in which a disease epidemic in host plants is linked with the insect vector population to describe the transmission process (Jeger *et al.*, 1998). This model has been used to compare the transmission characteristics of the four virus classes directly, and to explore the consequences for epidemic development and possible control options. Depending on the assumptions made about migration, it was possible to obtain an expression for the basic reproductive number (number of new infected plants resulting from an infected plant introduced into a susceptible plant population, R_0). Expressions were also obtained for equilibrium values for the host and vector population classes and a numerical analysis indicated that these equilibria were stable for known or reasonable estimates of the parameter values. R_0 was used to examine the relative contributions of key parameters in distinguishing the four virus disease classes using values and ranges taken directly from the literature or estimated indirectly. Pairwise plots of parameter values which satisfied the threshold criterion $R_0 = 1$ clearly separated the propagative class from the other categories. On holding the other parameters constant, a much larger vector population or vector activity was required to satisfy the epidemic threshold for propagative viruses. Similar conclusions were reached from plots of the healthy host and viruliferous vector populations against key parameters. The model framework was used to analyse the effectiveness of roguing (the removal and destruction) of diseased plants and/or reduction of the vector-population size, for example, by insecticide treatment or vegetation management. Roguing would only be effective for non-persistently transmitted viruses at relatively low vector-population densities. Roguing would usually only be needed for propagative viruses at very high population densities. There would be a clear advantage in reducing the vector-population density for propagative viruses, and control measures aimed at reducing populations would only be effective for these viruses.

Later, a continuous-time and deterministic model was developed by Madden *et al.* (2000) to characterize plant virus disease epidemics in relation to virus transmission mechanism and the population dynamics of the insect vectors. This model is a set of linked differential equations for healthy (virus-free), latently infected, infectious and removed (post-infectious) plant categories, and virus-free, latent and infective insects, with parameters based on the transmission classes, vector population dynamics, immigration/emigration rates and virus–plant interactions. The rate of change in diseased plants is expressed as a function of the density of infective insects, the number of plants visited per time and the probability of transmitting the virus per plant visit. Numerical solutions of the differential equations were used to determine transitional and steady-state levels of disease incidence. Disease incidence was also determined directly from the model parameters. Clear differences were found in disease development among the four transmission classes. The highest disease incidence was in the SP and CP classes relative to the others, especially at low insect density when there was no insect migration or when the vector status of emigrating insects was the same as that of immigrating ones. The PP and CP viruses were most affected by changes in vector longevity, rates of acquisition and inoculation of the virus by vectors, whereas the PP viruses were least affected by changes in insect mobility. When vector migration was considered, results depended on the fraction of infective insects in the immigration pool and the fraction of dying and emigrating vectors replaced by immigrants. The PP and CP viruses were most sensitive to changes in these factors. Based on model parameters, R_0 was derived for some circumstances and used to determine the steady-state level of disease incidence and an approximate exponential rate of disease increase early in the epidemic. This model is now being used to evaluate disease management strategies of various types.

Host Plant Infection by Helper-dependent Virus Complexes

Interactions between viruses in plants are common, and some viruses depend on such interactions for their survival. Frequently, a virus lacks some essential molecular function that another provides. A variety of non-circulatory transmitted viruses have evolved a vector transmission strategy that involves, in addition to virions, virus-encoded proteins that are not constituents of virions. These 'helpers' and the genes encoding them have been characterized for viruses in the genera Potyvirus and Caulimovirus. Several lines of evidence support the hypothesis that these helpers act by mediating retention of virions in regions of the vector's alimentary tract from which they subsequently can be egested to initiate an infection (Pirone and Blanc, 1996).

In 'helper-dependent' virus complexes, the helper virus is transmitted independently by a vector, whereas the dependent virus depends on molecular agents associated with the helper virus for transmission by a vector. A general mathematical model has been developed of the dynamics of host plant infection by a helper-dependent virus complex (Zhang *et al.*, 2000). Four categories of host plants were considered: healthy, infected with helper virus only, infected with dependent virus only and infected with both viruses. New planting of the host crop was constrained by a maximum abundance due to limitation of the cropping area. The ratio of infection rate to host loss rate due to infection is proposed as an important epidemiological quantity and was used as a measure of the mutual adaptation of the virus and host. A simple analysis of the distribution of the final equilibria illustrated that the dependent virus could affect the survival of the helper virus, so facilitation between the two can be reciprocal. The distribution of the final equilibria also indicate that a well-adapted helper virus increases the opportunity for a dependent virus to evolve and survive. The model, therefore, explains why infection with a helper virus usually causes no or little damage to plants, whereas infection with a dependent virus or mixed infection with both often causes very severe damage.

Vector Feeding Period Variability in Epidemiological Models of Persistent Plant Viruses

The probability of virus inoculation increases with the period of exposure of the host to the vector. In mathematical models of plant virus disease epidemics it is frequently assumed that virus transmission is a simple bilinear process, i.e. is proportional to the abundance of hosts, vectors and a constant 'contact rate' parameter. Thus, no account is taken of any minimum feeding period required for virus transmission or of how the feeding period duration affects the probability of transmission. A theoretical model has been developed to evaluate these effects (Grilli and Holt, 2000). The results of numerical simulation with two models, conventional and with variable feeding period, were compared. The conventional model was adequate when the mean feeding period by a vector on a plant (T) was greater than or equal to the average feeding period required for one inoculation event to occur (α). Particularly in pathosystems where the vectors are relatively inefficient virus transmitters the situation $T < \alpha$ can occur, leading to underestimation or overestimation of the inoculation rate when variability is ignored. Genetic changes in host or vector, e.g. associated with a new host plant variety, which result in an increase in the variability of the vector feeding period could give rise to unexpected changes in disease dynamics.

Examples of this are the cosmopolitan whitefly species, *Bemisia tabaci* and *Trialeurodes vaporariorum*. These have always been regarded as pests to a large range of worldwide crops. Both species are capable of transmitting plant viruses in a persistent manner, with *T. vaporariorum* being the vector of only a few 'clostero'-like viruses and *B. tabaci*, the vector of viruses in several groups. The largest group of viruses transmitted by *B. tabaci* are the Begomoviruses and *B. tabaci* is known to transmit around 60 members. Until recently, *B. tabaci* had been associated with only a limited range of host plants within any one region, although its total potential host range was large. Virus transmission was confined within the plant host range of each regional population of *B. tabaci*. The emergence of the polyphagous 'B' biotype of *B. tabaci*, with its increased host range of more than 600 plant species, has resulted in Begomoviruses infecting previously unaffected crops. As the 'B' biotype spreads further into Europe, European field and glasshouse crops have been shown to be susceptible to whitefly-transmitted viruses already endemic to other parts of the world. More than 20 colonies of *B. tabaci*, including both 'B' and non-'B' biotypes from disparate global locations have been compared for their ability to transmit more than 20 Begomoviruses (Markham *et al.*, 1994). All but two highly host-specific colonies were capable of transmitting most Begomoviruses tested. However, some viruses were transmitted more efficiently than others. The virus coat protein or capsid is essential for vector recognition and transmission. By comparing transmissible viruses at the molecular level to viruses that are no longer whitefly-transmissible, the active epitope on the virus coat protein could be identified for designing future virus control strategies (Markham *et al.*, 1994).

Competition Between Viruses in a Complex Plant–Pathogen Interaction

Interactions among viruses, vectors and host plants may influence the spread and success of plant viruses. Major factors include direct competition within host plants, direct competition within vectors, differences in transmission rates, and virus influences on vector behaviour and population dynamics. The aphid-transmitted barley yellow dwarf luteoviruses (BYDVs), which infect a broad range of grasses worldwide, represent a model system for addressing questions about the outcome of direct and indirect competition between viruses. Historical shifts in the relative prevalence of BYDV strains document the apparent displacement of one virus strain (PAV) by another (MAV) over 20 years (Power, 1996). In the BYDV system, transmission rate appears to play an important role in determining the outcome of competition between viruses. Moreover, the interaction between transmission rate and vector behaviour may be particularly important. PAV is the stronger competitor within hosts, where double infections occur more often than in insect vectors. PAV also has significant advantages due to higher overall transmission rates than MAV. In addition, vector aphids show a strong non-preference for PAV that may lead to greater rates of virus spread.

Interactions between BYDV and the aphid parasitoid, *Aphidius ervi*, have been investigated while sharing the vector/host, *Sitobion avenae* (Christiansen-Weniger *et al.*, 1998). Aphids which were parasitized during their second larval stage had access to virus-infected plants before, immediately after or several days after parasitoid attack. The larval development of *A. ervi* in *S. avenae* was significantly delayed when virus acquisition took place before or shortly after the parasitoid had hatched, but not when the parasitoid was at the second larval stage during virus acquisition. Similarly, the presence of BYDV led to a significantly higher aphid mortality when they acquired virus up to and including the time that *A. ervi* was at the first larval stage. Adult female parasitoids deposited fewer eggs in viruliferous aphids. Virus transmission was not reduced by parasitization, and aphids which were subjected to parasitoid attack transmitted BYDV more efficiently than unattacked insects.

Virus Spread and Vector Dynamics in Genetically Diverse Plant Populations

Little is known about the influence of genetic diversity in plant populations on the dynamics of plant viruses, particularly those transmitted by insects. For these viruses, plant genetic diversity may affect virus incidence through impacts on the population dynamics of the vector insects or through impacts on vector feeding behaviour, which determines transmission of the virus. Power (1991) explored the influence of plant genetic diversity on virus dispersal by aphid vectors and examined the biological mechanisms responsible for that influence. In a set of field experiments using the aphid-transmitted BYDV, the influence of genetic diversity in oat (*Avena sativa*) populations on the spread of the virus and on the population dynamics and movement behaviour of aphid vectors of the virus was examined. Only at relatively high aphid abundance were the densities of aphid vectors influenced by plant genetic diversity. In one year out of three, densities of the oat-bird cherry aphid, *Rhopalosiphum padi*, were significantly lower in the genetically diverse stand than in the genetically homogeneous stands. In no year were densities of the English grain aphid, *Sitobion avenae*, influenced by the host-plant population. Despite these weak or absent effects on vector abundance, the incidence of the virus was consistently lower in the genetically diverse oat populations.

Disease reduction in the diverse populations appears to depend upon changes in aphid movement behaviour that affect the efficiency of virus transmission. Mark–release experiments with *S. avenae* demonstrated that movement rates were significantly higher and plant tenure times were significantly lower in the genetically diverse oat populations. Because the BYDV requires several hours of aphid feeding for effective transmission, these reduced tenure times and increased travel time among plants led to a reduction of virus transmission. While plant genotype can clearly influence herbivorous insects dramatically, Power (1991) suggests that the effects on insects of genetic diversity in the host plant population are likely to be subtle and not easily detected using standard field sampling techniques, except at high insect densities. Yet even at low vector densities, behavioural responses to plant genetic diversity can lead to significant effects on the spread of pathogens.

Non-circulative Transmission of Plant Viruses by Leaf-feeding Beetles

Beetles can acquire virus very quickly, even after a single bite of an infected plant, but efficiency of transmission increases with longer feeding. Transmission by beetles that inflict gross wounding is incompletely understood. However, there are several leaf beetles (*Chrysomelidae*) and some weevils (*Curculionidae*) that are important sole vectors of bromo-, sobemo- and tymoviruses. The viruses soon appear in the body fluid (haemolymph) after feeding upon the virus source, and their retention in the beetles is for days or weeks. The viruses that are not transmitted by beetles are now known to be subject to specific inactivation by an enzyme (RNase) in the regurgitant fluid produced by the beetle during feeding. Beetle transmission seems to be accompanied by virus introduction into and transport through xylem. The movement of two beetle-transmissible viruses, southern bean mosaic comovirus (SBMV) and bean pod mottle comovirus (BPMV) and two non-beetle-transmissible viruses cowpea strain of tobacco masaic tobamovirus (CP-TMV) and tobacco ringspot nepovirus (TRSV) into the haemocoel of chrysomelid and coccinellid beetle vectors after ingestion has been studied to understand the mechanisms involved. None of the viruses was detected in the haemolymph of the Mexican bean beetle (*Epilachna varivestis*), an efficient plant virus vector in the family *Coccinellidae*, regardless of the acquisition source, type of virus or method of virus detection. The infectivity of viruses was not destroyed by the haemolymph of the Mexican bean beetle, as demonstrated by virus survival in haemolymph for up to 3 days after virus injection into the haemocoel. Only one beetle-transmissible and one non-beetle-transmissible virus tested were found in the haemocoel of the bean leaf beetle (*Cerotoma trifurcata*) and the spotted cucumber beetle (*Diabrotica undecimpunctata howardi*), both members of the family *Chrysomelidae* (Wang *et al.*, 1992). These results indicate that virus movement into the beetle haemocoel is determined by the nature of the interaction between the individual virus and beetle, and that some plant viruses which are non-circulative, such as BPMV, can be efficiently transmitted by beetle vectors.

Transmission of Viruses by Plant Nematodes

Transmission by nematodes resembles semi-persistent transmission by arthropods. Nematode-borne viruses are retained in the vector on the lining of the guide-sheath of the odontostyle in *Longidorus*, or in the lumen of the odontophore and the oesophagus in *Xiphinema*, or in the entire pharynx and oesophagus in *Trichodorus*. The virus particles are released during subsequent feeding at another site, for instance on uninfected plants. Transmission specificity, even within local populations of a single nematode species, is attributed to specific adsorption determined by properties of the viral coat protein since it is often correlated with serological specificity. There is no latency. Virus may be retained in these vectors for several months, especially at low temperature, but it is lost at moulting. Some nematodes have life cycles of about 2 years. It has been claimed that non-vector trichodorids that superficially wound root tissues may assist tobacco mosaic tobamovirus, and perhaps some other viruses such as tomato bushy stunt tombusvirus that persist vectorless in the soil, to infect plant roots. Many nematodes infest various crops and many wild plant species including weeds, as do many of the viruses they transmit. Eighteen species in the plant-parasitic nematode genera *Longidorus*, *Paralongidorus* and *Xiphinema* are vectors of 12 nepoviruses and in the genera *Paratrichodorus* and *Trichodorus* 13 species are vectors of all three tobraviruses. A characteristic of these vector nematode and virus associations is that serologically distinct nepovirus and virus strains are transmitted by different, but related, longidorid species. A specific association exists between vectors and their viruses which is a consequence of the nature, site and mechanism of virus particle retention within the vector. It is correlated with the serological properties of the virus coat protein and determined by the RNA-2 segment of the virus genome and by an inherited character of the vector. The virus coat protein is probably involved in the recognition process between vector and virus but is not the sole determinant for transmission of tobraviruses. Genetic changes made to proteins present in the RNA-2 segment of pea early browning tobravirus have been used to reveal the probable involvement of several proteins in vector transmission (Brown *et al.*, 1995). 'Protruding' C-terminal amino acid sequences of tobraviruses possibly link, with the aid of a viral determined helper factor, to the site of retention. The viruses are referred to as having 'specific' vector species, and the terminology has been adopted by researchers who refer to 'specificity' of transmission of viruses by vector nematodes. Further research has confirmed that specificity of transmission extends to populations of vector species and to minor serological variants of nepoviruses. It has also been shown to extend to tobraviruses and their vector species (Brown and Weischer, 1998).

Fungal Transmission of Plant Viruses

Fungi that transmit viruses in the soil were formerly called algal or lower fungi. Their taxonomic position and relationships remain a matter of debate. Only plasmodiophorids (class *Plasmodiophoromycetes*) and chytrids (class *Chytridiomycetes*) are known as virus vectors. They parasitize the roots of their plant host with a microscopic undifferentiated thallus or plasmodium, often a naked protoplast, inside the epidermal cells. Zoospores, which are biflagellate for the *Plasmodiophoromycetes* and uniflagellate for the *Chytridiomycetes*, are released from zoosporangia via an exit tube into the water around the root. These motile spores reinfect the same root closer to its developing tip, or move to other roots of adjacent plants. They externally attach, encyst while withdrawing their flagellae, and release their contents into epidermal cells or root hairs via an infection canal. The resting spores are thick-walled. How they are formed remains uncertain, but a process of zoospore conjugation may be involved. The resting spores occur either singly (*Olpidium* spp.) or in clusters (*Cystosori*: *Polymyxa* and *Spongospora* spp.). They remain in the soil when roots decay, to germinate and produce zoospores when water becomes available. The zoospores need water for movement (the fungi are aquatic) and are short-lived. In contrast, the resting spores can resist desiccation and may perennate in dry soils for many years. They can also be blown away in dry soil and be transported on tools, vehicles and planting materials, and in flowing water. Except for *Plasmodiophora brassicae*, causing clubroot of brassicas, and *Spongospora subterranea*, causing powdery scab on potato, very little is known about pathogenicity of the virus-transmitting zoosporic fungi. Thirty soil-borne viruses or virus-like agents are transmitted by five species of fungal vectors. Ten polyhedral viruses, of which nine are in the family *Tombusviridae*, are acquired in an *in vitro* manner and do not occur within the resting spores of their vectors, *Olpidium brassicae* and *O. bornovanus*. Fungal vectors for other viruses in the family should be sought even though tombusviruses are reputed to be soil transmitted without a vector. Eighteen rod-shaped viruses belonging to the furo- and bymovirus groups and to an unclassified group are acquired in the *in vivo* manner and survive within the resting spores of their vector, *O. brassicae*, *Polymyxa graminis*, *P. betae* and *Spongospora subterranea*. Non-persistent transmission by *Olpidium* spp. has been little studied in the last 20 years, but appears to depend on adsorption of virus to the outside of the fungal zoospores. The viral coat protein has an essential role in *in vitro* transmission. With *in vivo* transmission a site in the coat protein–read through protein (CP-RT) of beet necrotic yellow vein furovirus determines vector transmissibility as does a site in a similar 98-kDa polyprotein of barley mild mosaic bymovirus (Campbell, 1996). Such viruses are not transmitted in the fungal resting spores. The route by which the virus is transferred from the vector to the host may involve uptake into the zoospores and this merits further study.

Persistent transmission by *Olpidium*, *Polymyxa* and *Spongospora* spp. is less well characterized and some of the evidence used in support of this is inconclusive. The viruses are always carried inside zoospores, and they also persist in the fungal resting spores. Transmission depends on the genome of the vector and the virus, but not exclusively on the virus coat protein (Adams, 1991).

Concluding Remarks

In this chapter some common themes have emerged about the ways in which virus–vector interactions influence disease epidemics. Models that have been developed to describe the effects of virus–vector transmission on disease epidemics show clear differences in disease development among the four transmission classes for arthropod-transmitted viruses. For example, roguing was shown only to be effective for NP viruses, and only at low vector population densities, therefore indicating a clear advantage in reducing the vector population for control of viruses transmitted in this manner. In the case of viruses that are transmitted with the aid of 'helpers', the dependent virus was shown to affect survival of the helper, therefore a well-adapted helper is necessary for the survival and evolution of the dependent virus.

Transmission rate was shown to be important in the outcome of competition between viruses in the BYDV complex. BYDV transmission was not reduced by parasitization of the vector, but actually transmission rate increased in parasitized aphids. Research on the role of genetic diversity on disease epidemics shows an impact through vector feeding behaviour, but only at relatively high aphid populations. For example, the incidence of BYDV was lower in genetically diverse oat populations.

Concepts of transmission classes for viruses transmitted by beetles, nematodes and fungi are less well characterized and therefore no epidemiological models have yet been developed, but in the future such models could have an impact on disease management of these viruses.

References

Adams, M.J. (1991) Transmission of plant viruses by fungi. *Annals of Applied Biology* 118, 479–492.

Brown, D.J.F. and Weischer, B. (1998) Specificity, exclusivity and complementarity in the transmission of plant viruses by plant parasitic nematodes: an annotated terminology. *Fundamental and Applied Nematology* 21, 1–11.

Brown, D.J.F., Robertson, W.M. and Trudgill, D.L. (1995) Transmission of viruses by plant nematodes. *Annual Review of Phytopathology* 33, 223–249.

Campbell, R.N. (1996) Fungal transmission of plant viruses. *Annual Review of Phytopathology* 34, 87–108.

Christiansen-Weniger, P., Powell, C. and Hardie, J. (1998) Plant virus and parasitoid interactions in a shared insect vector/host. *Entomologia Experimentalis et Applicata* 8, 205–213.

Gray, S.M. and Banerjee, N. (1999) Mechanisms of arthropod transmission of plant and animal viruses. *Microbiology and Molecular Biology Reviews* 63, 128.

Grilli, M.P. and Holt, J. (2000) Vector feeding period variability in epidemiological models of persistent plant viruses. *Ecological Modelling* 126, 49–57.

Jeger, M.J., Van Den Bosch, F., Madden, L.V. and Holt, J. (1998) A model for analysing plant–virus transmission characteristics and epidemic development. *IMA Journal of Mathematics Applied in Medicine and Biology* 15, 1–18.

Madden, L.V., Jeger, M.J. and van den Bosch, F. (2000) A theoretical assessment of the affects of vector–virus transmission mechanism on plant virus disease epidemics. *Phytopathology* 90, 576–594.

Markham, P.G., Bedford, I.D., Liu, S.J. and Pinner, M.S. (1994) The transmission of geminiviruses by *Bemisia tabaci*. *Pesticide Science* 42, 123–128.

Nault, L.R. (1997) Arthropod transmission of plant viruses: a new synthesis. *Annals of the Entomological Society of America* 90, 521–541.

Pirone, T.P. and Blanc, S. (1996) Helper-dependent vector transmission of plant viruses. *Annual Review of Phytopathology* 34, 227–247.

Power, A.G. (1991) Virus spread and vector dynamics in genetically diverse plant-populations. *Ecology* 72, 232–241.

Power, A.G. (1996) Competition between viruses in a complex plant-pathogen. *Ecology* 77, 1004–1010.

Wang, R.Y., Gergerich, R.C. and Kim, K.S. (1992) Noncirculative transmission of plant-viruses by leaf-feeding beetles. *Phytopathology* 82, 946–950.

Zhang, X.S., Holt, J. and Colvin, J. (2000) Mathematical models of host plant infection by helper-dependent virus complexes: why are helper viruses always avirulent? *Phytopathology* 90, 85–93.

Functional Consequences and Maintenance of Vegetative Incompatibility in Fungal Populations

3

Rolf F. Hoekstra

Laboratory of Genetics, Department of Plant Sciences, Wageningen University, Dreijenlaan 2, NL-6703 HA, Wageningen, The Netherlands

Introduction

In a mycelial fungus, a growing hyphal tip may come into close contact with another hypha and this may result in a local hyphal fusion (anastomosis). As a consequence, a mycelium is a hyphal network rather than a bundle of branching but unconnected hyphae. When different conspecific mycelia meet each other, their hyphae may likewise come into physical contact. Sometimes intermycelial anastomosis follows and if this occurs at a sufficiently large scale along their common border and the established connections are stably maintained, a genetically chimaeric mycelium results containing a mixture of the parental nuclei (hence the name *heterokaryon*) and cytoplasms (*heteroplasmon*). However, whether or not a confrontation between two conspecific individuals results in a heterokaryon depends on their genotype at a number of so-called *het*-loci. Only allelic identity at all *het*-loci allows heterokaryon formation. Strains obeying this rule are said to be vegetatively compatible. Since the number of segregating *het*-loci (or *vic* loci, for vegetative incompatibility) in fungal populations often appears to be in the order of ten with generally two alleles per locus (Croft and Jinks, 1977; Cortesi and Milgroom, 1988; Perkins and Turner, 1988), this criterion for heterokaryon formation will in practice restrict heterokaryosis mainly to clonally related individuals or to close kin. Thus, vegetative incompatibility may in a loose sense be termed a self/non-self recognition system. In a confrontation between vegetatively incompatible strains the hyphal cells in the contact zone are destroyed. As a consequence heterokaryosis is prevented, but the incompatibility response does not always completely prevent heteroplasmy. Mitochondrial plasmids and dsRNA viruses

 Biotic Interactions in Plant–Pathogen Associations (eds M.J. Jeger and N.J. Spence)

have been shown to be able to cross the incompatibility barrier, although with very low probability (Debets *et al.*, 1994).

This chapter reviews general functional consequences of vegetative (in)compatibility in fungal populations and considers the question of to what extent these functional aspects may be responsible for the selective maintenance of the observed high levels of incompatibility. Special attention will be paid to the role of vegetative (in)compatibility in pathogenic fungi, especially in relation to biological control.

Vegetative Incompatibility as a Defence Against Deleterious Genetic Infection

Active maintenance by balancing selection of the tri-allelic polymorphism at the *het*-c locus in *Neurospora* and related genera has been suggested by Wu *et al.* (1998) on the basis of a molecular genetic analysis of these alleles. They concluded that this polymorphism is quite ancient and apparently stable, predating several speciation events. In this respect it resembles ancient polymorphisms in the vertebrate major histocompatibility complex (MHC). Their study, however, does not permit an elucidation of the putative selective forces involved. It has been suggested that a high level of vegetative incompatibility may have been selected to prevent infection by deleterious genetic elements (Caten, 1972; Hartl *et al.*, 1975). This would conceivably lead to balancing frequency-dependent selection, since common vegetative incompatibility types – because of a higher rate of compatible interactions – would suffer more from infection by deleterious elements than rare types. Suppressive nuclear (Pittenger and Brawner, 1961) and cytoplasmic (Griffiths *et al.*, 1990) factors are known and mechanisms preventing contamination with such elements should be selectively advantageous. Hartl *et al.* (1975) showed in a population genetic model analysis that under plausible conditions two alleles at a *het*-locus can be stably maintained in a fungal population when a deleterious suppressive nuclear factor is present in the population. Nauta and Hoekstra (1994) extended the analysis to a more general model allowing for more *het*-loci and not only considering nuclear but also cytoplasmic factors. They concluded that selection to prevent deleterious genetic infection is unlikely to be sufficient to explain the observed high numbers of vegetative incompatibility types. Selection for maintaining more than a few different types is too weak to be effective, mainly for the following reason. The maintenance (and spread) of a deleterious parasitic genetic element critically depends on the frequency of compatible interactions: only then anastomosis (followed by infection) will be sufficiently common to offset the selective elimination of this element due to the fitness loss it inflicts on its host. When the number of VC (vegetative compatibility) groups increases, the frequency of compatible interactions reduces, and quickly reaches the threshold below which a deleterious suppressive element can no longer be maintained. However, the model of Nauta and Hoekstra (1994) assumes a homogeneous and well-mixed population. It remains to be seen if their conclusions hold up in a (presumably more realistic) spatially structured population model. In comparing selection against a nuclear suppressive gene and against a cytoplasmic factor, Nauta and Hoekstra (1994) found selection by a cytoplasmic factor to be a more plausible explanation for the evolution of vegetative incompatibility than selection against a nuclear gene.

Milgroom and Cortesi (1999) investigated the genetic population structure of the chestnut blight fungus *Cryphonectria parasitica* with respect to vegetative incompatibility genotypes and looked for evidence in support of frequency-dependent balancing selection on the *het*-loci. The hypothesis of frequency-dependent selection was motivated by the occurrence of dsRNA viruses ('hypoviruses') in *Cryphonectria* which cause a reduction of virulence in this pathogenic fungus. Since virus transmission is severely restricted between incompatible genotypes, a rare incompatibility type might well be favoured by escaping virus infection. However, their data failed to support the frequency-dependent selection hypothesis.

Vegetative Incompatibility as a Protection of Genetic Identity

A fungal colony optimally adapted to its environment could be harmed by fusion with a less well-adapted colony. On the other hand, the latter colony would gain from a fusion by sapping resources from its more vigorous partner. It is thus conceivable that the preservation of a 'genetic identity' would provide a selection pressure contributing to the evolution and maintenance of vegetative incompatibility. A suggestion of this nature has been made by Todd and Rayner (1980). However, whether strains should in general avoid somatic fusion for this reason seems unclear on the basis of purely verbal arguments. It obviously will depend on how much on average a poorly adapted strain will gain from fusing with a more successful neighbour and how much on average the latter will lose from such a fusion. This problem has been quantitatively analysed by De Boer (1995). He considers models in which strains are characterized by an ecological preference and by a compatibility specificity. These strains grow in different ecological environments, where individuals of the same compatibility type fuse. The resulting colonies can be chimaeric with respect to the ecological preference of the strains that make up the colony. The growth of each strain depends on its match between preference and environment and on the total reproduction of the colony of which it is a member. The reproduction of the colony depends on its composition, i.e. on the average match between the preferences of its component strains and the environment. The theoretical analysis shows that the population evolves towards unique associations between the ecological preferences and compatibility specificities (i.e. a state in which strains do not fuse except with their own compatibility type) when the total reproduction in good conditions –

in which fusion would reduce fitness – is larger than reproduction in worse conditions – in which fusion would enhance fitness. Although this interesting model has been designed to explain slime mould aggregation and interactions between colonial tunicates, it is also applicable to asexual fungi. However, I doubt if the conclusions would survive the incorporation of explicit sexual genetics into the model. Sexual recombination would disrupt associations between genes coding for ecological preference and those coding for compatibility, making it hard to maintain specific associations between ecological preference and compatibility type.

Vegetative Incompatibility as a Defence Against Resource Plundering in the Sexual Phase

Debets and Griffiths (1998) have called attention to neglected aspects of the role of vegetative incompatibility in sexual crosses. The high frequency of polymorphism for vegetative incompatibility in fungal populations implies that generally the parents in a cross will be vegetatively incompatible. Whitehouse (1949) already noted that in fungi in which sex organ differentiation is required to complete the sexual cycle, vegetative heterokaryosis is restricted to nuclei of the same mating type. In these species (like *Neurospora crassa*) sexual crosses are only possible between parents of different mating type, who therefore are necessarily vegetatively incompatible. Particularly in heterothallic situations, when selfing is not possible, an established maternal protoperithecial colony has to await fertilization by incoming conidia from a culture of opposite mating type. Pheromone-induced inhibition of germination of conidia of opposite mating type that are close to protoperithecia (Bistis, 1981) helps the maternal culture to control the fertilization process by only allowing fusion between its trichogynes and the conidia. Because of the mating type-associated vegetative incompatibility, such conidia will be unable to fuse with the maternal somatic tissue. Paternal nuclei that have entered the protoperithecium during outcrossing cannot migrate into the vegetative hyphae (Dodge, 1935). Thus, several adaptations make use of the vegetative incompatibility mechanism to prevent incoming paternal conidia invading the maternal mycelium, presumably because this would allow them to plunder the maternal resources. In elegant experiments using the mutation a^{m33}, which is defective in the mating type-associated vegetative incompatibility, Debets and Griffiths (1998) showed that in the absence of vegetative incompatibility between the fertilizing conidia and the maternal culture, incoming conidia did actually fuse with the maternal tissue, and managed to obtain access to the maternal resources and initiate new fruiting bodies. It was even shown that when a mixture of conidia with different genetic markers (including the mutant a^{m33}) and of both mating types was put on a protoperithecial colony, some fruiting bodies actually produced ascospores that appeared to be offspring from two *conidial* parents, i.e. with no genes at all from the maternal colony that provides the resources for the production of the fruiting bodies and the ascospores. This is clearly a case of genetic parasitism, somewhat reminiscent of the germline parasitism (non-self cells entering the germline tissue and using this sexual route for their genetic propagation), found in fused colonies of the colonial invertebrate *Botryllus schlosseri* (Stoner and Weissman, 1996), which reinforces the interpretation of the self/non-self recognition system as a device to limit somatic cell and germ cell parasitism to closely related kin.

Consequences of Vegetative Incompatibility in Pathogenic Fungi

Vegetative incompatibility creates effective barriers within fungal populations for the horizontal transmission of genetic material. The barrier between two incompatible individuals is absolute for nuclear genes and not quite absolute (but still fairly effective) for cytoplasmic elements like dsRNA (Anagnostakis and Day, 1979) and mitochondrial plasmids (Debets *et al.*, 1994). A few cases are known where cytoplasmic RNA viruses debilitate their plant pathogenic fungal host. For example, in the chestnut blight fungus, *Cryphonectria parasitica*, a dsRNA virus is known to reduce the fungal virulence (Day *et al.*, 1977). Similarly, in the Dutch elm disease fungus, *Ophiostoma novo-ulmi*, mitochondrial virus-like RNA is present, probably associated with hypovirulence (Hong *et al.*, 1998). Thus, the endosymbiotic viruses can be viewed as 'hyperparasites' (Taylor *et al.*, 1998), their fungal host being a plant parasite. Because the viral hyperparasite weakens the plant parasitic fungus, it could be viewed as a mutualist of the host plant. However, it is also theoretically possible that the hyperparasite moves the transmission rate of the fungus closer to its optimum value (Michalakis *et al.*, 1992), in which case the hyperparasite is a mutualist of the fungal pathogen. Taylor *et al.* (1998) have analysed a model of a host pathogen system with three components: (plant) host, fungal pathogen and (viral) hyperparasite. They conclude that conditions for spread of a hyperparasite which reduces the pathogen virulence mainly depend on the functional relationship between the pathogen's rate of transmission and its virulence and on the horizontal transmission rate of the hyperparasite. In the context of a fungal pathogen infected by a hypovirus, the horizontal transmission of the virus will be strongly influenced by the extent of vegetative incompatibility in the fungal population. In general, a high rate of horizontal transmission of the (hyper-)parasite is expected to select for a high level of virulence, therefore for a strong debilitating effect on its fungal host (Andersen and May, 1982; Ewald, 1983; Frank, 1996). Thus, a low level of vegetative incompatibility in populations of a fungal pathogen will, in general, select for virulent viruses and thus contribute to reduction of the fungal virulence, while a high level of incompatibility will be much less effective in reducing the virulence of the fungal pathogen. These considerations imply that a high diversity of vegetative incompatibility in fungal populations will hamper biological control efforts of fungal disease by introducing hypoviruses. This conclusion is supported by the observation that in Europe the level of vegetative incompatibility in the chestnut blight fungus is lower than in North America, while the disease is milder in Europe than in North America (Anagnostakis *et al.*, 1986; Cortesi *et al.*, 1996).

Another functional aspect of vegetative incompatibility in pathogenic fungi may be related to within-host competition. With a high level of vegetative incompatibility it is likely that in the case of multiple infections, the different fungal genotypes present in a host plant will be incompatible. This will affect their mutual competition within the host for space and resources. At least part of their energy will be used up in these competitive interactions. Also, depending on the relationship between fungal virulence and the size of the fungal mycelium, a combination of several mutually incompatible fungal strains may well be less virulent than a single strain that can occupy its host plant uninhibited by other fungal competitors. I am not aware of relevant empirical data on this issue, but experimentation to test these ideas would be of interest.

Conclusions

Vegetative incompatibility is so widespread in natural populations of fungi that it effectively can be considered as a self/non-self recognition system which limits somatic fusions by anastomosis to clonally related individuals and close kin. Since the 1970s several relevant functional aspects of incompatibility have been proposed and discussed in the scientific literature, notably defence against parasitic genetic infections, protection of individual integrity to prevent introgression of less well-adapted genes, and prevention of parasitism of maternal resources during the sexual phase by paternal conidia. Although some quantitative model analysis has been applied to investigate to what extent these functions could generate sufficient selective force to maintain high levels of vegetative incompatibility, there is as yet no fully satisfactory quantitative explanation of the observed high diversity of incompatibility types. There is a need for more precise modelling as well as for experimental measurements of relevant parameters like rates of horizontal transfer of cytoplasmic elements, fitness reduction as a consequence of ecological competition between colonies, and extent of somatic anastomosis during the sexual interactions. A highly interesting question of practical importance is to what extent vegetative incompatibility interferes with biological control of pathogenic fungi using mycoviruses that reduce fungal virulence. Although several interesting relevant studies have appeared recently, a satisfactory answer to this question also requires more population genetic and epidemiological work, both empirical and theoretical.

References

Anagnostakis, S.L. and Day, P.R. (1979) Hypovirulence conversion in *Endothia parasitica*. *Phytopathology* 69, 1226–1229.

Anagnostakis, S.L., Hau, B. and Kranz, J. (1986) Diversity of vegetative compatibility groups of *Cryphonectria parasitica* in Connecticut and Europe. *Plant Disease* 70, 536–538.

Andersen, R.M. and May, R.M. (1982) Coevolution of hosts and parasites. *Parasitology* 85, 411–426.

Bistis, G.N. (1981) Chemotropic interactions between trichogynes and conidia of opposite mating type in *Neurospora crassa*. *Mycologia* 73, 959–975.

Caten, C.E. (1972) Vegetative incompatibility and cytoplasmic infection in fungi. *Journal of General Microbiology* 72, 221–229.

Cortesi, P. and Milgroom, M.G. (1988) Genetics of vegetative incompatibility in *Cryphonectria parasitica*. *Applied and Environmental Microbiology* 64, 2988–2994.

Cortesi, P., Milgroom, M.G. and Bisiach, M. (1996) Distribution and diversity of vegetative compatibility types in subpopulations of *Cryphonectria parasitica* in Italy. *Mycological Research* 100, 1087–1093.

Croft, J.H. and Jinks, J.L. (1977) Aspects of the population genetics of *Aspergillus nidulans*. In: Smith, J.E. and Pateman, J.A. (eds) *Genetics and Physiology of Aspergillus*. Academic Press, London, pp. 339–360.

Day, P.R., Dodds, J.A., Elliston, J.E., Jaynes, R.A. and Anagnostakis, S.L. (1977) Double-stranded RNA in *Endothia parasitica*. *Phytopathology* 67, 1393–1396.

Debets, A.J.M. and Griffiths, A.J.F. (1998) Polymorphism of *het*-genes prevents resource plundering in *Neurospora crassa*. *Mycological Research* 102, 1343–1349.

Debets, F., Yang, X. and Griffiths, A.J.F. (1994) Vegetative incompatibility in *Neurospora*: its effect on horizontal transfer of mitochondrial plasmids and senescence in natural populations. *Current Genetics* 26, 113–119.

De Boer, R.J. (1995) The evolution of polymorphic compatibility molecules. *Molecular Biology and Evolution* 12, 494–502.

Dodge, B.O. (1935) The mechanics of sexual reproduction in *Neurospora*. *Mycologia* 27, 418–438.

Ewald, P.W. (1983) Host–parasite relations, vectors and the evolution of disease severity. *Annual Review of Ecology and Systematics* 14, 465–485.

Frank, S. (1996) Models of parasite virulence. *Quarterly Review of Biology* 71, 37–78.

Griffiths, A.J.F., Kraus, S.R., Barton, R., Court, D.A., Myers, C.J. and Bertrand, H. (1990) Heterokaryotic transmission of senescence plasmid DNA in *Neurospora*. *Current Genetics* 17, 139–145.

Hartl, D., Dempster, E.R. and Brown, S.W. (1975) Adaptive significance of vegetative incompatibility in *Neurospora crassa*. *Genetics* 81, 553–569.

Hong, Y., Cole, T.E., Brasier, C.M. and Buck, K.W. (1998) Evolutionary relationships among putative RNA-dependent RNA polymerases encoded by a mitochondrial virus-like RNA in the Dutch elm disease fungus, *Ophiostoma novo-ulmi*, by other viruses and virus-like RNAs, and by the *Arabidopsis* mitochondrial genome. *Virology* 246, 158–169.

Michalakis, Y., Olivieri, I., Renaud, F. and Raymond, M. (1992) Pleiotropic action of parasites: how to be good for the host. *Trends in Ecology and Evolution* 7, 59–62.

Milgroom, M.G. and Cortesi, P. (1999) Analysis of population structure of the chestnut blight fungus based on vegetative incompatibility genotypes. *Proceedings of the National Academy of Sciences USA* 96, 10518–10523.

Nauta, M. and Hoekstra, R.F. (1994) Evolution of vegetative incompatibility in filamentous ascomycetes. I. Deterministic models. *Evolution* 48, 979–995.

Perkins, D.D. and Turner, B.C. (1988) Neurospora from natural populations: toward the population biology of a haploid eukaryote. *Experimental Mycology* 12, 91–131.

Pittenger, T.H. and Brawner, T.G. (1961) Genetic control of nuclear selection in *Neurospora* heterokaryons. *Genetics* 46, 1645–1663.

Stoner, D.S. and Weissman, J.L. (1996) Somatic and germ cell parasitism in a colonial ascidian: possible role for a highly polymorphic allorecognition system. *Proceedings of the National Academy of Sciences USA* 93, 15254–15259.

Taylor, D.R., Jarosz, A.M., Lenski, R.E. and Fullbright, D.W. (1998) The acquisition of hypovirulence in host–pathogen systems with three trophic levels. *American Naturalist* 151, 343–355.

Todd, T.K. and Rayner, A.D.M. (1980) Fungal individualism. *Science Progress (Oxford)* 66, 331–354.

Whitehouse, H.L.K. (1949) Heterothallism and sex in fungi. *Biological Reviews* 24, 411–447.

Wu, J., Saupe, S.J. and Glass, N.L. (1998) Evidence for balancing selection at the *het*-c heterokaryon incompatibility locus in a group of filamentous fungi. *Proceedings of the National Academy of Sciences USA* 95, 12398–12403.

Fungal Endophytes and Nematodes of Agricultural and Amenity Grasses

4

Roger Cook[1] and Graham C. Lewis[2]

[1]*Institute of Grassland and Environmental Research, Aberystwyth, Ceredigion SY23 3EB, UK;* [2]*Institute of Grassland and Environmental Research, North Wyke, Okehampton, Devon EX20 2SB, UK*

Introduction

Many fungi live as endophytes in plants, representing a continuum of relationships from antagonism to obligate mutualism (Saikkonen *et al.*, 1998). This continuum correlates with the way the fungus is transmitted. The mechanisms for this range from 'horizontal' spread via sexual spores to 'vertical' spread by clonal colonization of healthy seeds (Schardl, 1996). Fungal mutualists receive nutrition and protection from the host plant, which may benefit from the endophyte through improved competitiveness, and tolerance of biotic and abiotic stresses (Saikkonen *et al.*, 1998).

In grasses, one group of obligate mutualists has ecological and economic significance because of the impact of its secondary metabolites on herbivores. These endophytes infect leaves and stems of healthy plants but have no marked pathogenic effects. A range of insects, including sap-sucking aphids as well as biting herbivores, is affected by endophytes in tall fescue and perennial ryegrass (Breen, 1994). Endophyte-infected grasses can also cause toxicoses in grazing livestock. As well as protecting the plant from herbivory, these endophytes can increase plant yield, enhance root growth and modify water relations.

The fungal genotype determines both the type and quantity of secondary metabolites produced in endophyte-infected grasses. The symbiosis is affected by growing conditions, particularly by temperature. The host genotype may also directly affect the responses of herbivores to the phenotype presented by the fungus × grass interaction (Breen, 1994). The interactions of fungus and host genotype with the environment contribute to substantial variability in impact on herbivores.

Evidence for the effects of grass endophytes on root-feeding nematodes is equivocal, perhaps reflecting the variety of these interactions. None the less, there are striking examples of endophyte-infected grasses expressing very effective resistance to nematodes, which is not found in the endophyte-free host. These examples include nematodes of a number of crops in which natural genetic host resistance has proven elusive or difficult to manipulate. This chapter reviews relationships between grass endophytes of the genus *Neotyphodium*, and root parasitic and other nematodes associated with grasslands.

Grasses

The *Gramineae* is notable among plant families for the minimal incidence of its chemical defences against herbivory (Harborne, 1993). The few grasses that do produce toxins are C4 genera, including *Panicum*, which produce tryptamine and carboline alkaloids, and other tropical grasses that produce glycosides, oxalates and saponins, all of which may have toxic effects on grazing livestock (Cheeke, 1995).

Some grasses are so adapted to grazing by herbivores that their physiology is actually stimulated by regular cropping of their leaves. The grasses have other adaptations that favour survival in the ecological conditions between 'the forest and the fire', particularly basal meristems that allow re-generative growth from soil level and adventitious root systems that also can regenerate from the stem base. These adaptations also contribute to tolerance of above- and below-ground herbivory. It is considered that grasses rely on growth habit to survive defoliation and endophytic fungal toxins for defence (Cheeke, 1995).

These features account for the widespread occurrence and persistence of grazed natural grasslands, and the exploitation of grasses in managed and agricultural grasslands. Grasses are the essential basis for ruminant production systems whether by various grazing managements or by zero grazing of fresh or conserved grass. Grass is the bulk energy source for much of Europe's livestock production and is grown on 40% of the land area. In addition, the persistence of grasses under regular defoliation has made them attractive for amenity use, both for visual amenity as, for example, in lawns and road sides, as well as for providing surfaces suitable for a variety of sporting activities.

We review endophytes in grasses indigenous to temperate grasslands of Eurasia. The genus *Lolium* is native to Europe, temperate Asia and North Africa. *Festuca* species are native to temperate regions of the world, but tall fescue, *F. arundinacea*, is native to the same regions as *Lolium*. The more interesting phenomena involving endophytes are reported from countries where such grass species have been introduced, for example, from tall fescue in the USA and from perennial ryegrass, *L. perenne*, in New Zealand.

Grass Nematodes

Grazed grasslands have distinctive soil systems with abundant and diverse soil faunal and microbial communities, including soil-inhabiting nematodes (Bardgett and Cook, 1998). These microscopic worms are represented by families at all heterotrophic levels in soil food webs. Those that ingest decomposer organisms (bacterial and fungal feeders) contribute to enhanced nutrient cycling and this can have positive effects on plant growth. Other nematodes that feed on plant roots may have negative impacts upon plant production, but even these, through a series of interactions and feedback mechanisms, may contribute positively to grassland yields (Bardgett *et al.*, 1999; Yeates, 1999).

Realistic assessment of crop loss using treatments that kill nematodes is difficult to interpret since some nematodes do make positive contributions. None the less, damage to roots can be shown to reduce grassland productivity. Nematodes may either reduce plant growth rate and size, or debilitate the plant, reducing its tolerance and survival of other stresses (Cook and Yeates, 1993). Nematode feeding types have been categorized by Yeates *et al.* (1993). Root-feeding nematodes show considerable differences with respect to the site and mode of feeding as well as in the amount of damage they cause.

Most root feeding is by nematodes that pierce plant cells and withdraw cytoplasm. The response of the plant contributes to the impact that feeding may have on growth, and ranges from individual cell death, sometimes accompanied by a localized hypersensitive response, through various modifications of the cell that appear to allow it to survive nematode attack. In some cases, cells are modified so that the particular nematode species can feed for long periods at the same cell or group of modified cells (e.g. the root-knot and cyst-forming nematodes). These nematodes are highly adapted to parasitism, having lost the ability to migrate to new food sources, and compensating by increased rates of reproduction.

Nematodes recognized as pests of grasses and the situations in which damage may occur were reviewed by Cook and Yeates (1993) and, more recently, by Bernard *et al.* (1998). Briefly, some nematode populations reach densities such that penetration and feeding activities significantly damage root growth. This is exacerbated when grass is under other stress, including drought, or during re-growth after dormancy or re-seeding when many nematodes are available to attack the much reduced food resource.

Nematode responses to fungal endophytes in grasses will probably depend upon the particular way in which the nematode species feeds. The location of the nematode and its feeding sites can be used to classify root-feeding nematodes. Ectoparasites remain in the soil or on root surfaces, feeding either on outer cells (root hairs, epidermal cells, root caps) or inserting their stylet more deeply into roots, particularly in non-secondarily thickened roots. There are marked differences in the times for which morphologically similar nematode species may feed on individual cells or groups of cells. Endoparasites enter roots and feed on internal cells. There are intermediates in which partial penetration of roots allows nematodes to feed on some internal cells. The sedentary endoparasites include some of the more highly specialized parasites, inducing the development of modified plant cells as feeding sites for individual nematodes. Some ectoparasites that feed deep within the root tip also induce modifications to host cells.

The review of nematode–endophyte interactions is grouped according to the mode of parasitism of the nematode concerned (Tables 4.3–4.5).

Grass Endophytic Fungi

Systemic grass endophytes, until recently named *Acremonium* species, are now recognized as *Neotyphodium*. These host-specific fungi grow between host plant cells apparently without inducing defence responses and without symptoms. The sexual state, *Epichloë*, on occasion, causes the disease choke, that suppresses seed production by the host grass.

Distribution

In the field, *Neotyphodium* endophyte hyphae appear to be largely confined to leaf sheaths and stems, growing into flowers and the seed aleurone layer and relying on the grass seed for their dissemination. In embryos, hyphae have been seen in the root–shoot internode adjacent to vascular tissue. In practice, grass roots are generally considered to be free of endophyte hyphae although, *in vitro*, *N. coenophialum* has been isolated from the roots of tall fescue seedlings, particularly when infected seeds were germinated on agar. In sand, there were also substantial proportions of plants with up to half their root axes infected (Azevedo and Welty, 1995). The primary root axis was less often infected than were adventitious roots growing from the crown, and it was speculated that infection of new roots growing from the crown would help the continual colonization of the plant root system through the life of the plant. Using enzyme-linked immunosorbent assay (ELISA), Musgrave (1984) showed that 76–96% of endophyte mycelium was in the basal 3 cm of two perennial ryegrass cultivars in pastures. In glasshouse-grown plants, some 7% of total mycelial weight was below ground and in roots.

Endophyte infection is widespread in Europe. In a survey of 523 wild populations of *Lolium* spp. in 15 European countries (Lewis *et al.*, 1997), 38% had no infected plants, 48% had from 1 to 50% and 14% had from 51 to 100% infection. Other grass species in Europe may be infected by other endophyte species (e.g. Lewis, 1994; Eggestein *et al.*, 1996; Oliveira and Castro, 1997; Zabalgogeazcoa *et al.*, 1999; Table 4.1). In German perennial ryegrass pastures, incidence of *N. lolii* ranged from 1 to 30% of plants infected, with occasional populations with 80% infection (Oldenburg, 1997). In France, 70% of pastures had a low level of endophyte infection (Ravel *et al.*, 1997). In

Table 4.1. The main species of temperate grasses used for agriculture and amenity grass, their naturally occurring fungal endophyte species and the alkaloids produced. (After Siegel *et al.*, 1990; Bush *et al.*, 1997; Yue *et al.*, 1997.)

Grass species	Endophyte species	Alkaloids			
		Ergopeptines	Lolitrems	Peramine	Lolines
Perennial ryegrass	*Neotyphodium lolii*	Yes	Yes	Yes	No
Tall fescue	*N. coenophialum*	Yes	No	Yes	Yes
Meadow fescue	*N. uncinatum*	No	No	No	Yes
Fine fescues	*Epichlöe festucae*	Yes	Yes?	Yes	No?
Creeping bent	*Epichlöe* sp.	Yes	No	Yes	No?

Britain, infection is common, particularly in old pastures (Lewis, 1994). In natural fescue pasture in Arizona, USA, the number and type of alkaloids produced varied, and total alkaloid production was relatively limited in the wild. This diversity suggests that high alkaloid production in introduced species may not be representative of all grassland (Saikkonen *et al.*, 1998).

Secondary metabolites

Neotyphodium species produce a variety of metabolites, in plants and in culture. Some of these are the important toxins or anti-feedants effective against a range of herbivores, from grazing livestock to insect pests (Table 4.2). Alkaloids produced in endophyte-infected grasses are associated with toxicoses of grazing livestock. The impact of symbiosis also confers tolerance to abiotic stresses, particularly drought, as well as to herbivorous arthropods and to some pathogens (Bernard *et al.*, 1998). In the USA, tall fescue toxicosis affects cattle, sheep and horses: the disease syndrome reduces individual growth rate, reproduction and lactation. In New Zealand, livestock in many regions frequently suffer from a neuromuscular condition, ryegrass staggers, when grazing endophyte-infected ryegrasses. The risk is greater at the end of a prolonged dry spell in late summer, when grass growth has stopped and toxin level is elevated. Such conditions are infrequent in perennial ryegrass-growing regions of Europe, and reported cases of ryegrass staggers are sporadic (Lewis, 1997). One incident concerned staggers in horses fed hay from the aftermath of a seed crop, and flowering stems do have elevated toxin levels. The condition is usually reversible and symptoms disappear when animals are given clean feed.

Stress tolerance responses are only expressed in the presence of the stress, for example, differential root growth and production of sugars associated with drought tolerance in endophyte-infected plants is initiated only when drought begins. The morphological changes seen with endophyte infection in tall fescue are not solely associated with the presence of the fungus but also depend upon the host genotype (Bacon and Hill, 1996).

Table 4.2. Major secondary metabolites associated with fungal endophyte infection of fescues and perennial ryegrass and examples of activity against herbivores. (After sources in Table 4.1 and Oldenburg, 1997; Eerens *et al.*, 1998a).

Metabolites		Anti-herbivore activity	
Class	Compounds	Livestock	Insects
Saturated amino pyrrolizidines (loline alkaloids)		Fescue toxicosis	
	Perloline	+	+
	N-Formyl loline	++	+
	N-Acetyl loline	+	+: LD50 1 μg per insect
Ergopeptines (ergot alkaloids)		Heat stress	
	Ergonovine	+	+
	Ergocryptine	+	+: LC50 80 μg ml^{-1}
	Ergotamine	+	+: LC50 97 μg ml^{-1}
	Ergovaline	++	+: LC50 50 μg ml^{-1}
Pyrrolopyrazine			
	Peramine	–	+: anti-feedant at >1 ppm: synergism with other alkaloids
Indole diterpines (lolitrems)		Tremorgens: 'ryegrass staggers'	
	Paxilline (precursor of lolitrems)	+	Anti-feedant
	Lolitrem B	++: most abundant	Kills Argentine stem weevil at 5 ppm in diet

Loline alkaloids

The pyrrolozidine alkaloids (such as *N*-formyl loline) are active against invertebrate herbivores (Bush *et al.*, 1997) but have not been proven conclusively to be toxic to vertebrates. *N*-Formyl loline is usually the most abundant of the lolines, and is found in some 35% of symbiota (Bush *et al.*, 1997). The lolines have some activity against mammals and also are implicated in allelopathic effects of endophyte-infected tall fescue on other plants (Bush *et al.*, 1997). Lolines are potent insecticides with activity by contact and ingestion, although concentrations measured in roots are not usually high.

Ergot alkaloids

The ergot alkaloids are associated mainly with toxicity to vertebrates, although they have some effect on invertebrates. Ergot alkaloids, in particular ergovaline, are present and cause livestock toxicoses in both tall fescue and perennial ryegrass. Little is known of ergovaline concentrations in European grasses. In north-west Spain, concentrations of up to 0.55 mg kg^{-1} have been recorded in *L. perenne* plants used for breeding (Oliveira *et al.*, 1997). In Switzerland, ergovaline concentrations in *F. arundinacea* over 2 years averaged 0.7 with a maximum of 1.4 ppm dry matter (Bush and Schmidt, 1996). These concentrations were not significantly different from those in identical plant material grown in Kentucky, USA, at the same time.

Peramine

The pyrrolopyrazine, peramine, is active against invertebrate herbivores (Bush *et al.*, 1997) but has not been proven conclusively to be toxic to vertebrates. The best documented case of endophyte-induced deterrence of insect pests is that of Argentine stem weevil (*Listronotus bonariensis*), a major pest of perennial ryegrass in New Zealand (Popay *et al.*, 1990). Peramine is translocated to the leaves and is detected by adult female stem weevils during feeding but before egg-laying, with the result that the insect moves elsewhere. The efficacy of this biological control is such that a very large proportion of perennial ryegrass cultivars sown in New Zealand are endophyte-infected. Peramine is active against Argentine stem weevil at 10 ppm in artificial diets and concentrations exceed this in endophyte-infected leaf blade and sheath. Peramine seems to be mobile within the tiller, with evidence for movement from sheath to blade and from older to younger leaves (Keogh *et al.*, 1996). This may be a mechanism for protecting the most photosynthetically active and therefore more valuable leaves.

Lolitrems

In ryegrass, *N. lolii* distribution and lolitrem B concentration is greater in leaf sheaths than in leaf blades: concentrations vary in different parts of the plant and there is more lolitrem B in older than in younger leaf parts or leaves. A concentration of 2–2.5 ppm dry weight of lolitrem B in plants of *L. perenne* has been cited as the threshold for the induction of ryegrass staggers (di Menna *et al.*, 1992). Concentrations above this threshold have been recorded in Europe. In The Netherlands, samples of hay implicated in cases of ryegrass staggers contained concentrations of lolitrem B ranging from 1.82 to 6.06 ppm (Fink-Gremmels and Blom, 1994). Also, concentrations in plants of endophyte-infected *L. perenne* collected from UK grassland and grown in the glasshouse or field plots ranged from 0.08 to 5.5 ppm (Lewis, 1994). In contrast, maximum concentrations of only 0.68 and 1.4 ppm were detected in ecotypes of *L. perenne* from Ireland and Germany, respectively (Oldenburg, 1996; do Valle Ribeiro, 1996).

Insect responses

A wide range of insect pests is affected by endophyte infection of grasses (Clement *et al.*, 1994; Popay and Rowan, 1994). Endophyte infection of tall fescue and ryegrass leaves generally has an adverse effect on insect herbivores. Saikkonen *et al.* (1998) summarized tests with endophyte-infected leaves and herbivores: 66 and 71% showed adverse affects for tall fescue and ryegrass, respectively. The combinations included sap-sucking aphids. Endophyte-infected tall fescue affected infestation of the grass by barley yellow dwarf virus and by insects (West *et al.*, 1990). The root aphid, *Aploneura lentisci*, was not able to reproduce on *Festuca pratensis*, meadow fescue, infected by its specific endophyte *Neotyphodium uncinatum*. Two other aphids, feeding on leaves (*Rhopalomyzus padi*) and stem bases (*R. poae*), were also only prevalent on endophyte-free meadow fescue (Schmidt, 1993). Roots of tall fescue deterred herbivores, including nematodes (reviewed below) and root-feeding grubs (Potter *et al.*, 1992; Davison and Potter, 1995).

At the Institute of Grassland and Environmental Research, North Wyke, roots of grass plants grown in the glasshouse are often infected by the root aphid, *A. lentisci*. Perennial ryegrass plants are infested regardless of their endophyte status, but several *Festuca* species that are endophyte-infected remain completely free of root aphid. The likely reason for this distinction is that different alkaloids are produced in ryegrass and fescues (Table 4.1). Lolines, present in fescues, have been detected in roots (Bush *et al.*, 1993) whereas peramine in perennial ryegrass is almost undetectable in roots (Ball *et al.*, 1997).

Temperature accounts for seasonal differences with higher hyphal counts and anti-herbivory effects at 14–21°C than at either 7 or 28°C (Breen, 1994). In tall fescue pastures, Bernard *et al.* (1998) concluded that endophyte effects on invertebrates feeding on live or dead grass tissue were complex and usually species-specific. The effects are also subject to seasonal variations, perhaps related to variations in compounds produced by the grass–fungus symbiosis.

Endophytes increase phenotypic diversity of the grass and contribute to the fitness of all the plants in a population. Genetic diversity of the host is essential for survival of the endophyte which does not reproduce sexually. None the less, endophyte strains differ, and evolutionarily new combinations of grass and fungus have been formed (Saikkonen *et al.*, 1998). Genetic studies indicate that present-day endophytes may originate from hybridizations between the sexual stage *E. typhina* and *N. lolii* (Schardl *et al.*, 1994; Collett *et al.*, 1995). This makes it possible to consider developing new combinations of value to particular crop situations.

Endophyte infection is therefore both boon and bane in livestock systems (Joost, 1995). The losses of livestock sickness may to some extent be balanced by the gains from improved persistence and productivity of infected grasses. Careful attention to grazing management, especially in summer dry conditions can be an important factor in minimizing animal losses. The incorporation of herbage legumes in mixture with endophyte-infected grass can dilute the toxins. In contrast to forage grasses, endophyte is valuable in turf grasses for the endophyte-enhanced resistance to leaf-feeding pests and improved stress tolerance. Cultivars of perennial ryegrass and some turf fescues are already widely used, and suitable combinations in other important turf grasses are sought (Funk *et al.*, 1993; Funk and White, 1997). Some of these have modest improvements in resistance to root-feeding grubs.

None the less, the potentially harmful effects of endophyte-infections cannot be ignored. Much of southern and eastern Europe would likely be affected by animal toxicity in ryegrass and fescues, as are the warmer parts of New Zealand and the southern states of the USA. In both cases, because of the benefits of endophyte infection, it may be economic to manage and accept some risk of illness in livestock but the adverse effects on animal performance need to be balanced by consideration of the beneficial impacts on plant growth (Joost, 1995).

As well as exploiting the effects of endophytes to increase tolerance of stresses in grass, it is of course still possible to select endophyte-free grasses for genetic improvement of these characteristics. Improving the endophyte by selection for secondary metabolite production or by genetic engineering also has potential advantages (Latch, 1994; Latch and Fletcher, 1997; Bacon *et al.*, 1997). These approaches require further study of both stability of modified traits as well as further assessment of the routes of transmission of the fungi (Bacon *et al.*, 1997). The endophyte status of grass plants can be altered. Infection can be eliminated from seeds by fungicide or hot water treatment, or by storing the seed for several years. Plants can be rendered endophyte-free by fungicide treatment or by hydroponic culture (Lewis and Vaughan, 1995). Infection can be introduced to a seedling (Latch and Christensen, 1985), stem (Ravel *et al.*, 1994), callus culture (Johnson *et al.*, 1986), somatic embryo (Kearney *et al.*, 1991) and plantlets from meristems (O'Sullivan and Latch, 1993). The success rate of artificial inoculation varies according to the grass and endophyte species, for example, Naffaa *et al.* (1999) recorded success rates of infection of perennial ryegrass of 11–25% with the natural endophyte, *N. lolii*, and 90% with an isolate of *E. festucae* from a fine fescue. However, novel endophyte:grass associations are not always compatible (Christensen, 1995).

Endophyte Interactions with Nematodes

Evidence from the literature shows that studies have been focused on *Neotyphodium coenophialum* endophyte in tall fescue in the southern USA and on *N. lolii* on perennial ryegrass in New Zealand. It is worth emphasizing that these grasses have been introduced into the USA and New Zealand, respectively, becoming major forage species in these countries. There have been some studies of fescues and ryegrasses in Europe and of other grasses elsewhere. The conclusion is that endophyte–herbivore and/or stress interactions appear to be economically valuable in the USA and New Zealand, compensating to some extent for the disadvantages of endophyte infection. In the USA, at least some of this value is associated with improved persistence in summer droughts associated with control of root-feeding nematodes. There is some indication of control of other below-ground herbivores and for the presence of endophyte hyphae in roots in some circumstances (see previous sections). Here we review the evidence of effects of endophyte on nematodes, much of which is summarized by nematode feeding type in Tables 4.3–4.5.

Sedentary endoparasites

Root-knot nematodes

The effects of endophyte infection of tall fescue on the root-knot nematode, *Meloidogyne marylandi*, contrast with those of perennial ryegrass and red fescue on both *M. marylandi* and *M. naasi*, and of perennial ryegrass on *M. naasi* (Table 4.3). *M. marylandi* and *M. graminis* do not reproduce on tall fescue infected by endophyte: juvenile emergence from eggs, invasion of roots and female development were all less on E+ than E– grass (Kimmons *et al.*, 1989; Gwinn and Bernard, 1993). Four tall fescue clones with endophyte all controlled the nematode and one of these clones was a very good host when endophyte-free (Kirkpatrick *et al.*, 1990). Gwinn and Bernard (1993) showed that in endophyte-infected tall fescue cv. KY31 the inner walls of endodermal cells were substantially thicker than in endophyte-free plants.

In contrast, red fescue and perennial ryegrasses with their specific endophytes were as good hosts of *M. marylandi* as endophyte-free grasses (Gwinn and Bernard, 1993). In New Zealand, Stewart *et al.* (1993) reported that *Meloidogyne naasi* produced fewer galls (87 compared with 134 per plant) and females on perennial ryegrass with than without *N. lolii* infection. Root mass (dry weights) of E+ was less than that for E– plants, but not statistically significantly so. In pot experiments with a single clone of perennial ryegrass, endophyte had no affect on *M. naasi* invasion and multiplication (Cook *et al.*, 1991).

Ball *et al.* (1997) examined the response to *M. marylandi* of ryegrass plants (cv. Grasslands Nui) infected by one of six endophyte strains that produced mycotoxins different from usual natural associations. Two-week-old plants inoculated with 1000 eggs were assessed at 30 days: plants with endophyte had fewer females per gramme of root than endophyte-free grass: one combination had only 10% as many females as on endophyte-free plants. The resistant endophyte-infected plants produced both peramine and lolitrem B but not ergovaline nor other ergot alkaloids. However, there were different ryegrass genotypes in this experiment and the differences between these as nematode hosts were not assessed. Genotypes of perennial ryegrasses range from fully susceptible to resistant to *M. naasi* (Cook *et al.*, 1999) and tall fescue shows similar variation with respect to *M. marylandi* (Kirkpatrick *et al.*, 1990).

Table 4.3. Summary of the responses of root-knot nematodes (*Meloidogyne* spp.) to fungal endophytes of grasses in pot experiments in glasshouses.

	Grass		Nematode reproduction on plants infected (E+) or free (E–)			Assessments				
			Initial numbers of nematodes	Final numbers of nematodes						
Nematode	Species	Cultivar/ source	per plant	E+	E–	Final populations nematode	Days after inoculation or planting	Notes	Country and region	Reference
Meloidogyne marylandi	*Festuca arundinacea*	Selected seedlings from cv. KY31	3000	20	320 *	Egg masses per plant	56	Invasion in E+ < E–	USA, Tennessee	3
			5000	20	100 *					
		1 clone	5000	20	4200 *	Eggs per g root	63	Host genotype effect	USA, Arkansas	4
		3 clones		0–90	25–100 NS					
		KY31 clones	1000	0	280 *	Nematodes per plant	28	Root endodermis thicker in E+	USA, Tennessee	2
	F. rubra	Homologous seed lots	1000	170	130 NS	Nematodes per plant	28		USA, Tennessee	2
				55	105 NS	Egg masses per plant				
	Lolium perenne	Homologous seed lots	1000	520	430 NS	Nematodes per plant	28		USA, Tennessee	2
				359	288 NS	Egg masses per plant				
M. naasi	*L. perenne*	Single clone, P4	1000	250	260 NS	Per plant	10	Root weight E+ > E–	UK	1
				1170	1830 *	Per g root				
				140	180 NS	Per plant	35			
				170	260 NS	Per g root				
		cv. Nui	1000	87	134 *	Galls per plant	56	Root weight E+ > E–	New Zealand	5
				110	200 *	Females per plant				
				290	480 *	Females per g root				

References: 1, Cook *et al.*, 1991; 2, Gwinn and Bernard, 1993; 3, Kimmons *et al.*, 1990; 4, Kirkpatrick *et al.*, 1990; 5, Stewart *et al.*, 1993.
*, NS: E+ vs E– difference statistically significant or not.

Table 4.4. Summary of the responses of migratory root endoparasitic nematodes (*Pratylenchus* spp.) to fungal endophytes of grasses.

	Grass			Nematode reproduction on plants infected (E+) or free (E–)			Assessments					
					Final numbers of nematodes							
Nematode	Species	Cultivar/ source	Endophyte status	Initial numbers of nematodes	E+	E–	Final nematode population	Time after inoculation or planting	Experiment location	Notes	Country and region	Reference
Pratylenchus scribneri	*Festuca arundinacea*	KY31	75% E+ 100% E–	–	2	224 *	Per 100 cm^3 soil	1 year	Field plots		USA, Arkansas	8
P. scribneri	*F. arundinacea*	KY31	80% E+ 100% E–	–	Reduced	Pathogenic		2 years	Field plots		USA, Arkansas	9
P. scribneri	*F. arundinacea*	KY31		1000 per plant	80	390 *		63 days	Glasshouse	No reproduction on E+	USA, Tennessee	4
P. scribneri	*F. arundinacea*	KY31	Selected seedlings	1200 per pot 2000 per pot	<100 <100	1800 * 1200 *	Per pot	15 weeks	Glasshouse	Invade but no reproduction in E+	USA, Tennessee	4
P. scribneri	*F. arundinacea*	KY31	Clones	1000 per plant	30 30	80 * 300 *	In soil In roots	58 days	Glasshouse	Invaded but died in E+	USA, Tennessee	3
P. scribneri	*F. arundinacea*	KY31	E+ and E–	–	<1	0 NS	Per 100 cm^3 soil	1 year	Field plots		USA, Mississippi	5
					<1	0 NS		2 years				

P. sp.	*F. arundinacea*	Single clone	E+ and E–	370 per pot	1470 57	2160 NS 58 NS	In roots √n per pot	1 year	Pots of field soil		UK	2
P. sp.	*F. rubra*	Single clone	E+ and E–	370 per pot	400 56	160 NS 61	In roots √n per pot	1 year	Pots of field soil		UK	2
P. sp.	*F. pratensis*	Single clone	E+ and E–	370 per pot	2110 60	2250 NS 53 NS	In roots √n per pot	1 year	Pots of field soil		UK	2
P. pratensis *P. thornei*	*F. pratensis*	Cultivars	E+ and E–		34 0	94 * 19 *	Per 100 cm^3 soil		Field plots	(at one site: no effects at five others)	Germany	6
P. sp.	*Lolium perenne*	cv. Ellett cv. Nui	E+ E–		E+ < E–		Roots and soil	1 year	Field plots		New Zealand	7
P. sp.	*L. perenne*	Homologous population	E+ and E–		179 100	149 102	Per 100 cm^3 soil	1 year 2 years	Field plots		UK	1
P. sp.	*L. perenne*	Different clones	E+ and E–	370 per pot	460 39	2780 * 87 *	In roots √n per pot	1 year	Pots of field soil	Root weight E+ < E–	UK	2
P. sp.	*Holcus mollis*	Single clone	E+ and E –	370 per pot	1170 45	1080 NS 50 NS	In roots √n per pot	1 year	Pots of field soil		UK	2

References: 1, Cook *et al.*, 1991; 2, Cook and Lewis, unpublished; 3, Gwinn and Bernard, 1993; 4, Kimmons *et al.*, 1990; 5, O'Day *et al.*, 1993; 6, Schöberlein *et al.*, 1997; 7, Watson *et al.*, 1993; 8, West *et al.*, 1988; 9, West *et al.*, 1990.

Table 4.5. Summary of the responses of migratory root ectoparasitic nematodes to fungal endophytes of grasses.

Nematode	Grass: Species	Grass: Cultivar/ source	Endophyte status	Nematode reproduction on plants infected (E+) or free (E–): Initial numbers of nematodes per plant	Final numbers of nematodes: E+	Final numbers of nematodes: E–	Assessments	Time after inoculation or planting	Experiment location	Notes	Country and region	References
Stunt nematodes												
Tylenchorhynchus acutus	*Festuca arundinacea*	KY31	75% E+ 100% E–		76	306 *	Per 100 cm^3 soil	1 year	Field plots	Grass yield E+ 10% > E–	USA, Arkansas	7
T. acutus	*F. arundinacea*	KY31	80% E+ 100% E–		E+ < E–			2 years	Field plots	Nematode populations declined in all plots	USA, Arkansas	8
T. ewingi	*F. arundinacea*	KY31	80% E+ 100% E–		No difference			2 years	Field plots		USA, Arkansas	8
T. sp.	*F. arundinacea*	KY 31	E+ and E–		9 1	4 NS 2 NS		1 year 2 years	Field plots		USA, Mississippi	4
T. dubius	*F. pratensis*	Cultivars	E+ and E–		92	108 *	Per 300 g soil	1 year	Field plots	No effects at five other sites	Germany	6
Bitylenchus maximus					13	25 *						
Merlinius brevidens					33	100 *						
Spiral nematodes												
Helicotylenchus dihytera	*F. arundinacea*	KY31	98% E+ 100% E–	10 per 100 cm^3 soil	6 3	20 19	Per 5 g root Per 100 cm^3 soil	7 weeks	Pots of field soil	E+ bigger than E–	USA, Kentucky	5
H. multicinctus	*F. arundinacea*				Declined	Maintained					USA, Tennessee	3
H. pseudorobustus	*F. arundinacea*	KY31 selected seedlings	E+ and E–	1000 per pot	700 650	1000 NS 1200 NS	Per pot	8 weeks 10 weeks	Pots of compost		USA, Tennessee	3

H. sp.	*F. arundinacea*	KY31	E+ and E–		0 23	0 NS 6 NS	Per 100 cm^3 soil	1 year 2 years	Field plots		USA, Mississippi	4
H. sp.	*F. arundinacea*	Single clone	E+ and E–	1100 per pot	41	24 *	√ per pot	1 year	Pots of field soil		UK	1
H. sp.	*F. pratensis*	Single clone	E+ and E–	1100 per pot	27	22 NS	√ per pot	1 year	Pots of field soil		UK	1
H. sp.	*F. rubra*	Single clone	E+ and E–	1100 per pot	28	38 NS	√ per pot	1 year	Pots of field soil		UK	1
H. sp.	*Lolium multiflorum*	13 cvs	0–43% E+	10 per 100 cm^3 soil	Range 4–13 Range 7–42		Per 100cm^3 soil		Pots of field soil	No effects of E+	USA, Kentucky	5
H. sp.	*L. perenne*	Different clones	E+ and E–	1100 per pot	31	26 NS	√ per pot	1 year	Pots of field soil		UK	1
H. sp.	*Holcus mollis*	Single clone	E+ and E–	1100 per pot	22	45 *	√ per pot	1 year	Pots of field soil		UK	1
Pin nematodes												
Paratylenchus projectus	*F. arundinacea*	KY31	80% E+ 100% E–		Reduced	Pathogenic		2 years	Field plots	Increasing over time on E– only	USA, Arkansas	8
Females	*F. arundinacea*	KY31	E+ and E–		3 1	9 NS 5 NS	Per 100 cm^3 soil	1 year 2 years	Field plots	Effect on reproduction after 2 years	USA, Mississippi	4
Non-feeding stages					18 4	19 NS 24 *		1 year 2 years				
P. sp.	*F. arundinacea*	Single clone	E+ and E–		155	156 NS	√ per pot	1 year	Pots of field soil		UK	1
P. sp.	*F. pratensis*	Single clone	E+ and E–		21	22 NS	√ per pot	1 year	Pots of field soil		UK	1
P. goodeyi *P. microdorus*	*F. pratensis*	Cultivars	E+ and E–		2 13	77 * 40 *	Per 300 g soil	1 year	Field plots	No effects at five other sites	Germany	6
P. sp.	*F. rubra*	Single clone	E+ and E–		160	151 NS	√ per pot	1 year	Pots of field soil		UK	1

Table 4.5. *Continued.*

Nematode	Grass: Species	Grass: Cultivar/ source	Endophyte status	Nematode reproduction on plants infected (E+) or free (E–): Initial numbers of nematodes per plant	Final numbers of nematodes: E+	Final numbers of nematodes: E–	Assessments	Time after inoculation or planting	Experiment location	Notes	Country and region	References
P. sp.	*L. perenne*	3 cvs	E+ and E–		86	100	E+ as % E–		Pots of field soil		New Zealand	2
P. sp.	*L. perenne*	Different clones	E+ and E–		152	314 *	√ per pot	1 year	Pots of field soil	Root weight of E+ < E–	UK	1
P. sp.	*Holcus mollis*	Single clone	E+ and E–		333	320 NS	√ per pot	1 year	Pots of field soil		UK	1
Stubby root nematodes												
Paratrichodorus minor	*F. arundinacea*	KY31	98% E+	15 per 100 cm^3 soil	30	120	Per 5 g root		Pots of field soil	E+ grew better than E–	USA, Kentucky	5
			100% E–		94	106	Per 100 cm^3 soil					
Trichodorus primitivus	*F. pratensis*	Cultivars	E+ and E–		69	160 *	Per 300 g soil	1 year	Field plots	No effects at five other sites	Germany	6
P. minor	*L. multiflorum*	13 cvs	0–43% E+	15 per 100 cm^3 soil	Range 61–197 Range 16–72		Per 100 cm^3 soil		Pots of field soil	No effects of E+	USA, Kentucky	5

References: 1, Cook and Lewis, unpublished; 2, Eerens *et al.*, 1998b; 3, Kimmons *et al.*, 1990; 4, O'Day *et al.*, 1993; 5, Pedersen *et al.*, 1988; 6, Schöberlein *et al.*, 1997; 7, West *et al.*, 1988; 8, West *et al.*, 1990.

Migratory root parasites

Root lesion nematodes

The root lesion nematode, *Pratylenchus scribneri*, does not multiply on endophyte-infected tall fescue (Table 4.4). This has been demonstrated both with controlled inoculations of plants in pots and in naturally infested soil in pots and in field plot trials. In the field, control of *P. scribneri* has been related to the improved drought tolerance of endophyte-infected tall fescue (Kimmons *et al.*, 1990; Bernard *et al.*, 1998). In contrast, endophyte in KY31 and eight other tall fescue cultivars was not shown to control root lesion nematode in field plots over two seasons (O'Day *et al.*, 1993). In those trials, nematode populations were very small compared with those in experiments in which endophyte controlled the nematode (Table 4.4).

P. scribneri was not controlled by endophytes in either *F. rubra* or perennial ryegrass (Gwinn and Bernard, 1993). In glasshouse experiments with pots of naturally infested field soil and five species of grass, with or without respective endophytic fungi, *Pratylenchus* spp. were fewer only on endophyte-infected ryegrass (Cook and Lewis, unpublished; Table 4.4). In this example, the endophyte-infected plant root system was very much smaller than the endophyte-free control. In a field experiment with homologous ryegrass populations, there were no effects on *Pratylenchus* populations (Cook *et al.*, 1991). In field trials in New Zealand, fewer *Pratylenchus* spp. were recovered from endophyte-infected than from endophyte-free perennial ryegrass growing with white clover (Watson *et al.*, 1995). In one trial in Germany, two species of root lesion nematodes, *P. pratensis* and *P. thornei*, were both fewer under infected than endophyte-free meadow fescue (Schböerlein *et al.*, 1997).

Ectoparasites

Stunt nematodes of tall fescue illustrate the variation in nematode responses to endophyte (Table 4.5). *Tylenchorhynchus acutus* numbers on endophyte-infected tall fescue were only 25% of those on the uninfected (West *et al.*, 1988, 1990). A second species, *T. ewingi*, was not affected. Populations of other stunt nematodes, *T. (Bitylechus) dubius*, *T. maximus* and *Merlinius brevidens*, were reduced by endophytes in *F. pratensis* at one field trial site in Germany (Schöberlein *et al.*, 1997). At five other sites in Germany there was no effect. Different species of stunt nematodes feed differently on ryegrass roots, ranging from browsing for short periods on individual epidermal cells, to feeding for long periods as sedentary ectoparasites, and to species that feed as semi-endoparasites on cortical cells (Bridge and Hague, 1974). This would likely affect exposure to endophyte metabolites.

Soil population densities of other migratory nematodes are variously affected by endophytes (Table 4.5). There are more reports of spiral nematode control on endophyte-infected tall fescue than on other grasses. Even so, some species are not controlled by E+ tall fescue. These nematodes include migratory ectoparasites, some partial endoparasites and others that may be endoparasitic.

The observations on ectoparasitic pin nematodes, *Paratylenchus* spp., again suggest that endophyte-infected tall fescues control populations whereas other grasses do not (Table 4.5). In some experiments where *Paratylenchus* species were suppressed, other species of ectoparasitic nematodes appeared to be unaffected by endophyte status (West *et al.*, 1988; O'Day *et al.*, 1990; Bernard *et al.*, 1998). In New Zealand, a natural infestation of *Paratylenchus* sp. was allowed to develop, in a pot experiment, on three perennial ryegrass cultivars either free of endophyte or infected with a wild-type or selected endophyte. There were 16% more nematodes on endophyte-free plants, which were only slightly heavier (1% and 6% for roots and shoots, respectively) than endophyte-infected grasses. There was an indication of interactions between grass genotype and endophyte strain (Eerens *et al.*, 1998a,b). There is also variation between sites: Schöberlein *et al.* (1997) reported control by endophyte in *F. pratensis* at only one of six trial sites and O'Day *et al.* (1993) found no effect of endophyte status (Table 4.5).

In field soil, plant parasite nematode population densities were between two and ten times greater under infected than endophyte-free tall fescue (West *et al.*, 1989). Other ectoparasites including the economically significant stubby root nematodes are controlled by endophyte in tall fescues, and in meadow fescue at one site, but not at five others nor by ryegrass (Table 4.5). Of the six sites studied by Schöberlein *et al.* (1997), at only one were population densities of five plant parasites lower in plots of endophyte-infected than free *F. pratensis* cultivars. As with the root lesion nematodes, it seems that at these non-responsive sites other factors may have prevented nematode population increase. This may explain some cases of lack of a differential response to endophyte status. In field trials with perennial ryegrass, there were more plant nematodes under infected than endophyte-free grasses (Cook *et al.*, 1991). The differences were related directly to the better growth of E+ grass so that there was no difference between grasses in nematode numbers per gramme of root.

Non-plant parasitic nematodes

In pot trials, with natural grassland soil, after 1 year there were more *Aphelenchoides* associated with infected perennial ryegrass and tall fescue than with the endophyte-free clones: numbers on *F. pratensis*, *F. rubra* or *Holcus mollis* were not affected by endophyte presence or absence (Cook and Lewis, unpublished). Nematodes of this genus are frequently found in and on grass leaves and shoots, and are usually regarded as fungal feeders.

Entomopathogenic nematodes (*Heterorhabditis bacteriophora*) applied to beetle larvae fed with endophyte-infected or free tall fescue, were not harmed by the diet of the beetles. The nematodes killed more of the beetles that had fed on infected plants. Endophyte in beetle larvae had no effect on nematode reproduction (Grewal *et al.*, 1995).

Animal parasitic nematodes traverse grass leaves, and perhaps lodge in axils, before ingestion by grazing animals. It is conceivable that endophytes may have some impact upon these non-feeding stages. We are not aware of trials testing animal parasite survival on endophyte-infected grasses. Since these pre-infective juveniles do not feed, are resistant to other environmental stresses and retain a double cuticle, it is unlikely that they would be susceptible to endophytes in grasses. Lambs grazed on endophyte-infected ryegrass were not protected from parasites (Scales *et al.*, 1995).

Possible mechanisms

Plant morphology

In tall fescue, endoderm thickening was associated with effects on the root-knot nematode but not on *Pratylenchus* (Gwinn and Bernard, 1993). Invading juveniles of *Meloidogyne* migrate into the stele through undifferentiated tissue and so would not be affected during root penetration. However, a thickened endodermal layer would probably constrain growth of sedentary stages, reducing their growth and reproduction. *Pratylenchus* species migrate through and feed on cortical cells so would not be directly impeded by the thickened endodermis. Roberts *et al.* (1992) implicated fungal chitinases in the effects on nematodes. There was more chitinase activity in infected than endophyte-free clones of tall fescue cv. KY31, significantly so at 20 and 25 days after germination. In experiments with endophyte-infected KY31 and a fescue × ryegrass hybrid cv. Johnstone with low alkaloid production, infection of roots by *Meloidogyne marylandi* induced increased activity of chitinase in extracts from leaves but not from roots. The greatest chitinase activity was in the endophyte KY31 symbiosis. There was no direct link between chitinase activity and response to the nematode, but chitinase is a pathogenesis-related protein. There was no evidence of differences in nematode behaviour on the different grasses.

Plant size and quality

Where endophyte improves the quality or quantity of root, parasite populations may increase on endophyte-infected grasses, as in field trials with perennial ryegrasses (Cook *et al.*, 1991). Other benefits that preserve plant growth under stress may also act to make endophyte-infected plants support more rather then fewer nematodes. Plant size has to be taken into account when assessing impacts of endophyte on nematodes. For instance, Cook *et al.* (1991)

found 36% fewer *M. naasi* infective juveniles per unit of root mass in infected than in endophyte-free ryegrass clones at 10 and 35 days after inoculation, although totals per plant did not differ. In UK field trials, infected perennial ryegrass plants grew much better than controls, and this seemed to be the explanation for the increased populations of root parasites found by Cook *et al.* (1991). Conversely, the marked differences in root lesion nematode numbers on clones of ryegrass were related to the larger root mass of the endophyte-free clone.

Aphelenchoides may feed directly on the endophyte. There are differences in feeding preferences of other species of this fungal-feeding genus (Ruess *et al.*, 2000) but no indication as to the palatability of endophytes. These nematodes are frequently found in grass leaf tissues and in amongst leaf sheaths, in the same arena as the fungus.

Breen (1994) noted the complexity of interactions affecting endophytes. Watson *et al.* (1995) found complex interactions in which other rhizosphere organisms, notably arbuscular mycorrhizal fungi and root pathogens, influenced the effect of endophyte on the competitive interactions between ryegrass, white clover and invertebrate herbivores. They found that nematode species specific to white clover were more abundant when the companion grass was endophyte-free, probably because the clover had grown better than with infected grass which itself suffered less from stem-eating insects.

Toxins

Watson *et al.* (1993) observed that populations of root lesion nematodes in grass roots and rhizosphere soil were smaller in endophyte-infected grass. Noting that endophyte hyphae did not grow in roots, they suggested that toxins were translocated to roots in sufficient amounts to affect nematodes. In their assessments, Watson *et al.* (1993) consistently detected low concentrations of lolitrem B in E+ ryegrass (to 2 $\mu g\ g^{-1}$), but peramine (to 1.7 $\mu g\ g^{-1}$) was usually not detected. Although roots of endophyte-infected grasses do not generally bear hyphae, there remains the possibility that secondary metabolites produced in stem bases or leaves of infected plants may be transported to roots.

It seems clear that to be killed, nematodes must ingest toxin from endophyte-infected plants. Some, like the fungal feeder, *Aphelenchoides*, are probably tolerant, others may either be tolerant or may through their feeding habits or sites avoid exposure to translocated toxins. In some cases, the responses of plants to endophyte, such as wall thickening or anti-drought responses, may incidentally make them less good hosts of particular nematodes.

Alkaloids produced in endophyte-infected grasses may be more abundant at temperatures above those found in cool temperate grassland regions, for example, *N*-formyl loline production in ryegrass requires temperatures above

23°C. This may explain some of the variation in results. It may be that a relatively long period of exposure is required to control nematodes. For example, root lesion nematodes invaded roots but did not reproduce (Bernard *et al.*, 1998). In cool regions with transient toxin production, nematodes may survive in endophyte-infected plants.

Prospects for endophyte-induced resistance to nematodes

Grass endophyte associations are maintained in US and New Zealand grasses because of their beneficial features. In contrast, European-bred forage grass cultivars generally have little or no endophyte infection. The symbiotic interaction is very complex with a good deal of variability. This has good and bad aspects: good in that it may be possible from this diversity to identify useful combinations for exploitation but bad because practical field studies with existing material face too much uncontrolled variation.

There is clearly opportunity to exploit variation in endophytes to identify those most useful and to incorporate these into a proportion of grass parents. A prelude must be more studies with homogeneous populations or with genetically identical plants either infected or not by the endophyte.

Bernard *et al.* (1998) stress the importance of checking the endophyte status of grasses to the interpretation of reports of grass resistance or non-host status to nematodes. Cook *et al.* (1999) confirmed that perennial ryegrass clones resistant to *M. naasi* were all endophyte-free.

It is possible to distinguish between endophyte strains by selecting those that produce particular alkaloids. In New Zealand, the technique for artificially infecting seedlings has been used to produce cultivars of perennial ryegrass with strains of endophyte that do not induce formation of the animal toxins, lolitrem B and ergovaline (Latch and Fletcher, 1997). This is a rational approach to exploiting benefits and avoiding harmful effects (Latch, 1993, 1994) and will be the ideal way to solve the dilemma of tall fescue grazing systems in the hotter southern part of the USA and to remove occasional but unacceptable risks from other areas (Hoveland, 1993). A practical problem in maintaining endophyte-free pastures is that infected plants can establish where there is an extensive seed bank.

A useful start would be to select and define model grass and endophyte combinations that affect specific nematodes. These models can then be used for comparative chemical analyses of roots when grasses are grown under optimal conditions for toxin production and then to develop screening technique for best available combinations and to use these in controlled experiments to identify which nematode species can be affected.

The symbiosis offers unique opportunities to protect grasses from actual or potential damage in longer-lived pastures that seem to be essential for maintaining the economic efficiency of ruminant production from natural healthy grassland fodder. In any case the use of turf grasses with endophyte may have no disbenefits in some circumstances. Endophyte-infected cultivars of perennial ryegrasses and turf fescues and other species are produced and desirable (Funk *et al.*, 1993; Funk and White, 1997).

In conclusion, future studies of endophyte impacts on nematodes should take account of the variability in the interactions. This requires standardization of grass genotype and fungus strain as well as biochemical characterization of the combination. Environmental conditions need either to be controlled or at least measured to relate to effects on alkaloid production. It is also important to describe the feeding site and processes of the nematodes being studied. Control of these aspects will allow the development of experimental systems to test hypotheses about the relationship between parasite attributes and susceptibility to endophytes. Such studies have clear potential economic value, given that some combinations of grass and fungus provide fully effective resistance to nematodes not readily controlled either by natural genetic resistance or by other means. The use of selected endophytes on heterogenous grass populations may offer durable control of nematodes to the benefit both of forage and amenity grass and to rotational crops.

Acknowledgements

The Institute of Grassland and Environmental Research is an institute of the Biotechnology and Biological Sciences Research Council. Part of this work was funded by the UK Ministry of Agriculture, Fisheries and Food.

References

Azevedo, M.D. and Welty, R.E. (1995) A study of the fungal endophyte *Acremonium coenophialum* in the roots of tall fescue seedlings. *Mycologia* 87, 289–297.

Bacon, C.W. and Hill, N.S. (1996) Symptomless grass endophytes: products of coevolutionary symbioses and their role in the ecological adaptations of grasses. In: Redlin, S.C. and Carris, L.M. (eds) *Endophytic Fungi in Grasses and Woody Plants. Systematics, Ecology and Evolution.* American Phytopathological Society, St Paul, Minnesota, pp. 155–178.

Bacon, C.W., Richardson, M.D. and White, J.F. (1997) Modification and uses of endophyte-enhanced turfgrasses: a role for molecular technology. *Crop Science* 37, 1415–1425.

Ball, O.J.-P., Bernard, E.C. and Gwinn, K.D. (1997) Effect of selected *Neotyphodium lolii* isolates on root-knot nematode (*Meloidogyne marylandi*) numbers in perennial ryegrass. *Proceedings of the 50th Plant Protection Conference,* 1997, pp. 65–68.

Bardgett, R.D. and Cook, R. (1998) Functional aspects of soil animal diversity in agricultural grasslands. *Applied Soil Ecology* 10, 263–276.

Bardgett, R.D., Cook, R., Yeates, G.W. and Denton, C.S. (1999) The influence of nematodes on below-ground processes in grassland ecosystems. *Plant and Soil* 212, 23–33.

Bernard, E.C., Gwinn, K.D. and Griffin, G.D. (1998) Forage grasses. In: Barker, K.R., Pedersen, G.A. and Windham, G.L. (eds) *Plant Nematode Interactions. Agronomy Monograph No. 36*. American Society of Agronomy, Crop Science Society of America, Soil Science Society of America, Madison, Wisconsin, pp. 427–454.

Breen, J.P. (1994) *Acremonium* endophyte interactions with enhanced plant resistance to insects. *Annual Review of Entomology* 39, 401–423.

Bridge, J. and Hague, N.G.M. (1974) The feeding behaviour of *Tylenchorhynchus* and *Merlinius* species and their effect on growth of perennial ryegrass. *Nematologica* 20, 119–130.

Bush, L. and Schmidt, D. (1996) Alkaloid content of meadow fescue and tall fescue with their natural endophytes. *Proceedings of the 2nd International Conference on Harmful and Beneficial Microorganisms in Pastures, Turf and Grassland*, Paderborn, Germany, 1995, pp. 259–265.

Bush, L.P., Fannin, F.F., Siegel, M.R., Dahlman, D.L. and Burton, H.R. (1993) Chemistry, occurrence and biological effects of saturated pyrrolizidine alkaloids associated with endophyte-grass interactions. *Agriculture, Ecosystems and Environment* 44, 81–102.

Bush, L.P., Wilkinson, H.H. and Schardl, C.L. (1997) Bioprotective alkaloids of grass–fungal endophyte symbioses. *Plant Physiology* 114, 1–7.

Cheeke, P.R. (1995) Endogenous toxins and mycotoxins in forage grasses and their effects on livestock. *Journal of Animal Science* 73, 909–918.

Christensen, M.J. (1995) Variation in the ability of *Acremonium* endophytes of *Lolium perenne, Festuca arundinacea* and *F. pratensis* to form compatible associations in the three grasses. *Mycological Research* 99, 466–470.

Clement, S.R., Kaiser, W.J. and Eichenseer, H. (1994) *Acremonium* endophytes in germplasms of major grasses and their utilization for insect resistance. In: Bacon, C.W. and White, J.F. (eds) *Biotechnology of Endophytic Fungi of Grasses*. CRC Press, Boca Raton, Florida, pp. 185–200.

Collett, M.A., Bradshaw, R.E. and Scott, D.B. (1995) A mutualistic fungal symbiont of perennial ryegrass contains 2 different pyr4 genes, both expressing orotidine-5′-onphosphate decarboxylase. *Gene* 158, 31–39.

Cook, R., Lewis, G.C. and Mizen, K.A. (1991) Effects of plant-parasitic nematodes of infection of perennial ryegrass, *Lolium perenne*, by the endophytic fungus, *Acremonium lolii*. *Crop Protection* 10, 403–407.

Cook, R. and Yeates, G.W. (1993) Nematode pests of grassland and forage crops. In: Evans, K., Trudgill, D.L. and Webster, J.M. (eds) *Plant Parasitic Nematodes in Temperate Agriculture*. CAB International, Wallingford, UK, pp. 305–350.

Cook, R., Mizen, K.A. and Person-Dedryver, F. (1999) Resistance in ryegrasses *Lolium* spp. to three European populations of the root-knot nematode, *Meloidogyne naasi*. *Nematology* 1, 661–671.

Davison, A.W. and Potter, D.A. (1995) Responses of plant-feeding, predatory and soil-inhabiting invertebrates to *Acremonium* endophyte and nitrogen fertilization in tall fescue turf. *Journal of Economic Entomology* 88, 367–379.

Eerens, J.P.J., Lucas, R.J., Easton, H.S. and White, J.G.H. (1998a) Influence of the ryegrass endophyte (*Neotyphodium lolii*) in a cool moist environment. I Pasture production. *New Zealand Journal of Agricultural Research* 41, 39–48.

Eerens, J.P.J., Visker, M.H.P.W., Lucas, R.J., Easton, H.S. and White, J.G.H. (1998b) Influence of the ryegrass endophyte (*Neotyphodium lolii*) in a cool moist environment. IV Plant parasitic nematodes. *New Zealand Journal of Agricultural Research* 41, 209–217.

Eggestein, S.T., Pfannmöller, M, and Schöberlein, W. (1996) *Acremonium* spp. occurrence in cultivars and collected ecotypes of the genus *Festuca*. *Proceedings of the 2nd International Conference on Harmful and Beneficial Microorganisms in Pastures, Turf and Grassland*, Paderborn, Germany, 1995, pp. 161–167.

Fink-Gremmels, J. and Blom, M. (1994) Ryegrass staggers linked to lolitrem B: a case report. *Proceedings of the 6th Congress of European Association of Veterinary, Pharmacology and Toxicology*, Edinburgh, UK, 1994, pp. 129–130.

Funk, C.R. and White, J.F. (1997) Use of natural and transformed endophytes for turf improvement. In: Bacon, C.W. and Hill, N.S. (eds) Neotyphodium/*Grass Interactions*. Plenum Press, New York, pp. 229–239.

Funk, C.R., White, R.H. and Breen, J.P. (1993) Importance of *Acremonium* endophytes in turf-grass breeding and management. *Agriculture, Ecosystems and Environment* 44, 215–232.

Grewal, S.K., Grewal, P.S. and Gaugler, R. (1995) Endophytes of fescue grasses enhance susceptibility of *Popillia japonica* larvae to an entomopathogenic nematode. *Entomologia Experimentalis et Applicata* 74, 219–224.

Gwinn, K.D. and Bernard, E.C. (1993) Interactions of endophyte-infected grasses with the nematodes *Meloidogyne marylandii* and *Pratylenchus scribneri*. *Proceedings of the 2nd International Symposium on* Acremonium/*Grass Interactions*, pp. 156–160.

Harborne, J.B. (1993) *Introduction to Ecological Biochemistry*. Academic Press, London.

Hoveland, C.S. (1993) Importance and economic significance of the *Acremonium* endophytes to performance of animals and grass plant. *Agriculture, Ecosystems and Environment* 44, 3–12.

Johnson, M.C., Bush, L.P. and Siegel, M.R. (1986) Infection of tall fescue with *Acremonium coenophialum* by means of callus culture. *Plant Disease* 70, 380–382.

Joost, R.E. (1995) *Acremonium* in fescue and ryegrass – boon or bane – a review. *Journal of Animal Science* 73, 881–888.

Kearney, J.F., Parrott, W.A. and Hill, N.S. (1991) Infection of somatic embryos of tall fescue with *Acremonium coenophialum*. *Crop Science* 31, 979–984.

Keogh, R.G., Tapper, B.A. and Fletcher, R.H. (1996) Distributions of the fungal endophyte *Acremonium lolii*, and of the alkaloids lolitrem B and peramine, within perennial ryegrass. *New Zealand Journal of Agricultural Research* 39, 121 –127.

Kimmons, C.A., Gwinn, K.D. and Bernard, E.C. (1989) Reproduction of selected nematode species on endophyte-infected tall fescue. *Phytopathology* 79, 374.

Kimmons, C.A., Gwinn, K.D. and Bernard, E.C. (1990) Nematode reproduction on endophyte-infected and endophyte-free tall fescue. *Plant Disease* 74, 757–761.

Kirkpatrick, T.L., Barham, J.D. and Bateman, R.J. (1990) Host status for *Meloidogyne graminis* of tall fescue selections and clones with and without the endophyte *Acremonium coenophialum*. *Proceedings of the 1st International Symposium on* Acremonium/*Grass Interactions*, New Orleans, pp. 154–156.

Latch, G.C.M. (1993) Physiological interactions of endophytic fungi and their hosts – biotic stress tolerance imparted to grasses by endophytes. *Agriculture, Ecosystems and Environment* 44, 143–156.

Latch, G.C.M. (1994) Influence of *Acremonium* endophytes on perennial grass improvement. *New Zealand Journal of Agricultural Research* 37, 311–318.

Latch, G.C.M. and Christensen, M.J. (1985) Artificial infection of grasses with endophyte. *Annals of Applied Biology* 107, 17–24.

Latch, G.C.M. and Fletcher, L.R. (1997) Beneficial use of grasses artificially infected with endophytes, non-toxic to animals. *Proceedings of the XVIII International Grassland Congress*, Canada, pp. 11.7–11.8.

Lewis, G.C. (1994) Incidence of infection of grasses by endophytic fungi in the UK, and effects of infection on animal health, pest and disease damage and plant growth. *Proceedings of the 2nd International Conference on Harmful and Beneficial Microorganisms in Pastures, Turf and Grassland*, Paderborn, Germany, 1993, pp. 161–167.

Lewis, G.C. (1997) Significance of endophyte toxicosis and current practices for dealing with the problem in Europe. In: Bacon, C.W. and Hill, N.S. (eds) Neotyphodium/*Grass Interactions*. Plenum Press, New York, pp. 377–382.

Lewis, G.C. and Vaughan, B. (1995) Effect of hydroponic culture of ten genotypes of perennial ryegrass on infection by the fungal endophyte *Acremonium lolii*. Tests of Agrochemicals and Cultivars No. 16. *Annals of Applied Biology* 126 (Suppl.), 80–81.

Lewis, G.C., Ravel, C., Naafa, W., Astier, C. and Charmet, G. (1997) Occurrence of *Acremonium*-endophytes of wild populations of *Lolium* spp. in European countries and a relationship between level of infection and climate in France. *Annals of Applied Biology* 130, 227–238.

di Menna, M.E., Mortimer, P.H., Prestidge, R.A., Hawkes, A.D. and Sprosen, J.M. (1992) Lolitrem B concentrations, counts of *Acremonium lolii* hyphae, and the incidence of ryegrass staggers in lambs on plots of *A. lolii*-infected perennial ryegrass. *New Zealand Journal of Agricultural Research* 35, 211–217.

Musgrave, D.R. (1984) Detection of an endophytic fungus of *Lolium perenne* using enzyme-linked immunosorbent assay (ELISA). *New Zealand Journal of Agricultural Research* 27, 283–288.

Naffaa, W., Astier, C., Ravel, C. and Guillaumin, J.J. (1999) Creation of stable associations between perennial ryegrass or tall fescue and fungal endophytes. *Agronomie* 19, 133–144.

O'Day, M.H., Bailey, W.C. and Niblack, T.L. (1990) Phytonematode communities in tall fescue varieties with varying levels of endophyte infection. *Proceedings of the 1st International Symposium on* Acremonium/*Grass Interactions*, New Orleans, pp. 170–172.

O'Day, M.H., Niblack, T.L. and Bailey, W.C. (1993) Phytoparasitic nematode populations in *Festuca arundinacea* field plots in southwestern Missouri. *Journal of Nematology* 25, 900–906.

Oldenburg, E. (1996) Occurrence of the alkaloid lolitrem B in endophyte-infected *Lolium perenne*. *Proceedings of the 2nd International Conference on Harmful and Beneficial Microorganisms in Pastures, Turf and Grassland*, Paderborn, Germany, 1995, pp. 95–101.

Oldenburg, E. (1997) Endophytic fungi and alkaloid production in perennial ryegrass in Germany. *Grass and Forage Science* 52, 425–431.

Oliveira, J.A. and Castro, V. (1997) Incidence and viability of *Acremonium* endophytes in tall fescue accessions from North Spain. *Genetic Resources and Crop Evolution* 44, 519–522.

Oliveira, J.A., Rottinghaus, G.E., Collar, J. and Castro, P. (1997) The perennial ryegrass endophyte in Galicia, Northwest Spain. *Journal of Agricultural Science, Cambridge* 129, 173–177.

O'Sullivan, B.D. and Latch, G.C.M. (1993) Infection of plantlets, derived from ryegrass and tall fescue meristems, with *Acremonium* endophytes. *Proceedings of the 2nd International Symposium on* Acremonium/*Grass Interactions*, pp. 16–17.

Pedersen, J.F., Rodriguez-Kabana, R. and Shelby, R.A. (1988) Ryegrass cultivars and endophyte in tall fescue affect nematodes in grass and succeeding soybean. *Agronomy Journal* 80, 811–814.

Popay, A.J. and Rowan, D.D. (1994) Endophytic fungi as mediators of plant–insect interactions. In: Bernays, E.A. (ed.) *Insect–Plant Interactions*, Vol. V. CRC Press, Boca Raton, Florida, pp. 83–103.

Popay, A.J., Prestidge, R.A., Rowan, D.D. and Dymock, J.J. (1990) The role of *Acremonium lolii* mycotoxins in insect resistance of perennial ryegrass (*Lolium perenne*). *Proceedings of the 1st International Symposium on* Acremonium/*Grass Interactions*, New Orleans, pp. 44–48.

Potter, D.A., Patterson, C.G. and Redmond, C.T. (1992) Influence of turfgrass species and tall fescue endophyte on feeding ecology of Japanese beetle and southern masked chafer grubs (Coleoptera: Scarabaeidae). *Journal of Economic Entomology* 85, 900–909.

Ravel, C., Michalakis, Y. and Charmet, G. (1997) The effect of imperfect transmission on the frequency of mutualistic seed-borne endophytes in natural populations of grasses. *Oikos* 80, 18–24.

Ravel, C., Wartelle, D. and Charmet, G. (1994) Artificial infection of tillers from perennial ryegrass mature plants with *Acremonium* endophytes. *International Conference on Harmful and Beneficial Microorganisms in Grassland, Pastures and Turf*, Paderborn, Germany, pp. 123–125.

Roberts, C.A., Marek, S.M., Niblack, T.L. and Karr, A.L. (1992) Parasitic *Meloidogyne* and mutualistic *Acremonium* increase chitinase in tall fescue. *Journal of Chemical Ecology* 18, 1107–1116.

Ruess, L., Garcia Zapata, E.J. and Dighton, J. (2000) Food preferences of the fungal feeding nematode, *Aphelenchoides* sp. *Nematology* 2, 223–230.

Saikkonen, K., Faeth, S.H., Helander, M. and Sullivan, T.J. (1998) Fungal endophytes: a continuum of interactions with host plants. *Annual Review of Ecology and Systematics* 29, 319–344.

Scales, G.H., Knight, T.L. and Saville, D.J. (1995) Parasites and production performance of grazing lambs. *New Zealand Journal of Agricultural Research* 38, 237–247.

Schardl, C.L. (1996) *Epichloë* species: fungal symbionts of grasses. *Annual Review of Phytopathology* 34, 109–130.

Schardl, C.L., Leuchtann, A., Tsai, H.F., Collett, M.A., Watt, D.M. and Scott, D.B. (1994) Origin of a fungal symbiont of perennial ryegrass by interspecific hybridization of a mutualist with the ryegrass choke pathogen, *Epichloë typhina*. *Genetics* 136, 1307–1317.

Schmidt, D. (1993) Effects of *Acremonium uncinatum* and a *Phialophora*-like endophyte on the vigour, insect and disease resistance of meadow fescue. *Proceedings of the 2nd International Symposium on* Acremonium/*Grass Interactions*, pp. 185–188.

Schöberlein, W., Pfannmöller, M., Eggestein, S. and Szabová, M. (1997) Investigation of interactions between *Acremonium uncinatum* in *Festuca pratensis* and various nematode species in the soil. In: Bacon, C.W. and Hill, N.S. (eds) Neotyphodium/*Grass Interactions*. Plenum Press, New York, pp. 201–203.

Siegel, M.R., Latch, G.C.M., Bush, L.P., Fannin, F.F., Rowan, D.D., Tapper, B.A., Bacon, C.W. and Johnson, M.C. (1990) Fungal endophyte-infected grasses: alkaloid accumulation and aphid response. *Journal of Chemical Ecology* 16, 3301–3315.

Stewart, T.M., Mercer, C.F. and Grant, J.L. (1993) Development of *Meloidogyne naasi* on endophyte-infected and endophyte-free perennial ryegrass. *Australasian Plant Pathology* 22, 40–41.

do Valle Ribeiro, M.A.M. (1996) Ryegrass endophytes in Irish pastures. *Irish Farmer's Journal*, April 13, 34–35.

Watson, R.N., Prestidge, R.A. and Ball, O.J.-P. (1993) Suppression of white clover by ryegrass infected with *Acremonium* endophyte. *Proceedings of the 2nd International Symposium on* Acremonium/*Grass Interactions*, pp. 218–221.

Watson, R.N., Hume, D.E., Bell, N.L. and Neville, F.J. (1995) Plant-parasitic nematodes associated with perennial ryegrass and tall fescue with and without *Acremonium* endophyte. *Proceedings of the New Zealand Plant Protection Conference* 48, 199–203.

West, C.P., Izekor, E., Oosterhuis, D.M. and Robbins, R.T. (1988) The effect of *Acremonium coenophialum* on the growth and nematode infestation of tall fescue. *Plant and Soil* 112, 3–6.

West, C.P., Izekor, E., Elmi, A., Robbins, R.T. and Turner, K.E. (1989) Endophyte effects on drought tolerance, nematode infestation and persistence of tall fescue. *Proceedings of the Arkansas Fescue Toxicosis Conference, Arkansas Experimental Station Special Report* 140, 23–34.

West, C.P., Izekor, E., Robbins, R.T., Gergerich, R. and Mahmood, T. (1990) *Acremonium coenophialum* effects on infestations of barley yellow dwarf virus and soil-borne nematodes and insects in tall fescue. *Proceedings of the 1st International Symposium on* Acremonium/*Grass Interactions*, New Orleans, pp. 196–198.

Yeates, G.W. (1999) Effects of plants on nematode community structure. *Annual Review of Phytopathology* 37, 127–149.

Yeates, G.W., Bongers, T, de Goede, R.G.M., Freckman, D.W. and Georgieva, S.S. (1993) Feeding habits in soil nematode families and genera – an outline for soil ecologists. *Journal of Nematology* 25, 315–331.

Yue, Q., Logendra, S., Freehof, A. and Richardson, M.D. (1997) Alkaloids of turf-type fine fescue (*Festuca* spp.). In: Bacon, C.W. and Hill, N.S. (eds) Neotyphodium/*Grass Interactions*. Plenum Press, Oxford, pp. 285–287.

Zabalgogeazcoa, I., Vasquez de Aldana, B.R., Garcia Criado, B. and Garcia Ciudad, A. (1999) The infection of *Festuca rubra* by the endophyte *Epichlöe. festucae* in Mediterranean permanent grasslands. *Grass and Forage Science* 54, 91–95.

5

Feeding on Plant-pathogenic Fungi by Invertebrates: Comparison with Saprotrophic and Mycorrhizal Systems

T.P. McGonigle[1,2] and M. Hyakumachi[1]

[1]*Faculty of Agriculture, Gifu University, Gifu 501-1193, Japan;*
[2]*Department of Biological Sciences, Idaho State University, Campus Box 8044, ID 83209, USA*

Introduction

This chapter compares the effects of invertebrate grazing on pathogenic fungi to effects on saprotrophic and mycorrhizal fungi. Reviews on invertebrate grazing of saprotrophic (Reichle, 1977; Ingham, 1992; McGonigle, 1995) and mycorrhizal (Fitter and Sanders, 1992; McGonigle, 1995) fungi indicated a significant impact of fungivory on function. For saprotrophic fungi, function has been evaluated in terms of nutrient mineralization in an ecosystem context, whereas for mycorrhizal fungi, the focus has been on function in terms of nutrient acquisition by plants. Other reviews have discussed the role of grazing in determining the community structure of saprotrophic fungi (Visser, 1985; McGonigle, 1997). In contrast, reviews of grazing on plant-pathogenic fungi have focused on plant responses as seen in either disease suppression by reductions in fungal vigour (Curl, 1988; Curl and Harper, 1990) or as seen in disease exacerbation by plant wounding and inoculum vectoring (Kevan, 1965; Beute and Benson, 1979). With respect to pathogenic fungi, we limit ourselves here to consideration of the impact of grazing on the condition of the grazed fungus. For discussion of plant wounding and transfer of pathogen inoculum, the reader is referred to Beute and Benson (1979).

Broadly speaking, research on pathogenic, saprotrophic and mycorrhizal fungi has been done separately. The approach taken in this chapter is to summarize available information about the occurrence of grazing on plant-pathogenic fungi, and to evaluate this information in terms of the current understanding of the significance of grazing of saprotrophs and mycorrhizas. Our aim was to find similarities and differences in invertebrate grazing on these three types of fungi, and in this way to try to understand better the significance of invertebrate grazing.

The separation of the studies of pathogens, saprotrophs and mycorrhizas has made the evaluation of interactions among these co-occurring groups uncommon. Yet, such interactions are potentially important. Examples to show the role of such interactions in ecosystem processes can be found for pathogens and mycorrhizas (Newsham *et al.*, 1995a), for saprotrophs and mycorrhizas (Setala, 1995), and for pathogens and saprotrophs (Cook and Baker, 1983). The last section of this chapter focuses on these interactions and how they are affected by grazing.

Grazing on Soil-borne Pathogenic Fungi

Grazing on pathogenic fungi by microarthropods, nematodes and amoebae has been extensively studied. Modes of feeding differ among these animals. Springtails and oribatid mites have mouthparts to engulf particles of hyphae and spores, whereas some prostigmata and acarid mites have piercing mouthparts to suck out protoplasm (Moore *et al.*, 1988). Fungivorous nematodes pierce hyphae with a stylet and pump out fluid contents, leaving the hyphal wall mostly intact (Freckman and Baldwin, 1990). Amoebae cut and ingest a disc of wall from hyphae or sclerotia and withdraw the protoplasm, leaving grazed hyphae and sclerotia with a perforated appearance (Old and Chakraborty, 1986). Feeding on pathogenic fungi by dipterous larvae and by earthworms has sometimes been reported.

Microarthropods

Beginning with the work of Curl (1979), the collembola *Proisotoma minuta* and *Onychiurus encarpatus* were studied in relation to soil-borne diseases of cotton in Alabama, USA. In the top 20–25 cm of the profile, these collembola were up to three times more abundant directly alongside cotton plants compared with 20 cm away (Wiggins *et al.*, 1979). The collembola grazed various plant-pathogenic fungi such that fungal growth on agar was prevented (Curl *et al.*, 1985): the fungi included *Rhizoctonia solani* isolated locally and a virulent agent of cotton seedling disease and *Fusarium oxysporum* f. sp. *vasinfectum*, which causes cotton wilt. Further, these fungi along with other pathogens were eaten by *P. minuta* in preference to *Penicillium* and *Aspergillus* (Wiggins and Curl, 1979).

Experiments with *Rhizoctonia*-inoculated sterilized and non-sterilized soils in glass tubes and in pots (Curl, 1979; Lartey *et al.*, 1991) showed that these collembola can effectively suppress the disease caused by *Rhizoctonia* in cotton seedlings. However, the densities of collembola needed to effect this disease suppression under controlled conditions were typically an order of magnitude higher (Curl, 1979; Lartey *et al.*, 1991) than the maximum densities recorded in the field (Wiggins *et al.*, 1979; Rickerl *et al.*, 1989). This difference in population density suggests that the collembola in the field may constitute one of a number of agents that serve to restrict pathogen growth, but that they are not the main factor. Further, maximum densities in the field were detected under moderate or high soil moisture conditions, which did not reliably coincide with the timing of the greatest disease threat to seedlings (Rickerl *et al.*, 1989).

Nakamura *et al.* (1991a) tested the palatability of seven cultured pathogenic fungi in the genera *Fusarium*, *Rhizoctonia*, *Rosellinia*, *Cladosporium*, *Pythium* and *Helicobasidium* for four species of oribatid mites and seven species of collembola from crop fields. Among the collembola–fungus pairings, feeding was noted for 33 out of 49 combinations, with animal multiplication in 18 combinations (Nakamura *et al.*, 1991a). For the mites, feeding and multiplication were noted for only five and two out of 28 combinations, respectively (Nakamura *et al.*, 1991a). No feeding on these seven pathogenic fungi was noted for eight species of enchytraeids (Nakamura, 1993).

The collembola *Folsomia hidakana* was able to destroy mycelium of *R. solani* on agar and suppress disease occurrence in pot-grown seedlings of winter radish, cabbage, cucumber and burdock that had been inoculated with this pathogen (Nakamura *et al.*, 1991b). Similar results were noted for the collembola *Sinella curviseta* with *Fusarium oxysporum* f. sp. *cucumerinum* disease of cucumber (Nakamura *et al.*, 1992), and for the oribatid *Scheloribates azumaensis* with *R. solani* and winter radish (Enami and Nakamura, 1996). These studies in Japanese crop systems have shown a potential role for microarthropod grazing of pathogenic fungi to restrict disease severity in the field, but no animal population data were presented. Enami and Nakamura (1996) noted that *S. azumaensis* was tested in the laboratory at densities in excess of those found in the field, just as was the case for the pot studies on collembola in Alabama (Curl, 1979; Lartey *et al.*, 1991).

However, recent research has found microarthropod grazing can have an impact on pathogens at animal densities more similar to those in the field. Young sprouts from seed potatoes are attacked by *R. solani* in spring in The Netherlands. Various mites and springtails including *Folsomia fimetaria* successfully reduced disease severity in microcosms kept at cool temperatures representative of field conditions (Bollen *et al.*, 1991; Lootsma and Scholte, 1997a). Further experiments showed that the disease suppression was not sensitive to mild drought stress (Lootsma and Scholte, 1997b). At intermediate soil moisture levels and high pathogen inoculum density, *F. fimetaria* without other fauna reduced disease ratings from 75% to 54% at a population of 105 dm^{-3} (Lootsma and Scholte, 1997b). This collembola density converts to 16,000 m^{-2} for a 15-cm soil depth. Such a density is not atypical for collembola populations in the field, although agricultural management can restrict animal numbers (Butcher *et al.*, 1971).

Nematodes

The pot experiment of Rhodes and Linford (1959) showed that the fungivorous nematode *Aphelenchus avenae* could successfully control pathogenicity of *Pythium* for maize. Shortly thereafter, Mankau and Mankau (1963) and Townshend (1964) showed that *A. avenae* eats a wide range of pathogenic and saprotrophic fungi. Pathogenic fungi such as *Rhizoctonia*, *Verticillium*, *Fusarium* and *Cylindrocarpon* supported larger populations of *A. avenae* than saprotrophic fungi such as *Mucor*, *Penicillium* and *Rhizopus* (Mankau and Mankau, 1963; Townshend, 1964). Subsequent pot experiments showed similar suppressive action of *A. avenae* against *R. solani* on bean (Barker, 1964; Klink and Barker, 1968). *A. avenae* was also active against *Fusarium oxysporum* f. sp. *pisi* on *Pisum* (Klink and Barker, 1968), and against *R. solani* and *Fusarium solani* on *Medicago* (Barnes *et al.*, 1981). As frequently stated (Rhodes and Linford, 1959; Barker, 1964; Klink and Barker, 1968), the concentrations of nematodes in these pot experiments were in excess of those in the field, especially at times of the year with high disease risk. An example of a nematode population density from these experiments is 50 animals cm^{-3}, which brought about a 50% reduction in disease index (Barker, 1964). In contrast, the nematode community in the top 5 cm of a short-grass prairie increased from 20 g^{-1} dry soil in spring to a seasonal maximum of 80 g^{-1} dry soil in autumn (Ingham *et al.*, 1986), but with 65% of these nematodes being bacterivores (Ingham *et al.*, 1985). Across greater depths, the density of the total nematode community can be even less; Freckman (1988) gives a range of $0.3–9.0 \times 10^6$ m^{-2}. Klink and Barker (1968) calculated that 4000–6000 nematodes were needed per millilitre of laboratory culture of fungal inoculum to suppress soil-borne disease fungi in pot experiments.

Nematicides were found to increase disease severity of *R. solani* on sprouting seed potatoes in The Netherlands (Hofman *et al.*, 1991). This effect indicates that mycophagous nematodes normally keep the disease partially suppressed, which is an interpretation that is consistent with the field occurrence of these fungivores in the region (Hofman and Jacob, 1989). Microcosm experiments were able to show disease reductions by putting into the microcosms the mycophagous nematode *A. avenae* isolated locally from potato fields (Bollen *et al.*, 1991; Lootsma and Scholte, 1997a). In addition, nematodes were not restricted by mild water-stress conditions. Instead, migration of nematodes to roots in a drying soil caused more effective disease control (Lootsma and Scholte, 1997b). Migration of bacterivorous nematodes to roots has been shown in response to a burst of bacterial growth on decaying roots (Griffiths and Caul, 1993). The effects of collembola and nematodes appear to be additive (Lootsma and Scholte, 1997b). In the experiment of Lootsma and Scholte (1997b) conducted at 0.07 g H_2O g^{-1} dry soil and under high pathogen inoculum density, disease rating was reduced from 72% to 41% in systems with nematode populations that ranged from 19,000 to 46,000 dm^{-3} through time. Although nematode densities in potato fields as sampled by auger were an order of magnitude lower than this, rhizosphere densities were considerably higher, falling in the range 50–500 g^{-1} of fresh root and adhering soil combined (Hofman and Jacob, 1989).

Amoebae

Old (1967) found that 2–4 μm holes appeared in pigmented conidia of the small-grain root-rot fungus *Cochliobolus sativus* after incubation in moist soil. After 10 years of investigation, Old (1977) was able to report that the agent causing these holes was a mycophagous amoeba. Reports followed of observations and isolations of similar and various other mycophagous amoebae from soils of Ontario, Canada (Anderson and Patrick, 1978), France (Alabouvette *et al.*, 1979), Washington State, USA (Homma *et al.*, 1979), South Australia (Chakraborty *et al.*, 1983) and Japan (Homma and Ishii, 1984). In all cases, the amoebae ate soil-borne plant-pathogenic fungi.

The review by Old and Chakraborty (1986) listed 12 genera of mycophagous amoebae, with each species of amoeba typically able to consume a variety of pathogenic fungi. In addition to *Cochliobolus*, susceptible fungi contain several genera including *Fusarium*, *Gaeumannomyces*, *Phytophthora*, *Pythium*, *Rhizoctonia* and *Verticillium*. Both hyphae and spores are eaten.

Based on their occurrence, mycophagous amoebae appear to be one among several factors (Hornby, 1990) contributing to the ability of suppressive soils to limit take-all disease of wheat by *Gaeumannomyces graminis tritici* (Chakraborty and Warcup, 1983, 1984). Populations of mycophagous amoebae were consistently higher in take-all suppressive compared to non-suppressive Australian soils throughout two wheat crops with an intervening fallow (Chakraborty, 1983). In a greenhouse experiment, adding mycophagous amoebae was as successful as adding suppressive soil to restrict take-all development (Chakraborty and Warcup, 1985). However, mycophagous amoebae may not always be involved in disease decline. For sugar beet at three sites in Japan with root rot caused by *Rhizoctonia solani* AG-2–2, densities of mycophagous amoebae varied from place to place within each site. Yet these densities were unrelated to the occurrence of patches in the field with healthy plants, patches with diseased plants and patches with healthy plants where previously disease was severe (Hyakumachi *et al.*, 1982).

Other fauna

Grazing on the sclerotia of *R. solani* AG-2–2 by larvae of the fungus gnat *Pnyxia scabiei* (Diptera: Sciaridae) was evaluated in several studies in the 1980s (Naito, 1988; Naito and Sugimoto, 1988; Naito *et al.*, 1988), which were later summarized by Naito and Makino (1995). Various laboratory and field experiments revealed that although this larva occurs widely but at low population densities among crop fields in the region, populations increase in response to local *Rhizoctonia* disease development. As well as hyphal grazing, fragments of sclerotia were ingested, with no fungal vitality after gut passage. Additions of either larvae or residues of diseased tissues with larvae were successful in controlling damping off in pot studies (Naito and Makino, 1995). The pattern that emerges from this work is that the populations of fly larvae follow one step behind the disease development, increasing in response to an abundant food resource (Fig. 5.1). In this way, grazing by fly larvae may help to limit the severity of disease. However, the major factor in *Rhizoctonia* disease decline is thought to be a change in the population of soil microbes (Hyakumachi, 1996).

Conversion of wheat cultivation to reduced tillage in Australia was found to increase bare-patch disease caused by *R. solani*. In pot trials, Stephens *et al.* (1993, 1994) found that earthworms were just as effective in reducing damage caused by *Rhizoctonia* as was soil disturbance. To explain this effect, feeding of the earthworms on the fungi was suggested as one possible mechanism among several, including nitrogen mineralization and reduced substrate availability for saprotrophic nutrition (Stephens *et al.*, 1993). Similarly, earthworms reduced the severity of club-root galls of cabbage caused by *Plasmodiophora brassicae* in pot experiments (Nakamura, 1996; Nakamura *et al.*, 1995). Densities of resting spores of this myxomycete were unchanged by earthworm activity, but passage through the gut was thought to have rendered them ineffective (Nakamura, 1996; Nakamura *et al.*, 1995). Using worm compost, Szczech *et al.* (1993) reduced disease severity of *Phytophthora nicotianae* var. *nicotianae* and *Fusarium oxysporum* f. sp. *lycopersici* on tomato. The direct involvement of grazing by earthworms was not established.

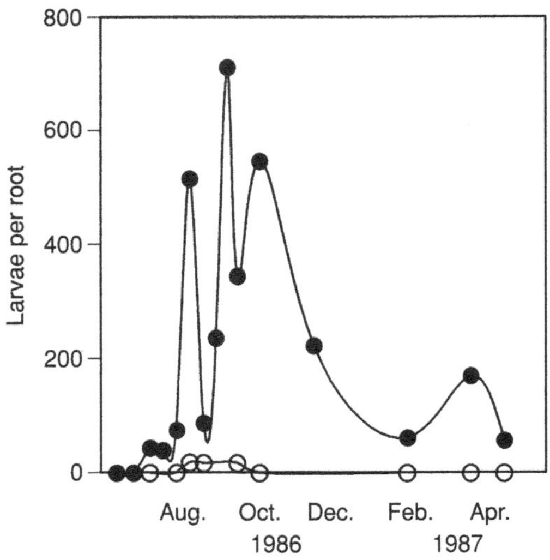

Fig. 5.1. Numbers of larvae of the fungus gnat *Pnyxia scabiei* on diseased (closed circles) and healthy (open circles) roots of sugar beet in Japan. Disease ratings for *Rhizoctonia solani* were high in 1986 from August to October, inclusive. (Modified from Naito and Makino, 1995.)

Grazing on Plant-pathogenic Fungi Above Ground

Although reports of fungivore grazing on plant-pathogenic fungi on shoots are rare, Kessler (1990) described one clear case involving reductions in leaf-spot (*Gnomonia leptostyla*) disease incidence for black walnut *Juglans nigra* when grown in association with the woody shrub *Elaeagnus umbellata*. Experiments revealed that microarthropods at high densities in the *E. umbellata* litter grazed heavily on perithecia of *G. leptostyla* on the surface of the walnut leaves, thereby restricting persistence of disease inoculum over the winter and reducing disease the next year (Kessler, 1990).

Batra and Stavely (1994) described migration of spider mites to uredinia of *Uromyces appendiculatus* on *Phaseolus vulgaris*, causing two- to sixfold greater mite numbers on diseased leaves. The mites carried uredinospores to rust-free plants. However, ingestion of fungal material by the mites was not shown; these fluid feeders were imbibing pale yellow liquid droplets that formed on uredinia under near-saturated moisture conditions (Batra and Stavely, 1994).

Although we could not find any similar examples in the literature, the unequivocal nature of the study of Kessler (1990) leads us to speculate that many other case histories of grazing on plant-pathogenic fungi above ground are waiting to be discovered in nature.

Grazing in Saprotrophic Systems

The direct effect of grazing on the function of saprotrophic fungi involves the excretion of nitrogen from the bodies of the grazers. In this way, a contribution is made to community function as the nitrogen released is in a mineralized or easily mineralized form. Mites excrete guanine or uric acid by means of tubules that lead into the posterior of the alimentary canal (Evans, 1992). The main excretory product for nematodes is ammonia, which can pass through the body wall (Freckman, 1988). For protozoa, ammonia and urea are released (Anderson *et al.*, 1981).

McGonigle (1995) reviewed papers for the direct contribution of fungivore grazing to nitrogen mineralization. Fungivores typically comprise less than half of the fauna, with bacterivores as the other main group. There are also carnivores, herbivores and saprovores. Ecosystem nitrogen budget studies for pine forest (Persson *et al.*, 1980; Persson, 1983), short-grass prairie (Hunt *et al.*, 1987) and crop fields (Paustian *et al.*, 1990) estimated that 10–50% of nitrogen mineralization is attributable to excretion by the soil fauna. However, when the fungivores were considered alone, only 2–13% of ecosystem nitrogen mineralization was estimated to come from mycophage excretion (Persson, 1983; Hunt *et al.*, 1987).

Indirect effects of grazing are comminution and mixing of substrate, and changes in vigour of the grazed fungus, all of which can modify mineralization or immobilization by the microbes. Although there is consensus that these indirect effects are important (Crossley, 1977; Visser, 1985; Setala and Huhta, 1991), the quantification of them remains intractable. At the very least, we can say that if 10–50% of mineralization is attributable to excretion from fauna, then indirect effects of grazing must play a part in somewhere between none and all of the remaining mineralization by the microbes. As revealed by biocide experiments in the field, the combination of the direct and indirect effects of grazing has a major influence on nutrient mineralization and immobilization (Beare *et al.*, 1992).

Laboratory experiments with saprotrophic fungi have shown clear effects of grazing on fungal community structure. To have any such effects, grazing must be selective (Crawley, 1983). Preferences among fungal food sources are often noted for fungivorous microarthropods (Shaw, 1988; Chen *et al.*, 1995; Kaneko *et al.*, 1995), although some species are more generalist fungal feeders (Chen *et al.*, 1995).

Different outcomes are expected if the grazed fungus is dominant or subordinate in the community in the absence of grazing (McGonigle, 1997). Suppression of an otherwise dominant fungus by grazing will promote greater diversity in the fungal community. Newell (1984a) found that across a range of temperatures *Marasmius androsaceous* grew faster than the sympatric basidiomycete *Mycena galopus*. Thereby, in ungrazed systems *M. androsaceous* more extensively colonized needles than *M. galopus*. However, under grazing pressure from collembola with a preference for *M. androsaceous*, substrate colonization by *M. galopus* was more extensive (Newell, 1984b).

In contrast, where two saprotrophic fungi are co-dominant, selective grazing on one of them will act to polarize the community, and so reduce diversity. Parkinson *et al.* (1979) found that two basidiomycete aspen-litter saprotrophs were equally abundant in the absence of grazing, either when grown alone or together. Yet, when the fungi were grown together, collembola strongly suppressed the growth of the palatable fungus, so that the less-grazed fungus came to dominate the microcosms (Parkinson *et al.*, 1979).

Although grazing on a dominant species can increase diversity by making the fungi in the community be of more equal abundances, high intensities of grazing can eliminate sensitive species altogether, thereby reducing the species richness. Wicklow and Yocum (1982) varied grazing intensity by adding different numbers of mycophagous sciarid fly larvae to dung microcosms. From 14 saprotrophic species isolated in the ungrazed dung, increments of grazing pressure reduced the number of fungal species that were isolated. Above 10 larvae g^{-1} dung, the community richness remained within six and nine grazing-tolerant species (Wicklow and Yocum, 1982).

A further effect of grazing on communities is to speed the transition from early- to late-successional stages where such successions are in progress. Klironomos *et al.* (1992) showed that collembola were able to speed up the transition from primary to secondary fungal saprotrophs on conifer needles by several weeks, as compared to ungrazed microcosms.

Grazing in Mycorrhizal Systems

Grazing on arbuscular mycorrhizal (AM) fungi is thought to act by severing the connections (Fig. 5.2) between the mycelium that is inside and outside the roots (Fitter and Sanders, 1992). These connections occur at a frequency of about 1 mm^{-1} (Fitter, 1991). Where these connections are damaged, the fungus in the root retains the capacity to withdraw carbon from the host, but the extraradical mycelium has a reduced ability to transfer phosphorus from soil to plant. This effect of grazing is relevant to arbuscular mycorrhizas but not ectomycorrhizas, because the latter have extensive hyphal contact between root and extraradical mycelium.

The capacity of AM fungi to contribute to plant nutrition was restricted by fungivore grazing in pots (Warnock *et al.*, 1982; Finlay, 1985) and in field experiments (McGonigle and Fitter, 1988). However, Klironomos and Kendrick (1996) found that the springtail *Folsomia candida* had a clear preference for pigmented saprotrophic hyphae over hyaline AM hyphae. This preference suggests that grazing on AM fungi in the field may not occur widely enough to produce the impact on plant growth that has been seen in laboratory experiments (Warnock *et al.*, 1982; Finlay, 1985). Even though they do not feed extensively, however, high densities of collembola can cause physical damage to AM extraradical mycelium (Klironomos and Ursic, 1998).

Riffle (1971) investigated the grazing by the nematode *Aphelencoides cibolensis* on 43 species of fungi on agar, almost all of which were

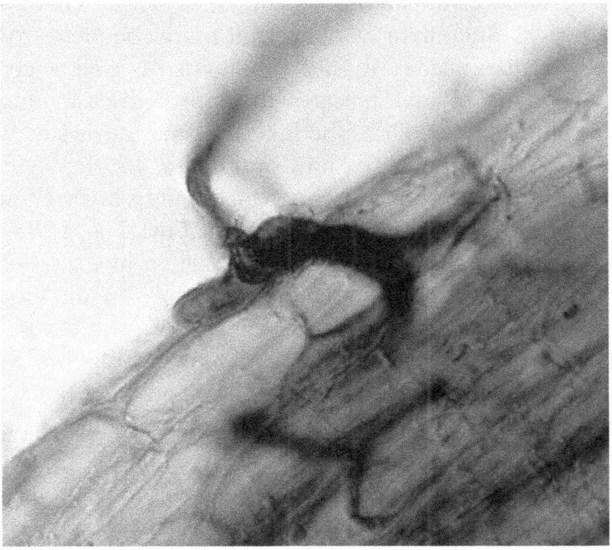

Fig. 5.2. A connection between the mycelium of an arbuscular mycorrhizal fungus inside and outside of a root. The swelling on the root surface is an appressorium.

ectomycorrhizal. The only pathogenic species included was *Armillaria mellea*. There was considerable variation in the responses of the nematode and of the fungi to the grazing (Riffle, 1971). In 11 species, the fungus supported good nematode development yet grew faster than the rate of consumption, yielding both healthy mycelia and dense nematode populations. In a separate group of 12 species, the fungi were not eaten and the growth and viability of the mycelium was good. Riffle (1971) suspected that the fungi produced substances which were toxic to the nematodes. In the third group of a further 13 species, the fungus was heavily grazed by the nematode, leading to destruction of the fungus and a large nematode population. *A. mellea* was in this third group. For the remaining seven species, both fungus and nematode did poorly due to consumption of aerial hyphae and slow growth of hyphae. The wide range of responses of fungus and nematode seen by Riffle (1971) means that generalizations about grazing on ectomycorrhizal fungi are difficult to make.

Plant-pathogenic Fungi Compared with Saprotrophic and Mycorrhizal Fungi

Effects of grazing on function of pathogenic fungi

Caution should be taken before concluding that disease damage in the field is tempered to a large extent by faunal grazing on pathogenic fungi. As the examples given above show, a clearer understanding of the significance of grazing will need more data for fauna population densities in association with disease in the field. Such data would permit more confident extrapolation of grazing experiments under controlled conditions. As well as concerns about effective populations of invertebrates to control disease fungi, there are two further complications. First, the invertebrates can act as vectors of the pathogenic fungi (Kevan, 1965; Shaw and Beute, 1979; Wiggins and Curl, 1979). Second, when fungi are scarce as a food source, invertebrates can cause damage to roots directly with their mouthparts (Curl *et al.*, 1988).

The effect of grazing to enhance mineralization by excretion of nitrogen from the bodies of the invertebrates must apply to all fungi alike. For saprotrophic fungi, this enhancement is viewed as a stimulation of function, because the mineralization process is evaluated within the context of the function of the ecosystem (Crossley, 1977). However, the functioning of plant-pathogenic fungi is usually seen in the context of disease occurrence on the plant. In one microcosm study with an emphasis on mineralization, *Fusarium oxysporum* caused the release of 20 μg N g^{-1} dry soil as ammonium from chitin (Trofymow and Coleman, 1982). This release was increased to 26 μg N g^{-1} dry soil by the presence of *A. avenae* in the microcosms (Trofymow and Coleman, 1982).

Pathogenic fungi are susceptible to grazing only when they are outside the host. Thereby, grazing could affect pathogen persistence outside the host, contact with and entry into the host, and the exit and dispersal from the host. In this regard, plant-pathogenic fungi resemble mycorrhizal fungi. The situation is different for saprotrophic fungi, because grazing is directly on to the absorptive mycelium. Exceptions are when access to the absorptive mycelium is restricted, such as with wood decomposition (Dyer *et al.*, 1992). For a pathogen outside the living host, Garrett (1976) distinguished between dormant and saprotrophic survival. In dormant survival, the fungus relies on mycelial reserves. In saprotrophic survival, the pathogen utilizes necrotrophic substrates in the period before entering a host and starting to consume biotrophically. Grazing would be expected to have a negative impact on a pathogen in a state of dormant survival or in the transition from a period of dormant survival to a biotrophic phase. This impact on the condition of the fungus would be negative, because grazing would constitute a drain on the resources of the fungus that are needed to overcome the host defences and establish biotrophic nutrition. Using the terminology of Garrett (1981), pathogenic root-infecting fungi can be divided into two groups. There are specialized parasites such as *Gauemannomyces* that overcome host defences but have weak saprotrophic capability, and there are unspecialized parasites such as *Pythium* and *Rhizoctonia* that have a stronger saprotrophic capability, but for which development in healthy roots is kept at low levels by plant defences. Specialized parasites in a state of dormant survival would be expected to be especially vulnerable to the impact of grazing. The carbon source for a plant-pathogenic fungus in a saprotrophic survival mode will transfer from outside to inside the root during the development of biotrophic nutrition. This transition occurs similarly for ectomycorrhizal fungi. Specialized parasites are more similar to AM fungi, with respect to this change of carbon resource, because of their very limited potential for carbon utilization outside the host. Specialized parasites such as *Gauemannomyces* make limited use of carbon sources within residues. In contrast, the strategy of AM fungi is to use the extensive lipid reserves in their large spores, and to simply grow from root to root using lipid reserves in vesicles. A lack of host specificity and a low palatability to grazers help AM fungi to persist outside the host as a mycelium.

Positive effects of grazing on fungal activity have been noted, although the mechanisms involved can differ between regular saprotrophs and pathogenic fungi undergoing saprotrophic nutrition. Collembola grazing was able to stimulate respiration of the pathogen *Botrytis cinerea* on agar at the time when the fungus was in a senescent phase following saprotrophic growth (Hanlon, 1981). In soil, low densities of *F. candida* increased the length-density of active hyphae of *F. oxysporum* from 3.2 to 4.8 m g^{-1}, although the mechanism of this stimulation was unclear (Moore, 1988). Two different mechanisms not related to the removal of senescent hyphae have been noted for saprotrophic fungi. These mechanisms are a change in fungal growth mode and an improvement in substrate quality. Respiration of the soil mould *Mortierella isabellina* on agar was stimulated by collembola (Bengtsson and Rundgren, 1983), but the increase was recorded in the period of days following cessation of a grazing episode. This response was later found to be due to the fungus switching from a slow-growing prostrate form to a faster growing more aerial form, which was associated with greater respiration (Hedlund *et al.*, 1991). The change in form was thought to be a strategy to try to escape grazing (Hedlund *et al.*, 1991). Hedlund and Augustsson (1995) saw a stimulation of respiration of *M. isabellina* in response to adding enchytraeid worms.

A stimulation of respiration was seen for saprotrophic fungi on oak-leaf litter following introduction of woodlice, but this was probably due to comminution of substrate, leading to greater saprotrophic activity (Hanlon and Anderson, 1980). Grazing on a saprotrophic fungus can also lead to stimulation of respiration by other microbes. Collembola feeding on the basidiomycete *Coriolus versicolor* in leaf-litter microcosms caused increased microbial respiration, but this was in association with a replacement of fungal with bacterial biomass (Hanlon and Anderson, 1979). This stimulation of respiration was caused by the collembola improving the condition of the substrate for bacteria. Ineson *et al.* (1982) also found increased fungal activity with collembola grazing in litter microcosms. Low densities of collembola increased hyphal mass from 20 to 32 mg g^{-1} litter, whereas high collembola densities reduced the fungi from 24 to 8 mg g^{-1} litter (Ineson *et al.*, 1982). However, it is possible to have a direct stimulation of the fungus that is not related to changes in the condition of the substrate. Growth and respiration of fungi coming out from millipede pellets increased in response to collembola grazing (van der Drift and Jansen, 1977). This stimulation of respiration was not caused by modification of the substrate as in the studies of Hanlon and Anderson (1979, 1980). The mechanism was unclear.

Stimulatory effects on function have also been noted for AM fungi. Greenhouse experiments showed a negative effect of grazing on function of AM fungi, although at relatively high fauna densities in excess of 100 dm^{-3} (Warnock *et al.*, 1982; Finlay, 1985). However, at low animal densities a stimulation of mycorrhizal function can occur (Finlay, 1985; Harris and Boerner, 1990). Shoot-P content of mycorrhizal *Allium porrum* increased from 0.2 to 0.5 mg P per shoot in response to low densities of collembola, although shoot-P content decreased with higher collembola populations (Finlay, 1985). A similar effect was seen for *Geranium robertianum*. In a greenhouse potting mix in pots, shoot-P content increased from 3.4 to 4.8 mg P per shoot in response to low densities of collembola, but under higher collembola densities a shoot-P content of 4.3 mg P per shoot was found (Harris and Boerner, 1990). The stimulation of shoot-P content for *G. robertianum* was associated with an increased concentration of plant-available soil-P (Harris and Boerner, 1990). Thus, low collembola densities appear to stimulate P mineralization without impeding mycorrhizal function.

Effects of grazing on communities of pathogenic fungi

When grazing on saprotrophic fungi affects the fungal community, those fungi are present in the same place and are utilizing the same or similar substrates for decomposition (Parkinson *et al.*, 1979; Wicklow and Yocum, 1982; Newell, 1984b; Klironomos *et al.*, 1992). Selective grazing on certain fungi can then lead to a relatively greater ability of the ungrazed fungi to use the resource at the exclusion of the sympatric and trophically similar species. Specificity in plant–pathogen associations would be expected to prevent such events. If a heavily grazed fungus were specific in its pathogenicity to one type of plant, then a separate and lightly grazed fungal pathogen specific to a different plant would be unaffected.

The effect of grazing on the community of plant-pathogenic fungi would be more similar to that seen for saprotrophic fungi where the pathogenic fungi are of high saprotrophic capability and are engaged in saprotrophic survival outside the host. Mycorrhizal fungi have low host specificity, and so the communities of these fungi should behave in a similar manner to those of saprotrophic fungi.

Grazing and Interactions Among Pathogens, Saprotrophs and Mycorrhizas

Interactions have been described between AM fungi and pathogenic fungi. Newsham *et al.* (1995a) inoculated seedlings of the annual grass *Vulpia ciliata* with *F. oxysporum* and with an AM species, and transplanted them into the field. The isolates for both types of inocula had been obtained from the same field site. Mycorrhizal colonization conferred on the grass partial resistance from both infection and disease losses related to *Fusarium*. Infection from the plant-pathogenic fungus *Embellisia chlamydospora* from field inocula was also suppressed (Newsham *et al.*, 1995a). The carbon cost of mycorrhizal colonization that is given up by a plant when that plant does not benefit from infection in terms of phosphorus nutrition can thereby be explained in evolutionary terms (Newsham *et al.*, 1995b). Various examples of similar interactions (Read, 1999) are now known. Azcon-Aguilar and Barea (1996) reviewed the possible mechanisms of these types of interactions between AM fungi and plant-pathogenic fungi. A host-mediated response to compensate for pathogen losses is possible, such as by improved phosphorus nutrition. However, this was shown not to be the case in the study of Newsham *et al.* (1995a). A more likely possibility is deletion competition between the two types of fungi for photosynthate or space to occupy in the roots. Elicitation of plant defence mechanisms by the AM fungus as the means by which disease resistance is brought about does not seem to occur (Azcon-Aguilar and Barea, 1996; Dassi *et al.*, 1998).

Setala (1995) noted an interaction between saprotrophic and ectomycorrhizal fungi in the field. When the soil fauna was manipulated to high levels, ectomycorrhizal biomass was reduced, but nitrogen nutrition of pine and birch was enhanced. This effect was explained (Setala, 1995) as a stimulation of nitrogen mineralization by saprotrophic microbes in response to greater grazing. The mycorrhizas were reduced to a lower yet functional level that worked in concert with the released nitrogen to enhance plant growth (Setala, 1995).

Interactions involving AM fungi also occur. McAllister *et al.* (1997) found that inoculation of pots separately with the saprotrophic fungi *Alternaria alternata* and *Fusarium equiseti* caused a reduction in colonization of maize and lettuce by the AM fungus *Glomus mosseae*. This response was seen only when the saprotrophs were added prior to the AM fungus (McAllister *et al.*, 1997), suggesting a similar mechanism operates as with interactions between AM and pathogenic fungi (Azcon-Aguilar and Barea, 1996).

Interactions between pathogenic fungi and saprotrophic fungi form a significant part of the biological control literature (Cook, 1993). The mechanisms of interaction are antibiosis, hyperparasitism, and competition for nutrients and infection sites (Cook and Baker, 1983).

We expect that the effect of grazing on interactions between plant-pathogenic, saprotrophic and mycorrhizal fungi will be determined by the selectivity of the grazing. If one fungus suffers heavier losses to grazing than another, the outcome of a competitive interaction between those two fungi for a resource, such as a root to colonize, will doubtless be affected. In this regard, it is noteworthy that pathogenic fungi are usually more palatable than saprotrophic fungi (Mankau and Mankau, 1963; Townshend, 1964; Wiggins and Curl, 1979), which in turn are preferred more than AM fungi (Klironomos and Kendrick, 1996). Ectomycorrhizal fungi are sometimes selected by microarthropods but are sometimes not eaten. For example, Shaw (1985) found that for *Onychiurus armatus* the ectomycorrhizal fungus *Lactarius rufus* was the most preferred food among the ectomycorrhizal and saprotrophic fungi that were tested, whereas the ectomycorrhizal fungus *Paxillus involutus* was not palatable.

The collembola *P. minuta* and *O. encarpatus* readily ate *Rhizoctonia* and *Fusarium*, but the disease-antagonist fungi *Gliocladium virens*, *Trichoderma harzianum* and *Laetisaria arvalis* (Curl *et al.*, 1988; Lartey *et al.*, 1989) are not palatable to these collembola. Lartey *et al.* (1994) found that a combined addition of *P. minuta* and these antagonist fungi to pots of non-sterilized soil inoculated with *R. solani* reduced disease severity more effectively in cotton than did the antagonist fungi alone.

Palatability of fungi to microarthropods is related to chemicals produced by the fungi. Bengtsson *et al.* (1988) found that *O. armatus* could locate fungal food sources by their odour. In addition, different chemical odours are produced by different fungi, and the chemicals from palatable species can serve as attractants in the absence of fungi (Bengtsson *et al.*, 1991). However, repugnant chemicals may also be involved for non-palatable fungal species. Fungi such as *G. virens* and *T. harzianum* produce antifungal enzymes (Jeffries, 1997), and it is tempting to speculate that the lack of palatability of these fungi is linked to the production of some extruded materials. Shaw (1985) suggested that toxins, where they occur in sporophores of mycorrhizal fungi, are incidental, and that their main function is to restrict feeding on the vegetative mycelium. If fungi produce toxins to protect the absorptive mycelium from grazing, then wood-degrading fungi would have no need of such chemicals. The reason is that the hyphae are ramified into the wood substrate, and so they are out of reach of the grazers. In keeping with the interpretation that toxins are to protect the vegetative mycelium from grazing, Shaw (1985) noted that toxin production was not found in a survey among wood-decomposing fungi, and that lack of edibility of those sporophores for humans where it occurred was due to their toughness.

Conclusions

Grazing occurs on plant-pathogenic fungi by invertebrates such as microarthropods, nematodes and protozoa. Grazing on soil-borne fungal pathogens has often been reported. Reports of invertebrates feeding on pathogenic fungi on plants above ground are rare, probably because this grazing has been mostly overlooked. For saprotrophic fungi, grazing can affect ecosystem function and fungal community structure. Grazing can reduce nutrient transfer to plants by mycorrhizal fungi. For pathogens, emphasis has been on the action by fauna to promote disease by wounding plants and to transfer inoculum among host individuals.

Similarities and differences were noted in a comparison of grazing by invertebrate fauna on plant-pathogenic, saprotrophic and mycorrhizal fungi. The direct effect of grazing to mineralize nitrogen by excretion from the bodies of the grazers is the same for pathogenic, saprotrophic and mycorrhizal fungi. For saprotrophic fungi, this process is often viewed as a contribution to ecosystem function. Mycorrhizal and pathogenic fungi are grazed outside the host and away from the site of biotrophic carbon absorption. In contrast, saprotrophic fungi are grazed directly at the site of carbon acquisition, unless access is restricted such as with wood decomposers. For systems with specificity between pathogen and host, we expect that effects of grazing on the structure of communities of pathogenic fungi would be less marked than for saprotrophs. For plant-pathogenic fungi, removal of senescent hyphae by grazing can cause an increase in microbe growth. For saprotrophic fungi, an alternate and more active fungal growth mode is sometimes adopted in response to grazing. Alternatively, grazing of saprotrophic fungi can enhance decomposition by improving the condition of the substrate. For mycorrhizas, fungivore grazing at low levels seems to enhance plant growth by improving phosphorus availability.

Feeding preferences are an important part of any effects of grazing. Pathogenic fungi are eaten in preference to saprotrophs, which in turn are more palatable than arbuscular mycorrhizal fungi. The palatability of ectomycorrhizal fungi is variable. Invariably, chemicals produced by the fungi seem to dictate this palatability.

Pathogenic, saprotrophic and mycorrhizal fungi typically co-occur in the rhizosphere and associated soil regions. The interactions between these fungi and the significance of grazing to modify those interactions need to be more thoroughly studied in order to develop a better understanding of the activities of these fungi in the field.

References

Alabouvette, C., Pussard, M. and Pons, R. (1979) Isolation and characterization of a mycophagous amoeba: *Thecamoeba granifera* subsp. *minor*. In: Schippers, B. and Gams, W. (eds) *Soil-borne Plant Pathogens*. Academic Press, London, pp. 629–633.

Anderson, R.V., Coleman, D.C. and Cole, C.V. (1981) Effects of saprotrophic grazing on net mineralization. *Ecological Bulletin* 33, 201–216.

Anderson, T.R. and Patrick, Z.A. (1978) Mycophagous amoeboid organisms from soil that perforate spores of *Thielaviopsis basicola* and *Cochliobolus sativus*. *Phytopathology* 68, 1618–1626.

Azcon-Aguilar, C. and Barea, J.M. (1996) Arbuscular mycorrhizas and biological control of soil-borne plant pathogens – an overview of the mechanisms involved. *Mycorrhiza* 6, 457–464.

Barker, K.R. (1964) On the disease reduction and reproduction of the nematode *Aphelenchus avenae* on isolates of *Rhizoctonia solani*. *Plant Disease Reporter* 48, 428–432.

Barnes, G.L., Russell, C.C. and Foster, W.D. (1981) *Aphelenchus avenae*, a potential biological control agent for root rot fungi. *Plant Disease* 65, 423–424.

Batra, L.R. and Stavely, J.R. (1994) Attraction of twospotted spider mite to bean rust uredinia. *Plant Disease* 78, 282–284.

Beare, M.H., Parmelee, R.W., Hendrix, P.F., Cheng, W., Coleman, D.C. and Crossley, D.A. (1992) Microbial interactions and effects on litter nitrogen and decomposition in agroecosytems. *Ecological Monographs* 62, 569–591.

Bengtsson, G. and Rundgren, S. (1983) Respiration and growth of a fungus, *Mortierella isabellina*, in response to grazing by *Onychiurus armatus* (Collembola). *Soil Biology and Biochemistry* 15, 469–473.

Bengtsson, G., Erlandsson, A. and Rundgren, S. (1988) Soil odour attracts soil Collembola. *Soil Biology and Biochemistry* 20, 25–30.

Bengtsson, G., Hedlund, K. and Rundgren, S. (1991) Selective odor perception in the soil Collembola by *Onychiurus armatus*. *Journal of Chemical Ecology* 17, 2113–2125.

Beute, M.K. and Benson, D.M. (1979) Relation of small soil fauna to plant disease. *Annual Review of Phytopathology* 17, 485–502.

Bollen, G.J., Middlekoop, J. and Hofman, T.W. (1991) Effects of soil fauna on infection of potato sprouts by *Rhizoctonia solani*. In: Beemster, A.B.R. (ed.) *Biotic Interactions and Soil-borne Diseases*. Elsevier, Amsterdam, pp. 27–34.

Butcher, J.W., Snider, R. and Snider, R.J. (1971) Bioecology of edaphic collembola and acarina. *Annual Review of Entomology* 16, 249–288.

Chakraborty, S. (1983) Population dynamics of mycophagous and other amoebae in soils suppressive to wheat take-all. *Soil Biology and Biochemistry* 15, 661–664.

Chakraborty, S. and Warcup, J.H. (1983) Soil amoebae and saprophytic survival of *Gaeumannomyces graminis tritici* in a suppressive pasture soil. *Soil Biology and Biochemistry* 15, 181–185.

Chakraborty S., and Warcup J.H. (1984) Populations of mycophagous and other amoebae in take-all suppressive soils. *Soil Biology and Biochemistry* 16, 197–199.

Chakraborty, S. and Warcup, J.H. (1985) Reduction of take-all by mycophagous amoebae in pot bioassays. In: Parker, C.A., Rovira, A.D., Moore, K.J., Wong, P.T.W. and Kollmorgan, J.F. (eds) *Ecology and Management of Soil-borne Plant Pathogens.* American Phytopathological Society, St Paul, Minnesota, pp. 107–109.

Chakraborty, S., Old, K.M. and Warcup, J.H. (1983) Amoebae from a take-all suppressive soil which feed on *Gaeumannomyces graminis tritici* and other soil fungi. *Soil Biology and Biochemistry* 15, 17–24.

Chen, B., Snider, R.J. and Snider, R.M. (1995) Food preference and effects of food type on the life history of some soil collembola. *Pedobiologia* 39, 496–505.

Cook, R.J. (1993) Making greater use of introduced microorganisms for biological control of plant pathogens. *Annual Review of Phytopatholgy* 31, 53–80.

Cook, R.J. and Baker, K.F. (1983) *The Nature and Practice of Biological Control of Plant Pathogens.* American Phytopathological Society, St Paul, Minnesota.

Crawley, M.J. (1983) *Herbivory.* Blackwell Scientific, Oxford.

Crossley, E.A. (1977) The role of terrestrial saprophagous arthropods in forest soils: current status of concepts. In: Mattson, W.J. (ed.) *The Role of Arthropods in Forest Ecosystems.* Springer-Verlag, New York, pp. 49–56.

Curl, E.A. (1979) Effects of mycophagous collembola on *Rhizoctonia solani* and cotton-seedling disease. In: Schippers, B. and Gams, W. (eds) *Soil-borne Plant Pathogens.* Academic Press, London, pp. 253–269.

Curl, E.A. (1988) The role of soil microfauna in plant-disease suppression. *Critical Reviews in Plant Science* 7, 175–196.

Curl, E.A. and Harper J.D. (1990) Fauna-microflora interactions. In: Lynch, J.M. (ed.) *The Rhizosphere.* John Wiley & Sons, Chichester, pp. 369–388.

Curl, E.A., Gudauskas, R.T., Harper, J.D. and Peterson, C.M. (1985) Effects of soil insects on populations and germination of fungal propagules. In: Parker, C.A., Rovira, A.D., Moore, K.J., Wong, P.T.W. and Kollmorgan, J.F. (eds) *Ecology and Management of Soil-borne Plant Pathogens.* American Phytopathological Society, St Paul, Minnesota, pp. 20–23.

Curl, E.A., Lartey, R. and Peterson, C.M. (1988) Interactions between root pathogens and soil microarthropods. *Agriculture, Ecosystems and Environment* 24, 249–261.

Dassi, B., Dumas-Gaudot, E. and Gianinazzi, S. (1998) Do pathogenesis-related (PR) proteins play a role in bioprotection of tomato roots towards *Phytophthora parasitica*? *Physiological and Molecular Plant Pathology* 52, 167–183.

Dyer, H.C., Boddy, L. and Preston-Meek, C.M. (1992) Effect of the nematode *Panagrellus redivivus* on growth and enzyme production by *Phanerochaeta velutina* and *Stereum hirsutum*. *Mycological Research* 96, 1019–1028.

Enami, Y. and Nakamura, Y. (1996) Influence of *Scheloribates azumaensis* (Acari: Oribatida) on *Rhizoctonia solani*, the cause of radish root rot. *Pedobiologia* 40, 251–254.

Evans, G.O. (1992) *Principles of Acarology*. CAB International, Wallingford.

Finlay, R.D. (1985) Interactions between soil micro-arthropods and endomycorrhizal associations of higher plants. In: Fitter, A.H. (ed.) *Ecological Interactions in Soil*. Blackwell Scientific, Oxford, pp. 319–331.

Fitter, A.H. (1991) Costs and benefits of mycorrhizas: implications for functioning under natural conditions. *Experientia* 47, 350–355.

Fitter, A.H. and Sanders, I.R. (1992) Interactions with the soil fauna. In: Allen, M.F. (ed.) *Mycorrhizal Functioning*. Chapman and Hall, New York, pp. 333–354.

Freckman, D.W. (1988) Bacterivorous nematodes and organic-matter decomposition. *Agriculture, Ecosystems and Environment* 24, 195–217.

Freckman, D.W. and Baldwin, J.G. (1990) Nematoda. In: Dindal, D.L. (ed.) *Soil Biology Guide*. John Wiley & Sons, New York, pp. 155–200.

Garrett, S.D. (1976) Influence of nitrogen on cellulolysis rate and saprophytic survival in soil of some cereal root-rot fungi. *Soil Biology and Biochemistry* 8, 229–234.

Garrett, S.D. (1981) *Soil Fungi and Soil Fertility*. Pergamon Press, Oxford.

Griffiths, B.S. and Caul, S. (1993) Migration of bacteria-feeding nematodes, but not protozoa, to decomposing grass residues. *Biology and Fertility of Soils* 15, 201–207.

Hanlon, R.G.D. (1981) Influence of grazing by collembola on the activity of senescent fungal colonies grown on media of different nutrient concentration. *Oikos* 36, 362–367.

Hanlon, R.G.D. and Anderson, J.M. (1979) The effects of collembola grazing on microbial activity in decomposing leaf litter. *Oecologia* 38, 93–99.

Hanlon, R.G.D. and Anderson, J.M. (1980) Influence of macroarthropod feeding activities on microflora in decomposing oak leaves. *Soil Biology and Biochemistry* 12, 225–261.

Harris, K.K. and Boerner, R.E.J. (1990) Effects of belowground grazing by collembola on growth, mycorrhizal infection, and P uptake of *Geranium robertianum*. *Plant and Soil* 129, 203–210.

Hedlund, K. and Augustsson, A. (1995) Effects of enchytraeid grazing on fungal growth and respiration. *Soil Biology and Biochemistry* 27, 905–909.

Hedlund, K., Boddy, L. and Peterson, C.M. (1991) Mycelial responses of the soil fungus *Mortierella isabellina*, to grazing by *Onychiurus armatus* (Collembola). *Soil Biology and Biochemistry* 23, 361–366.

Hofman, T.W. and Jacob, J.J. (1989) Distribution and dynamics of mycophagous and microbivorous nematodes in potato fields and their relationship to some food sources. *Annals of Applied Biology* 115, 291–298.

Hofman, T.W., Middlekoop, J. and Bollen, G.J. (1991) Causes of the increased incidence of *Rhizoctonia solani* in potato crops treated with nematicides. In: Beemster, A.B.R. (ed.) *Biotic Interactions and Soil-borne Diseases*. Elsevier, Amsterdam, pp. 21–26.

Homma, Y. and Ishii, M. (1984) Perforation of hyphae of sclerotia of *Rhizoctonia solani* Kuhn by mycophagous soil amoebae from vegetable field soils in Japan. *Annals of the Phytopathological Society of Japan* 50, 229–240.

Homma, Y., Sitton, J.W., Cook, R.J. and Old K.M. (1979) Perforation and destruction of pigmented hyphae of *Gaeumannomyces graminis* by vampyrellid amoebae from Pacific Northwest field soils. *Phytopathology* 69, 1118–1122.

Hornby, D. (1990) Root diseases. In: Lynch, J.M. (ed.) *The Rhizosphere*. John Wiley & Sons, Chichester, pp. 233–258.

Hunt, H.W., Coleman, D.C., Ingham, E.R., Ingham, R.E., Moore, J.C., Rose, S.L., Reid, C.P.P. and Morley, C.R. (1987) The detrital food web in a shortgrass prairie. *Biology and Fertility of Soils* 3, 57–68.

Hyakumachi, M. (1996) Mechanisms involved in disease decline. In: Sneh, B., Jabaji-Hare, S., Neate, S. and Dijst, G. (eds) Rhizoctonia *Species: Taxonomy, Molecular Biology, Ecology, Pathology and Disease Control.* Kluwer, Dordrecht, pp. 227–235.

Hyakumachi, M., Homma, Y. and Ui, T. (1982) Presence of vampyrellid amoebae in the soils of sugarbeet monoculture fields, and its relation with sugarbeet root rot decline. *Memoirs of the Faculty of Agriculture of Hokkaido University* 13, 441–444.

Ineson, P., Leonard, M.A. and Anderson, J.M. (1982) Effect of collembolan grazing upon nitrogen and cation leaching from decomposing leaf litter. *Soil Biology and Biochemistry* 14, 601–605.

Ingham, E.R., Anderson, R.V., Gould, W.D. and Coleman, D.C. (1985) Vertical distribution of nematodes in a short grass prairie. *Pedobiologia* 28, 155–160.

Ingham, E.R., Trofymow, J.A., Ames, R.N., Hunt, H.W., Morley, C.R., Moore, J.C. and Coleman, D.C. (1986) Trophic interactions and nitrogen cycling in a semi-arid grassland soil. I. Seasonal dynamics of the natural populations, their interactions and effects on nitrogen cycling. *Journal of Applied Ecology* 23, 597–614.

Ingham, R.E. (1992) Interactions between invertebrates and fungi: effects on nutrient availability. In: Caroll, G.C. and Wicklow, D.T. (eds) *The Fungal Community*, 2nd edn. Marcel Dekker, New York, pp. 669–690.

Jeffries, P. (1997) Mycoparasitism. In: Wicklow, D.T. and Söderström B.E. (eds) *The Mycota. IV. Environmental and Microbial Relationships.* Springer-Verlag, Berlin, pp. 149–164.

Kaneko, N., McLean, M.A. and Parkinson, D. (1995) Grazing preference of *Onychiurus subtenuis* (Collembola) and *Oppiella nova* (Oribatei) for fungal species inoculated on pine needles. *Pedobiologia* 39, 538–546.

Kessler, K.J. (1990) Destruction of *Gnomonia leptostyla* perithecia on *Juglans nigra* leaves by microarthropods associated with *Elaeagnus umbellata* litter. *Mycologia* 82, 387–390.

Kevan, D.K.McE. (1965) The soil fauna – its nature and biology. In: Baker, K.F. and Synder, W.C. (eds) *Ecology of Soil-borne Plant Pathogens.* University of California Press, Berkeley, pp. 33–51.

Klink, J.W. and Barker, K.R. (1968) Effect of *Aphelenchus avenae* on the survival and pathogenic activity of root-rotting fungi. *Phytopathology* 58, 228–232.

Klironomos, J.N. and Kendrick, W.B. (1996) Palatability of microfungi to soil arthropods in relation to the functioning of arbuscular mycorrhizae. *Biology and Fertility of Soils* 21, 43–52.

Klironomos, J.N. and Ursic, M. (1998) Density-dependent grazing on the extraradical hyphal network of the arbuscular mycorrhizal fungus, *Glomus intraradices*, by the collembolan, *Folsomia candida*. *Biology and Fertility of Soils* 26, 250–523.

Klironomos, J.N., Widden, P. and Deslandes, I. (1992) Feeding preferences of the collembolan *Folsomia candida* in relation to microfungal successions on decaying litter. *Soil Biology and Biochemistry* 24, 685–692.

Lartey, R.T., Curl, E.A., Peterson, C.M. and Harper, J.D. (1989) Mycophagous grazing and food preference of *Proisotoma minuta* (Collembola: Isotomidae) and *Onychiurus encarpatus* (Collembola: Onychiuridae). *Environmental Entomology* 18, 334–337.

Lartey, R.T., Curl, E.A., Peterson, C.M. and Williams, J.C. (1991) Control of *Rhizoctonia solani* and cotton seedling disease by *Laetisaria arvalis* and a mycophagous insect *Proisotoma minuta* (Collembola). *Journal of Phytopathology* 133, 89–98.

Lartey, R.T., Curl, E.A. and Peterson, C.M. (1994) Interactions of mycophagous collembola and biological control fungi in the suppression of *Rhizoctonia solani*. *Soil Biology and Biochemistry* 26, 81–88.

Lootsma, M. and Scholte, K. (1997a) Effects of the springtail *Folsomia fimetaria* and the nematode *Aphelenchus avenae* on *Rhizoctonia solani* stem infection of potato at temperatures of 10 and 15°C. *Plant Pathology* 46, 203–208.

Lootsma, M. and Scholte, K. (1997b) Effects of soil moisture content on the suppression of *Rhizoctonia* stem canker on potato by the nematode *Aphelenchus avenae* and the springtail *Folsomia fimetaria*. *Plant Pathology* 46, 209–215.

Mankau, R. and Mankau, S.K. (1963) The role of mycophagous nematodes in the soil. I. The relationships of *Aphelenchus avenae* to phytopathogenic fungi. In: Doeskin, J. and van der Drift, F. (eds) *Soil Fauna, Soil Microflora and their Relationships*. North Holland, Amsterdam, pp. 271–280.

McAllister, C.B., Garcia-Garrido, J.M., Garcia-Romera, I., Godeas, A. and Ocampo, J.A. (1997) Interaction between *Alternaria alternata* or *Fusarium equiseti* and *Glomus mosseae* and its effects on plant growth. *Biology and Fertility of Soils* 24, 301–305.

McGonigle, T.P. (1995) The significance of grazing on fungi in nutrient cycling. *Canadian Journal of Botany* 73(Suppl. 1), S1370-S1376.

McGonigle, T.P. (1997) Fungivores. In: Wicklow, D.T. and Söderström, B. (eds) *The Mycota*, Vol. IV, *Environmental and Microbial Relationships*. Springer-Verlag, Berlin, pp. 237–248.

McGonigle, T.P. and Fitter, A.H. (1988) Ecological consequences of arthropod grazing on VA mycorrhizal fungi. *Proceedings of the Royal Society of Edinburgh* (Section B) 94, 25–32.

Moore, J.C. (1988) The influence of microarthropods on symbiotic and non-symbiotic mutualism in detrital-based below-ground food webs. *Agriculture, Ecosystems and Environment* 24, 147–159.

Moore, J.C., Walter, D.E. and Hunt, H.W. (1988) Arthropod regulation of micro- and mesobiota in below-ground detrital food webs. *Annual Review of Entomology* 33, 419–439.

Naito, S. (1988) Control of soil-borne pathogens by mycophagous soil animals. A case of the fungus gnat *Pnyxia scabiei* (Hopkins), feeding sclerotia of *Rhizoctonia solani* Kuhn. *Plant Protection* 42, 255–258.

Naito, S. and Makino, S. (1995) Control of sclerotia of *Rhizoctonia solani* by a sciarid fly, *Pnyxia scabiei*, in soil. *Japan Agricultural Research Quarterly* 29, 31–37.

Naito, S. and Sugimoto, T. (1988) Control of sugarbeet damping off *Rhizoctonia solani* (Kuhn) by mycophagous *Pnyxia scabiei* (Hopkins). *Annals of the Phytopathological Society of Japan* 54, 317–318.

Naito, S., Makino, S. and Sugimoto, T. (1988) *Pnyxia scabiei* (Hopkins) feeding on sclerotia of *Rhizoctonia solani* Kuhn and its population changes in sugarbeet root rot field. *Annals of the Phytopathological Society of Japan* 54, 52–59.

Nakamura, Y. (1993) Relation between soil microorganisms and soil animals – biological control of soil borne pathogen by collembola, oribatids and enchytraeids. *Soil Microorganisms* 42, 43–59.

Nakamura, Y. (1996) Interactions between earthworms and microorganisms in biological control of plant root pathogens. *Farming Japan* 30(6), 37–43.

Nakamura, Y., Itakura, J. and Matsuzaki, I. (1991a) Mycophagous meso soil animals from cropfields in Fukushima Pref. *Edaphologia* 45, 49–54.

Nakamura, Y., Matsuzaki, I. and Itakura, J. (1991b) Effects of mycophagous collembola on *Rhizoctonia solani* Kuhn causing radish, cucumber, cabbage and burdock seedling diseases. *Edaphologia* 47, 41–45.

Nakamura, Y., Matsuzaki, I. and Itakura, J. (1992) Effect of grazing by *Sinella curviseta* (Collembola) on *Fusarium oxysporum* f. sp. *cucumerinum* causing cucumber disease. *Pedobiologia* 36, 168–171.

Nakamura, Y., Itakura, J. and Matsuzaki, I. (1995) Influence of the earthworm *Pheretima hilgendorfi* (Megascolecidae) on *Plasmodiophora brassicae* clubroot galls of cabbage seedlings in pot. *Edaphologia* 54, 39–41.

Newsham, K.K., Fitter, A.H. and Watkinson, A.R. (1995a). Arbuscular mycorrhiza protect an annual grass from root pathogenic fungi in the field. *Journal of Ecology* 83, 991–1000.

Newsham, K.K., Fitter, A.H. and Watkinson, A.R. (1995b). Multi-functionality and biodiversity in arbuscular mycorrhizas. *Trends in Ecology and Evolution* 10, 407–411.

Newell, K. (1984a) Interaction between two decomposer basidiomycetes and a collembolan under sitka spruce: distribution, abundance and selective grazing. *Soil Biology and Biochemistry* 16, 227–233.

Newell, K. (1984b) Interaction between two decomposer basidiomycetes and a collembolan under sitka spruce: grazing and its potential effects on fungal distribution and litter decomposition. *Soil Biology and Biochemistry* 16, 235–239.

Old, K.M. (1967) Effect of natural soil on survival of *Cochliobolus sativus*. *Transactions of the British Mycological Society* 50, 615–624.

Old, K.M. (1977) Giant soil amoebae cause perforation of conidia of *Cochliobolus sativus*. *Transactions of the British Mycological Society* 68, 277–281.

Old, K.M. and Chakraborty, S. (1986) Mycophagous soil amoebae: their biology and significance in the ecology of soil-borne plant pathogens. *Progress in Protistology* 1, 163–194.

Parkinson, D., Visser, S. and Whittaker, J.B. (1979) Effects of collembolan grazing on fungal colonization of leaf litter. *Soil Biology and Biochemistry* 11, 529–535.

Paustian, K., Andren, O., Clarholm, M., Hansson, A.C., Johansson, G., Lagerlof, J., Lindberg, T., Pettersson, R. and Sohlenhius, B. (1990) Carbon and nitrogen budgets of four agro-ecosystems with annual and perennial crops, with and without nitrogen fertilization. *Journal of Applied Ecology* 27, 60–84.

Persson, T. (1983) Influence of soil animals on nitrogen mineralization in a northern Scots pine forest. In: Lebrun, P., Andre, H.M., de Medts, C., Gregoire-Wibo and Wanthy, G. (eds) *New Trends in Biology*. Dien-Brichart, Louvain-la-Neuve, pp. 117–126.

Persson, T., Baath, E., Clarholm, M., Lundkvist, H., Söderström, B.E. and Sohlenhius, B. (1980) Trophic structure, biomass dynamics and carbon metabolism of soil organisms in a Scots pine forest. *Ecological Bulletin* 32, 419–459.

Read, D.J. (1999) Mycorrhiza – the state of the art. In: Varma, A. and Hock, B. (eds) *Mycorrhiza*. Springer-Verlag, Berlin, pp. 3–34.

Reichle, D.E. (1977) The role of soil invertebrates in nutrient cycling. *Ecological Bulletin* 25, 145–156.

Rhodes, H.L. and Linford, M.B. (1959) Control of *Pythium* root rot by the nematode *Aphelenchus avenae*. *Plant Disease Reporter* 43, 323–328.

Rickerl, D.H., Curl, E.A. and Touchton, J.T. (1989) Tillage and rotation effects on collembola populations and *Rhizoctonia* infestations. *Soil Tillage Research* 15, 41–49.

Riffle, J.W. (1971) Effect of nematodes on root-inhabiting fungi. In: Hacskaylo, E. (ed.) *Mycorrhizae*. USDA Forest Service Miscellaneous Publication 1189. United States Government Printing Service, Washington, DC, pp. 97–113.

Setala, H. (1995) Growth of birch and pine seedlings in relation to grazing by soil fauna on ectomycorrhizal fungi. *Ecology* 76, 1844–1851.

Setala, H. and Huhta, V. (1991) Soil fauna increase *Betula pendula* growth: laboratory experiments with coniferous forest floor. *Ecology* 72, 665–671.

Shaw, P.J.A. (1985) Grazing preferences of *Onychiurus armatus* (Insecta: Collembola) for mycorrhizal and saprotrophic fungi of pine plantations. In: Fitter, A.H., Atkinson, D., Read, D.J. and Usher, M.B. (eds) *Ecological Interactions in Soil*. Blackwell Scientific, Oxford, pp. 333–337.

Shaw, P.J.A. (1988) A consistent hierarchy in the feeding preferences of the collembola *Onychiurus armatus*. *Pedobiologia* 31, 179–187.

Shaw, H.D. and Beute, M.K. (1979) Evidence for the involvement of soilborne mites in *Pythium* pod rot of peanut. *Phytopathology* 69, 204–207.

Stephens, P.M., Davoren, C.W., Doube, B.M., Ryder, M.H., Benger, A.M. and Neate, S.M. (1993) Reduced severity of *Rhizoctonia solani* disease on wheat seedlings associated with the presence of the earthworm *Aporrectodea trapezoides* (Lumbricidae). *Soil Biology and Biochemistry* 25, 1477–1484.

Stephens, P.M., Davoren, C.W., Ryder, M.H. and Doube, B.M. (1994) Influence of the earthworms *Aporrectodea rosea* and *Aporrectodea trapezoides* on *Rhizoctonia solani* disease of wheat seedlings and the interaction with a surface mulch of cereal-pea straw. *Soil Biology and Biochemistry* 26, 1285–1287.

Szczech, M., Rondomanski, W., Brzeski, M.W., Smolinska, U. and Kotowski, J.F. (1993) Suppressive effect of a commercial earthworm compost on some root infecting pathogens of cabbage and tomato. *Biological and Agricultural Horticulture* 10, 47–52.

Townshend, J.L. (1964) Fungus hosts of *Aphelenchus avenae* Bastian, 1865 and *Bursaphelenchus fungivorus* Franklin & Hooper, 1962 and their attractiveness to these nematode species. *Canadian Journal of Microbiology* 10, 727–737.

Trofymow, J.A. and Coleman, D.C. (1982) The role of bacterivorous and fungivorous nematodes in cellulose and chitin decomposition in the context of a root/rhizosphere/soil conceptual model. In: Freckman, D.W. (ed.) *Nematodes in Ecosystems*. University of Texas Press, Austin, pp. 117–137.

van der Drift, J. and Jansen, E. (1977) Grazing of springtails on hyphal mats and its influence on fungal growth and respiration. *Ecological Bulletin* 25, 203–209.

Visser, S. (1985) The role of soil invertebrates in determining the composition of soil microbial communities. In: Fitter, A.H., Atkinson, D., Read, D.J. and Usher, M.B. (eds) *Ecological Interactions in Soil*. Blackwell Scientific, Oxford, pp. 297–317.

Warnock A.J., Fitter, A.H. and Usher, M.B. (1982) The influence of a springtail *Folsomia candida* (Insecta: Collembola) on mycorrhizal association of leek *Allium porrum* and the vesicular-arbuscular mycorrhizal endophyte *Glomus fasciculatus*. *New Phytologist* 90, 285–292.

Wicklow, D.T. and Yocum, D.H. (1982) Effects of larval grazing by *Lycoriella mali* (Diptera: Sciaridae) on species abundance of coprophilous fungi. *Transactions of the British Mycological Society* 78, 29–32.

Wiggins, E.A. and Curl, E.A. (1979) Interactions of collembola and microflora of cotton rhizosphere. *Phytopathology* 69, 244–249.
Wiggins, E.A., Curl, E.A. and Harper, J.D. (1979) Effects of fertility and cotton rhizospheres on populations of collembola. *Pedobiologia* 19, 75–82.

Plant Interactions with Endophytic Bacteria

6

Johannes Hallmann

Institut for Plant Diseases, University of Bonn, Nußallee 9, 53115 Bonn, Germany

Introduction

The first reports regarding the existence of bacteria residing in the internal tissue of non-symbiotic plants date back to the 1870s and were reviewed by Smith in 1911 and Hollis in 1951. Since then numerous articles have described endophytic communities in a broad spectrum of plant species and in various plant organs. However, recently increased interest in endophytic bacteria has taken place especially for those bacteria having commercial features such as plant growth promotion and stimulation of plant defence mechanisms. With all the work published within the last decade the question arises: what do we really know about plant interactions with endophytic bacteria and how do these interactions influence plant health? The best characterized plant–endophytic bacteria interactions are those of nitrogen-fixing bacteria, like the *Rhizobium*–legume symbiosis or the symbiosis between free-living N-fixing bacteria with several grass species. These endophytic bacteria improve the plant's nitrogen status, but do not directly affect plant health, except that a stronger plant can better resist attack by plant pathogens. The work on N-fixing endophytic bacteria has been excellently reviewed over the past few years by Döbereiner and Pedrosa (1987), Hecht-Buchholz (1998) and others and will therefore be only covered marginally within this chapter. However, a second group of endophytic bacteria has captured the plant pathologist's major interest, those endophytic bacteria capable of improving plant health. It is the objective of the current review to summarize recent work on endophytic bacteria and plant health with special emphasis on the interactions between the plant and the bacterial endophytes. The topics being covered will range from the source of endophytic bacteria, precolonization as well as postcolonization interactions, all the way to specificity as related to plant and bacterial species. The chapter will raise several questions such as: how do endophytic bacteria improve plant health, what are the major biotic and abiotic factors influencing plant–endophyte interactions, what is known about recognition and communication between plant and endophytic bacteria, and how does this relate to disease suppression? Several of these research areas have not yet been explored and answers in these cases can only be speculative. The author's primary goal is to stimulate discussion as well as research interest in this exciting and still wide open research area. It is hoped that this review will help to prepare the ground to use endophytic bacteria for plant health promotion in modern agriculture systems.

Terminology

Two key terms of this chapter need to be defined from the start: 'endophytic bacteria' and 'interaction'. With respect to the several definitions for 'endophytic bacteria' (reviewed by Hallmann *et al.*, 1997b), all of which have their legitimate place in science, this review will use the term 'endophytic bacteria' to describe those bacteria colonizing the plant internally without doing substantive harm to the plant. This definition deliberately excludes pathogenic bacteria of internal plant tissue to express the epidemiological and physiological differences of those two bacterial groups. Endophytic bacteria are commonly isolated from internal plant tissue either directly by centrifugation or pressure bomb extraction as well as indirectly following disinfestation of the plant surface (Hallmann *et al.*, 1997a). Crucial for all these procedures is a proper sterility check to avoid false positive endophytes such as colonizers of the rhizoplane/phyllosphere or contaminants from the surrounding environment. However, all these techniques are limited in that they do not consider obligate colonizers or dormant stages of endophytic colonizers. Although of scientific interest, their non-culturability on artificial nutrient media currently excludes them from being preferred candidates for plant growth promotion and biocontrol.

The second term 'interaction' is used in plant pathology quantitatively and qualitatively in describing the interrelationships between two or more factors involved in plant diseases (Sikora and Carter, 1987). This definition creates problems since the association between the plant and endophytic bacteria has not yet been and probably will not be clearly defined in the future. Therefore, 'interaction' and 'interrelationship' will be used as synonyms to describe plant–bacterial endophyte associations. Based on the above definition for endophytic bacteria, which stresses the fact that the bacteria do not cause any deleterious affects, only those types of interactions are acceptable where none of the two partners are affected in a negative manner. Considering this definition three different types of interactions can be described: (i) neutralism –

where none of the partners affect each other; (ii) symbiosis – where both partners benefit from each other; and (iii) commensalism – where one partner benefits from the interactions and the other remains unaffected. Which of these three interactions best describes the plant–endophyte association depends on several factors that will be discussed in this chapter.

Plant–Endophyte Associations

Endophytic bacteria colonize a broad spectrum of plant species and plant parts as summarized in Table 6.1. Although far from complete, the list clearly demonstrates the ubiquitous presence of endophytic bacteria within the plant kingdom covering annual as well as perennial crops. Additional information on the omnipresence of endophytic bacteria in nature is gained from the work with tissue-cultured plants, where the often observed difficulties in producing microbial-free plants indicates the strong association endophytic bacteria have developed with their host plant. It has often been stated that all plants in nature harbour endophytic bacteria. However, this has not been verified due to a lack of plants which possibly provide less favourable conditions for endophytic growth such as xerophytes and halophytes, or plants containing high amounts of alkaloids with antimicrobial properties within their apoplast. Isolation of very low numbers of bacteria presents a limiting factor for such plants. In addition, the literature usually covers only the detection of endophytic bacteria and does not explicitly refer to endophyte-free situations. If absence of endophytic bacteria is observed, it is usually attributed to inappropriate isolation methods, population densities below detection levels or non-culturability of the bacteria. All these factors underline the importance of improving detection methods and the development of highly sensitive as well as specific markers for bacterial endophytes. Molecular methods like BOX-PCR and denaturing gradient gel electrophoresis (DGGE) are promising tools for exploration of this group of non-culturable endophytic bacteria.

Endophytic Community Densities

The total population density of endophytic bacteria found in plants depends on plant species, plant genotype, plant tissue, growth stage and environmental conditions. The bacterial spectrum, conversely, seems to be a consequence of niche specialization, differences in colonization pathway or some form of mutual exclusion operating between different bacterial populations (Sturz *et al.*, 1997). Overall, population densities of endophytic bacteria are found to be highest in the root tissue with approximately log 5 cfu g^{-1} fresh weight. This density is significantly lower than that reported for pathogenic bacteria which can range from log 8 to log 10 cfu g^{-1} fresh weight.

Table 6.1. Plant species with their associative endophytic bacteria and potential interactions.

Plant species	Plant part	Bacterial taxa[a]	Plant interactions	Reference
Cabbage	Leaves	*Kluyvera, Alcaligenes*	Control of *Xanthomonas campestris* pv. *campestris*	Assis *et al.*, 1998
Clover	Leaves, stem, root, nodules	*Pantoea, Agrobacterium, Rhizobium, Bacillus, Curtobacterium*	Growth promotion	Sturz *et al.*, 1997
Cotton	Roots, stem	*Aureobacterium, Bacillus, Phyllobacterium, Pseudomonas, Burkholderia*	Control of *Fusarium oxysporum* f. sp. *vasinfectum, Rhizoctonia solani*	Chen *et al.*, 1995
Cotton	Seeds	*Bacillus*	Control of *Rhizoctonia solani* and *Sclerotium rolfsii*	Pleban *et al.*, 1995
Cotton	Root, radicle, stem, boll, flowers	*Erwinia, Bacillus, Clavibacter, Xanthomonas*		Mishagi and Donndelinger, 1990
Cotton	Root, stem	*Bacillus, Agrobacterium, Burkholderia, Serratia*		McInroy and Kloepper, 1995a
Cotton	Root	*Burkholderia, Phyllobacterium*	Control of *Meloidogyne incognita*	Hallmann *et al.*, 1999
Cucumber	Fruit	*Pseudomonadaceae, Enterobacteriaceae, Achromobacteriaceae, Micrococcaceae*		Samish *et al.*, 1961
Cucumber	Root	*Pseudomonas, Bacillus, Enterobacter, Agrobacterium, Chryseobacterium, Burkholderia*		Mahaffee and Kloepper, 1997
Lucerne	Root	*Pseudomonas, Erwinia*-like bacteria		Gagné *et al.*, 1987
Maize	Root	*Bacillus, Pseudomonas, Corynebacterium*		Lalande *et al.*, 1989
Maize	Root, stem	*Burkholderia, Enterobacter, Bacillus*		McInroy and Kloepper, 1995a
Maize	Stem	*Enterobacter, Klebsiella, Pseudomonas*		Fisher *et al.*, 1992
Potato	Tuber	*Bacillus*		Hollis, 1951
Potato	Tuber	*Micrococcus, Pseudomonas, Bacillus, Flavobacterium, Xanthomonas, Agrobacterium,* coryneforms		De Boer and Copeman, 1974

Table 6.1. *Continued.*

Plant species	Plant part	Bacterial taxa[a]	Plant interactions	Reference
Potato	Tuber	*Curtobacterium, Pantoea*	Control of *Erwinia carotovora* var. *atroseptica*	Sturz and Metheson, 1996
Potato	Tuber	*Curtobacterium, Pseudomonas*	Antibiosis against *Fusarium* spp., *Phytophthora infestans*	Sturz *et al.*, 1999
Rape	Root	*Bacillus, Flavobacterium, Micrococcus, Rathayibacter*		Germida *et al.*, 1998
Red clover	Root, stem, leaves	*Pantoea, Agrobacterium, Bacillus, Pseudomonas, Curtobacterium*	Growth promotion	Sturz *et al.*, 1997
Rice	Stem	*Pseudomonas*	Control of *Rhizoctonia solani*	Krishnamurthy and Gnana-manickam, 1997
Sugar beet	Root	*Bacillus, Erwinia, Pseudomonas, Corynebacterium, Lactobacillus, Xanthomonas*		Jacobs *et al.*, 1985
Wheat	Root	*Bacillus, Flavobacterium, Micrococcus, Rathayibacter*		Germida *et al.*, 1998
Div. species	Seed	*Bacillus, Erwinia, Flavobacterium, Pseudomonas*		Mundt and Hinkle, 1976
Grapevine	Stem	*Enterobacter, Pseudomonas, Pantoea, Rhodococcus*		Bell *et al.*, 1995
Rough lemon	Root	*Pseudomonas, Enterobacter, Bacillus, Corynebacterium*		Gardner *et al.*, 1982
Lodgepole pine	Root	*Bacillus*	N-fixation	Chanway, 1998
Oak	Stem	*Bacillus, Pseudomonas*	Control of *Ceratocystis fagacearum*	Brooks *et al.*, 1994

[a]This table summarizes those bacterial taxa isolated most frequently from the host plant and/or showing important traits regarding pathogen control or plant growth promotion.

The population densities in aerial parts usually average around log 4 cfu g^{-1} fresh stem weight and log 3 cfu g^{-1} in the leaf tissue (Quadt-Hallmann and Kloepper, 1996; Hallmann *et al.*, 1997b). The lowest and at times absence of endophytic colonization is found in generative organs like flowers, fruits and seeds. The reason for this gradient of bacterial numbers from the root to the seed gives rise to some speculation. The root is thought to be the preferred site of bacterial entrance into the plant which would explain high bacterial numbers in the root at early growth stages. However, endophytic bacteria also enter plant tissue via stomata, hydathodes and micropores of above-ground plant tissue. Furthermore, they can move systemically throughout the plant which should theoretically lead to an equilibrium in bacterial density within the plant. The fact that rather low endophytic population densities are found in above-ground plant parts could also be an indicator of less favourable conditions in that tissue caused by high amount and/or large daily fluctuations in temperature, water content, nutrient availability and UV radiation. In contrast, the root system seems to provide a more buffered habitat with regards to water availability and temperature changes. In addition, root exudates provide a rich and continuous flow of nutrients available to the bacteria. The total endophyte density within a given type of plant tissue is not constant and changes over time. Major factors influencing total population densities include plant age as well as various biotic and abiotic environmental factors. Intensive studies regarding the population dynamics of endophytic bacteria in roots and stems of cotton and sweetcorn over two successive cropping seasons were reported by McInroy and Kloepper (1995b). The same authors also reported that internal population densities commonly increased within the first 3 weeks and then remained at almost constant levels for the rest of the growth period.

Source of Endophytic Colonization

The source of endophytic colonization is diverse and can range from transmission via seeds and vegetative planting material to entrance from the surrounding environment such as the rhizosphere and phyllosphere. Bacterial transmission via vegetative planting material has been demonstrated for *Acetobacter diazotrophicus* for two successive cuttings of sugarcane (Dong *et al.*, 1994). Whereas occurrence of endophytic bacteria in healthy seeds has been reported for some plants (Mundt and Hinkle, 1976; Mishagi and Donndelinger, 1990), no such colonization was found for others. Surface disinfestation of cucumber, cotton or bean seeds usually showed them to be free of endophytic colonizers (Hallmann *et al.*, unpublished). These results are in contrast to those of McInroy and Kloepper (1995a) who recovered bacterial endophytes from surface-disinfested cotton and sweetcorn seeds. These authors also discussed survival of bacterial endophytes on the surface following chemical disinfestation procedures in recesses created by the seed coat of the imbibing seed. The fact that bacterial recovery from surface-disinfested cotton seeds (log 3–5 cfu g^{-1} seed) was higher than on sweetcorn (<log 1 cfu per seed) was attributed to the rough surface of the cotton seed. The fine cracks on the seed coat may allow bacterial survival whereas in the case of sweetcorn the surface is smooth and intact, providing an effective barrier to bacterial entrance (McInroy and Kloepper, 1995a).

However, if seed transmission does occur, the bacterium has to be an excellent systemic colonizer, able to grow internally within the plant and colonize the ovules. To this author's knowledge, the complete passage from seed to seed has not yet been reported for any endophytic bacterium. The main source of endophytic colonization is believed to be the rhizosphere and phyllosphere as indicated by the high similarity between bacteria found internally in the root with those found in the rhizosphere or phyllosphere. In greenhouse experiments with cotton, 82% of the endophytic species identified from root tissue were also isolated from the rhizosphere (Hallmann *et al.*, 1999), confirming that endophytic bacteria represent a segment of the rhizosphere microbial community. Similar results were reported for field-grown cucumber with 94% correspondence between endorhiza and rhizosphere communities (Mahaffee and Kloepper, 1997). In this study, common genera of the endorhiza such as *Agrobacterium*, *Bacillus*, *Chryseobacterium*, *Enterobacter* and *Pseudomonas* also predominated the rhizosphere community. In contrast to the rhizosphere/endorhiza habitat few reports exist covering comparative community studies of the phyllosphere/endosphere microcosm. For pathogenic bacteria, Beattie and Lindow (1995) proposed that an active exchange exists between individual cells of one pathogen colonizing the leaf externally and internally. This mechanism is probably also common to endophytic bacteria. However, the same authors assume that only phytopathogenic bacteria are capable of endophytic growth. The presence of non-pathogenic isolates of the genera *Pasteurella*, *Escherichia*, *Curtobacterium* and *Xanthomonas* in the leaves but not in the roots of red clover clearly indicates endophytic colonization through the leaves (Sturz *et al.*, 1997). The results obtained with the phyllosphere seem to indicate that the rhizosphere is the predominant source for endophytic colonization. There are several reasons to support this view: (i) the root is exposed to bacteria long before the seedling emerges and therefore before leaves can be colonized by phyllosphere organisms; and (ii) bacteria colonizing the root at early stages in plant growth might be able to colonize the plant systemically and pre-occupy stems, cotyledons and leaves. Assuming a first-come first-served strategy, phyllosphere bacteria would have only a small chance to colonize those preoccupied internal niches to become endophytic. Such competition for colonization sites has been shown for the rhizosphere. When potato roots were inoculated with *Rhizobium etli* G12 3 days ahead of inoculation with *Bacillus sphaericus* B43, and roots were studied 7 days after the second bacterial treatment, *R. etli* G12 was the dominant colonizer of the root surface, whereas in the opposite case *R. etli* G12 was almost non-detectable (Hallmann *et al.*, unpublished). When both bacteria were inoculated simultaneously, they both colonized the root surface, but at lower densities. A final reason favouring the rhizosphere over the phyllosphere as the predominant source for endophytic colonization might be attributed to the leaf surfaces. The thick-walled epidermal cells and depositions of cuticula and waxes on the leaves represent a stronger barrier to microbial entrance than does the thin unwaxed root surface with its many site roots, root hairs and wounds.

Competition between microorganisms within the rhizosphere or phyllosphere probably has significant effects on the spectrum of bacteria becoming endophytic. As shown by Quadt-Hallmann *et al.* (1997b), the total population density of endophytic *Enterobacter asburiae* JM22 was not affected when JM22 was co-inoculated with the pure rhizosphere colonizer *Micrococcus agilis*. However, when *E. asburiae* JM22 was applied simultaneously with the endophytic bacterium *Paenibacillus macerans*, the population densities of both endophytic bacteria were lower compared to individual application. Inter- and intraspecific competition between microbial organisms is a common phenomenon within any microbial community. Besides, the competiveness of a single isolate might also be affected by biotic and abiotic factors characteristic for a given growth substrate. How do different plant growth substrates varying in their indigenous microbial community affect plant colonization of an introduced endophytic bacterium? Appropriate data to answer this question are not yet available and it can only be speculated that biotic and abiotic factors characteristic for a given plant growth substrate such as pH, water content, organic matter, binding capacity and microorganisms will not only affect survival of an introduced endophyte but also its ability to colonize a plant. Comparing the recovery of *E. asburiae* JM22 from different growth substrates, the endophyte was reisolated at high concentrations from sand, loamy sand and ground clay, less from sandy loam, and only at low concentrations from peat-based Pro-Mix (Quadt-Hallmann and Kloepper, 1996). The low level of bacterial isolation from the latter might be related to the high concentration of humic substances in peat, which not only affects bacterial survival but also leads to bacterial adsorption to soil particles. However, further studies are necessary to improve our understanding of how plant growth substrates affect the colonization potential of endophytic bacteria. In conclusion, the rhizosphere seems to be the major source for endophytic colonization and any parameter influencing the indigenous bacterial community within the rhizosphere will consequently affect the spectrum of bacteria found internally in the plant tissue.

Precolonization Interactions

Strong interactions between the host plant and endophytic bacteria already appear prior to endophytic colonization and seem to be a significant prerequisite for successful establishment of the plant–endophytic bacterial association. With the exception of endophytic bacteria transmitted by seed or vegetative propagation material, successful endophytic colonization goes through several important stages including host finding, recognition, colonization of the plant surface and entrance into the internal plant tissue. With respect to the external and internal colonization phases characteristic for most endophytic bacteria, their association with the plant can be subdivided into precolonization and postcolonization interactions. Precolonization interactions will include bacterial movement towards the root, bacterial attachment to the root surface, plant–bacterial recognition processes at the root surface and finally root penetration by the bacterium (Fig. 6.1), whereas postcolonization

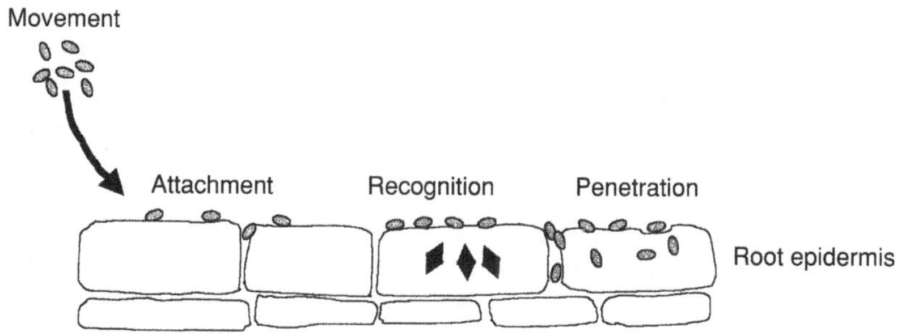

Fig. 6.1. Stages of precolonization interactions occurring at the root surface between the host plant and endophytic bacteria.

interactions will consider bacterial multiplication and localization within the root tissue, including potential plant beneficial effects. The importance of each of these mechanisms is still unclear, and only some of them might be essential to establish a successful plant–endophyte association.

Movement

Endophytic bacteria probably find their host by chemotaxis, accidental encounter or a mixture of the two. Root exudates released by the plant initiate a nutrient gradient which is believed to direct potential endophytic bacteria to the root surface. For example, migration of the beneficial plant-associated bacteria *Pseudomonas fluorescens* and *Azospirillum brasilense* towards wheat roots was stimulated by various wheat genotypes and by synthetic attractants (Bashan, 1986). Random contact with the root might be an important and often neglected prerequisite for penetration, especially when considering the effects that abiotic factors like rainfall can have on bacterial movement within the soil.

Attachment

The importance of attachment in plant–bacteria interactions has been reviewed excellently by Romantchuk (1992) for plant pathogenic bacteria. In incompatible interactions between the plant and plant-pathogenic bacteria, those bacteria which attach to the host cell wall induce structural damage to the plasma membrane, resulting in the release of electrolytes and the subsequent death of the host cell. During this process, toxic phenolic compounds are also released from the host cell to kill the pathogenic bacteria in the intercellular space. However, in compatible interactions bacterial attachment to plant cells can trigger the release of nutrients or stimulants for bacterial growth through mild degeneration of the host cell membrane. From the *Rhizobium*–legume symbiosis we know that attachment of the rhizobia to the rhizoplane is essential for bacterial penetration (Vance, 1983). The attachment, mediated by bacteria surface polysaccharides with root surface lectins, is very specific and only those bacteria bound to the root surface can become symbionts. Similar processes for endophytic bacteria can only be assumed and only limited data support this hypothesis. For example, Duijff *et al.* (1997) reported adherance of *Pseudomonas fluorescens* strain WCS417r to the root surface, which occasionally was associated with exocellular structures likely to be involved in polar attachment. The authors especially highlight the involvement of the O-antigenic side chain of the outer membrane lipopolysaccharides of WCS417r as playing a significant role in attachment. The role of lipopolysaccharides in plant interactions with endophytic bacteria will be discussed in more detail in the section on modes of action of biological control strains.

Recognition

Is recognition an essential attribute required for a compatible plant–endophyte association? When endophytic bacteria are in close proximity to the plant surface questions arise as to the presence of bacterial mechanisms for recognition of a suitable host or plant mechanisms for recognition of an appropriate endophyte. If recognition is involved, then a form of specific contact between an elicitor released by the bacterium and a corresponding receptor of the host plant has to be established. According to Vance (1983), this contact may occur extracellularly as an early event within the plant–endophyte association or may occur later at the intercellular or intracellular level. Furthermore, the result of recognition can be either positive, resulting in successful establishment of the plant–endophyte association or negative, culminating in plant responses like hypersensitivity, induced resistance, phytoalexin accumulation, cell wall depositions and papillae formation that exclude the bacteria from entrance into the plant. Whereas the importance of such recognition processes as a prerequisite for bacterial entrance is well studied for the *Rhizobium*–legume interaction and plant pathogens, we can only assume that similar processes apply to the plant–endophytic bacteria association. The presence of a recognition process would explain why only selected bacteria of the soil, rhizosphere or phyllosphere environment can become endophytic and not others. Under normal conditions with intact plant surfaces, extracellular recognition processes are probably the preferred mechanism. However, what if these recognition sites are destroyed due to wounding caused by either pathogen attack, soil disturbances or unfavourable weather conditions such as hail – will this lead to establishment of a completely different spectrum of endophytic bacteria? When the endophytic community of cotton roots 7 days after nematode infestation was compared with a non-infested control, bacterial species richness and diversity only changed marginally whereas total endophytic populations in the roots had increased sixfold and those of stems twofold following nematode infestation (Hallmann, unpublished). The most common endophytic genera isolated from nematode-free and nematode-infested plants were *Agrobacterium* and *Pseudomonas* accounting for 23% and 19% of the total population in the case of *Agrobacterium* and for 19% each in the case of *Pseudomonas*. The fact that the endophytic bacterial species spectrum was not altered significantly due to wounding caused by nematodes might indicate that extracellular recognition processes were probably replaced by inter- or intracellular recognition processes or that this form of biotic wounding has no selective effect on bacterial entrance.

However, recognition of endophytic bacteria by the plant can also result in stimulation of plant defence mechanisms such as hypersensitive cell collapse or accumulation of antimicrobial substances which then further inhibit endophytic colonization. Gardner *et al.* (1982) reported that most bacteria do not exist freely within xylem vessels, but become entrapped in fibrous occlusions which might be part of a plant defence response. It has been shown that avirulent plant pathogens and saprophytic bacteria are bound to plant cell walls immediately after penetration or they become agglutinated (reviewed by Vance, 1983). Both these plant responses prevent further multiplication and spread of the bacteria in the plant tissue. Conversely, endophytic bacteria are known to move systemically within the plant and even colonize different host tissues, indicating no such plant-mediated immobilization mechanisms are present. It can be concluded that if recognition processes between plants and endophytic bacteria appear, they will not necessarily result in plant defence responses or the endophytic bacteria are capable of inhibiting or bypassing such defence mechanisms. Overall, the importance of recognition processes in plant–endophytic bacteria associations is still inconclusive and requires further research input for a better understanding.

Penetration

The main routes of entry for endophytic bacteria are: (i) natural openings such as hydathodes, stomata and lenticels; (ii) wounds caused by abrasion with soil particles, pathogen damage, formation of lateral roots; (iii) micropores; and (iv) abiotic mechanical damage, e.g. hail. However, the first and probably most important entry for endophytic bacteria is thought to be through wounds and micropores present in the early stages of root development. Young root tissue is usually weak and undifferentiated and a Casparian strip has not yet been formed to prevent endophytic bacteria from movement into deeper layers of the root tissue. In addition, plant pathogens can facilitate bacterial entrance as has been demonstrated for plant parasitic nematodes (Hallmann *et al.*, 1998) and the fungal pathogen *Rhizoctonia solani* (Mahaffee and Kloepper, 1997). For example, an increase in nematode infestation was positively correlated with increasing total population densities of endophytic bacteria. Besides causing wounds that serve as entry points for the bacteria, the nematode juveniles also carried individual bacterial cells adhering to their cuticle (Hallmann *et al.*, 1998). Increased endophytic bacterial populations as a consequence of fungal pathogen attack has been reported for *R. solani* infection of *Phaseolus vulgaris* (Mahaffee and Kloepper, 1997). As a consequence of *Rhizoctonia* damage, the two endophytes *Enterobacter asburiae* JM22 and *Pseudomonas fluorescens* 89B-27 extensively colonized the fungal necrotic tissue, from where they expanded 2–4 cm into healthy tissue.

For the above-mentioned modes of entry, the bacteria take advantage of natural or artificial openings present at the plant surface. However, the question still remains as to whether bacterial penetration is passive or a more active process mediated by the endophytic bacterium itself and/or the plant. A pure passive penetration can be assumed for natural openings such as hydathodes and stomata, where entrance is supported by an undisturbed water film extending from the leaf surface into the hydathode or stoma. Bacterial cells which have reached the hydathode and stoma can easily colonize the intercellular space of the leaf. However, for wounds no such uptake has yet been reported. Addressing this subject, Quadt-Hallmann *et al.* (1997a) inoculated roots and leaves of bean plants with cells of *Enterobacter asburiae* JM22 previously killed in a 1.0% glutaraldehyde solution and found no bacterial uptake by cotton roots and leaves, although living cells of the same bacterium did colonize cotton roots. Similarly, seedlings grown with minimal disturbance in liquid media or on water agar were penetrated by endophytic bacteria long before lateral roots were formed (Levanony and Bashan, 1989; Quadt-Hallmann *et al.*, 1997a). More recently it was demonstrated that at least some bacteria do produce cellulolytic enzymes in close proximity with epidermal cells which might enable them to actively gain entrance (Benhamou *et al.*, 1996). Interestingly, these enzymatic reactions were not observed after the bacteria had established in the intercellular spaces of the host plant. This finding seems logical, since continued cell wall degradation would destroy the plant tissue and the bacterium by definition would be considered a pathogen. Does this mean that endophytic bacteria selectively produce hydrolytic enzymes to enter the plant and stop expressing them after successful colonization? If this is the case, are the genes expressing cellulolytic and pectinolytic activity regulated by the host plant or the endophyte? To date, involvement of cellulolytic and pectinolytic enzymes in bacterial entry have only been reported for plant pathogens and *Rhizobium* species. However, besides being of bacterial origin, cell wall degrading enzymes are also released by the plant in response to specific bacterial polysaccharides which pass through the cell wall and interact with the host plant nucleus to stimulate enzyme production. Whereas cellulolytic enzymes are required for plant entry, bacterial metabolites are believed to support bacterial entry, although the role of bacterial metabolites in the infection process of endophytic bacteria is still questionable. For *Rhizobium trifolii* it has been shown that cell-free bacterial exudates applied prior to *Rhizobium* inoculation increased the susceptibility of white clover root hairs to infection (Beringer *et al.* in Vance, 1983). Furthermore, Sturz *et al.*

(1997) reported that bacterial metabolites of either endophytic bacteria or rhizobacteria can stimulate *Rhizobium* to colonize non-host legumes. However, all these studies were conducted under laboratory conditions with metabolites produced in nutrient-rich media. Therefore, it can only be speculated as to the existence of similar processes in the rhizosphere. Also of great interest is the density of rhizosphere bacteria that actually become endophytic. You and Zhou (1989) estimated that for *Alcaligenes faecalis* artificially inoculated at 5×10^5 cells ml^{-1} only 10% of the root surface population actually entered the root tissue. Under field conditions this level of colonization will probably be far less. In conclusion, we know very little about the mechanisms behind the fundamental importance of bacterial penetration processes needed for establishment of the plant–endophytic bacteria association.

Postcolonization Interactions

Following penetration of the plant tissue, endophytic bacteria have to multiply and colonize the plant tissue to establish a successful plant–endophytic bacteria association. Endophytic bacteria either colonize special parts of the plant extensively, become systemic colonizers or remain latent in the tissue where penetration occurred. Thus, the plant–endophytic bacteria association can be either neutral to the plant or positive when plant growth and/or health are stimulated. The following sections discuss important aspects of endophytic colonization: multiplication, localization, biocontrol potential, growth promotion and modes of action.

Multiplication

Endophytic population densities are generally low and seldom exceed log 3–5 cfu g^{-1} fresh plant tissue. This makes demonstration of bacterial multiplication very difficult. When leaves were infiltrated with a bacterial suspension, no multiplication of these bacteria in the leaf intercellular spaces was observed (summarized in Beattie and Lindow, 1995). In contrast, Hurek *et al.* (1994) reported inter- and intracellular multiplication of *Azoarcus* sp. in the root cortex of rice and Kaller grass as shown by electron microscopy. Similarly, Gardner *et al.* (1982), studying the long-term survival of several pseudomonads in citrus, suggested a compatible and dynamic association of the bacteria with the host plant. Lamb *et al.* (1996) reported persistence of *Pseudomonas aureofaciens* in maize plants due to continual bacterial invasion of newly formed lateral and crown roots. However, the common appearance of endophytic bacteria, their increase to a certain equilibrium density and the availability of nutrients within the plant tissues are all strong arguments favouring bacterial multiplication. Additional support in favour of endophytic bacteria multiplication in tissue is given by the fact that those intercellular

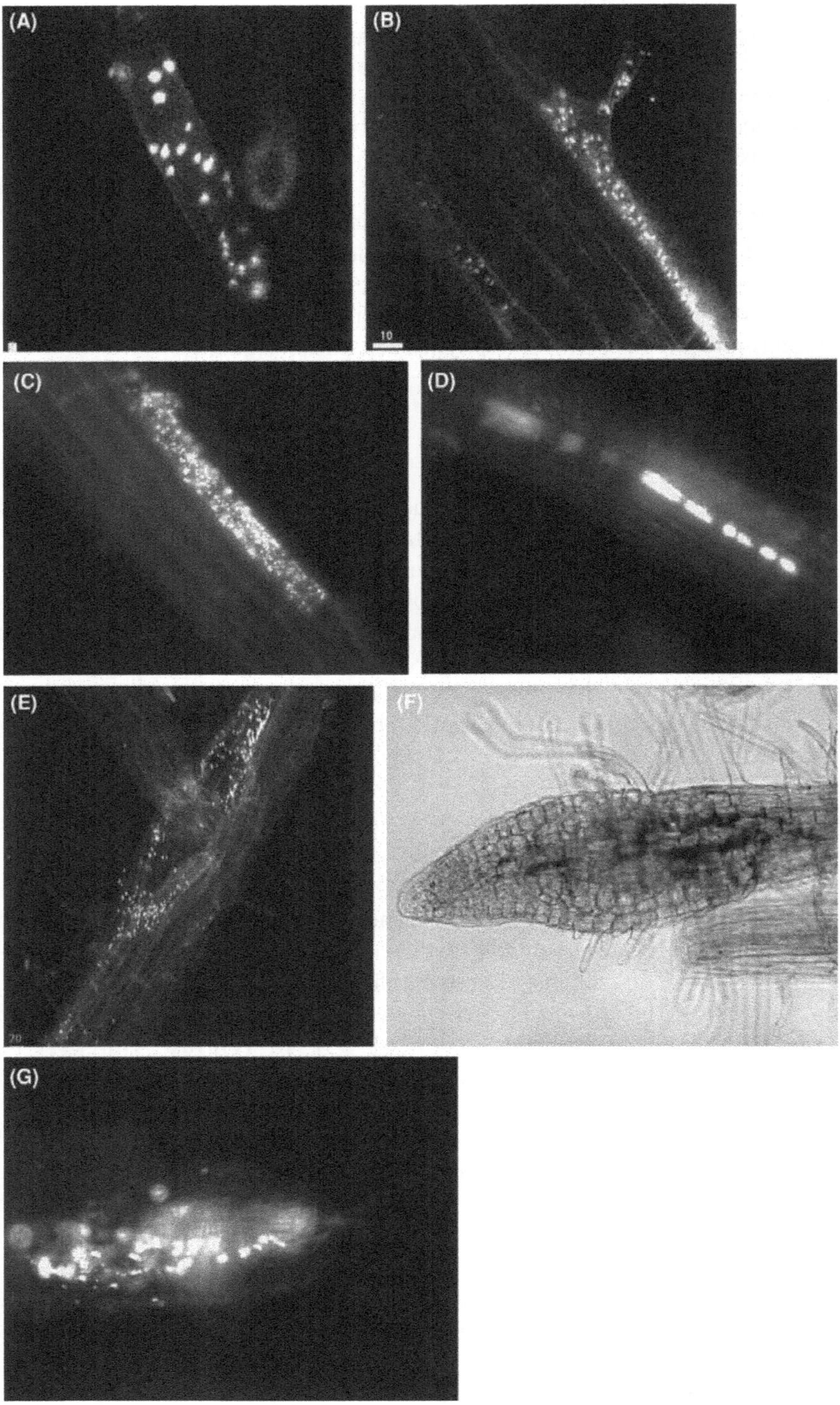
(A)
(B)
10
(C)
(D)
(E)
(F)
(G)

spaces or epidermal cells colonized by endophytic bacteria are usually packed with bacteria as demonstrated by Quadt-Hallmann and Kloepper (1996). Exactly how such high bacterial population densities can be established and maintained within plant microsites will be discussed in the following section.

Localization

Endophytic bacteria have been reported to colonize various plant parts such as roots, tubers, stems, leaves, fruits and seeds (Table 6.1). Even immature flower buds have been reported to contain endophytic bacteria (Mishagi and Donndelinger, 1990). In most cases these bacteria were isolated from surface disinfested tissue. With modern detection methods endophytic bacteria can be studied in selected plant tissues or even individual plant cells. For example, immuno-gold labelling in combination with transmission electron microscopy (Quadt-Hallmann and Kloepper, 1996) or bacterial marking with green fluorescent protein (GFP) (Hallmann *et al.*, 2001) have been successfully used in the past. *Enterobacter asburiae* JM22 and *Rhizobium etli* G12, for example, were localized in intercellular spaces of the root cortex, inside epidermal cells including root hairs and even in areas of conducting elements using GFP-marked strains (Fig. 6.2). However, bacterial colonization of the vascular system still needs further attention to clarify whether endophytic bacteria primarily colonize the xylem, the phloem, both or just the intercellular spaces surrounding the conducting elements. As mentioned earlier, bacterial colonization of the host plant can be differentiated into an external and internal phase. Externally, endophytic bacteria are found either randomly over the root surface, under collapsed epidermal cells or concentrated in grooves between epidermal cells as well as associated with wounds and lateral root formation (Fig. 6.3). Internal colonization can be intra- or intercellular, whereas the latter represents the major niche for bacterial endophytes. Although endophytic bacteria are reported to occur in the intercellular spaces throughout the

Fig. 6.2. Microscopic visualization of preferred colonization sites of *gfp*-marked endophytic bacteria on various host plants using epifluorescence (EM) and confocal laser scanning microscopy (CLSM). *Enterobacter asburiae* JM22 (pGT-kan), an endophytic bacterium found to colonize plants systemically, preferably colonizes individual epidermal cells as being shown for root hairs of *Phaseolus vulgaris* (A) and *Arabidopsis thaliana* (B). Epidermal cells colonized by *E. asburiae* JM22 (pGT-kan) were tightly packed with the bacterium (C). Epifluorescence images of *E. asburiae* JM22 (pGT-kan) forming an uneven string of microcolonies within the centre of an *A. thaliana* root (D). Visualization of the nematode-antagonistic rhizobacterium *Rhizobium etli* G12 (pGT-trp) around the lateral root base of *A. thaliana* (E). Bright field (F) and epifluorescence (G) image of a nematode gall on *Arabidopsis* caused by *Meloidogyne incognita* showing strong GFP fluorescence of *R. etli* G12 (pGT-trp) within the galled tissue. The magnification is indicated by the length of the bars or squares expressed in micrometres.

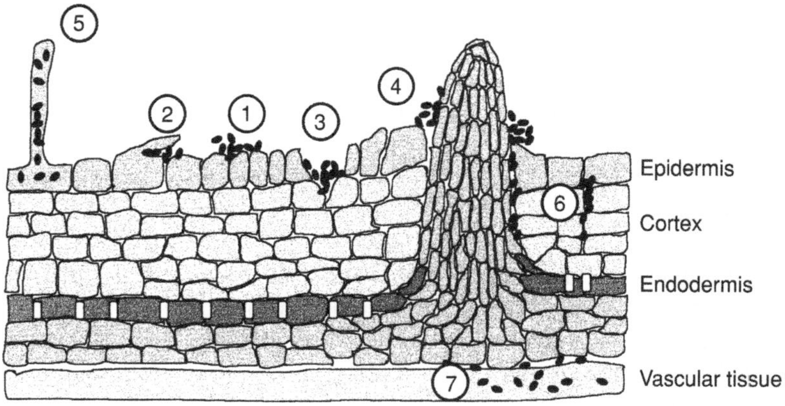

Fig. 6.3. External and internal colonization of plant roots by endophytic bacteria can be either: 1) random over the root surface, or 2) more specific below collapsed epidermal cells, 3) in association with wounds and 4) at sites of lateral root formation. Internal bacterial colonization occurs 5) intracellulary in root epidermal cells including root hairs, 6) intercellular within the root cortex or 7) in association with the conducting elements.

plant, most endophytic bacteria are probably found in the intercellular spaces of the root cortex where they can appear in high densities. High numbers of endophytic bacteria have also been found in close association with the conducting elements, and intracellularly in root epidermal cells including root hairs. The following section will focus on abiotic factors favouring endophytic colonization of certain plant tissues.

In general, successful bacterial colonization requires the availability of free plant tissue that supplies nutrients for bacterial metabolism. Very little is known about the internal space available for endophytic colonization. Bacteria colonizing the apoplast can basically select from a large network of connecting intercellular spaces, parts of the vascular tissue and the leaf's parenchyma tissue. Beattie and Lindow (1995) suggested that bacteria entering a leaf through a stoma or hydathode may have access to the intercellular space and a vast array of interconnected spaces within the leaf's mesophyll and parenchyma tissue. The size of the mesophyll air space can range from 23% to over 50% of the total leaf volume depending on plant species and leaf maturity, which would provide favourable conditions for endophytic bacteria. Within the root and stem tissue plant cells are usually more compact, therefore less space will be available for the endophytic bacteria in these regions. However, certain plant species provide intercellular spaces filled with air like the aerenchyma of rice plants and some other grass species that are potential colonization sites for endophytic bacteria (Reinhold *et al.*, 1987). By using cryo-scanning electron microscopy, intercellular spaces of maize roots expected to contain air were shown to be filled with fluid (Canny, 1995). The fluid was not just water but contained minerals at concentrations similar to those in the cortical cells surrounding them. However, not all the spaces contained fluid all the time, and the proportion of dry versus wet spaces appeared to depend on the age and water status of the root. For sugarcane, the intercellular space of stem internodes was found to contain a 10–12% solution of sucrose indicating the potential role of the intercellular space as storage for photoassimilates. This sugar solution can even be colonized by endophytic bacteria as has been shown for the N_2-fixing bacterium *Acetobacter diazotrophicus* (Dong *et al.*, 1994). Although fluid-filled intercellular spaces seem to be fairly widespread in stem and root tissue of various plant species, they do not appear in leaves (Canny, 1995).

Since space does not seem to be a limiting factor for endophytic colonizers, it should be expected that endophytic bacteria are more or less homogenously distributed within the plant tissue. However, results obtained to date on bacterial localization often present a completely different picture. Intercellular spaces or even individual epidermal cells colonized by endophytic bacteria are usually packed with bacterial cells, whereas adjacent areas are bacteria-free. High bacterial numbers found in such localized tissue can only be explained by a continuous and unlimited supply of nutrients from the plant at this site.

Knowledge concerning nutrient availability in the apoplast and bacterial demand for nutrients is scant. Nutrient concentrations in intercellular spaces are given as 2–3 mM amino acids, 3–6 mM glucose and 0.1–100 mM sucrose. However, these concentrations are highly variable depending on plant tissue examined and sampling time. It is doubtful that the amount of nutrients initially present within the intercellular spaces is sufficient to support bacterial survival in high numbers. If additional nutrients are provided by the plant, the questions still remain as to where do these nutrients come from and how are they translocated. It is known that pathogenic bacteria produce toxins which break down the selective permeability of the host cell membranes and cause efflux of soluble cell nutrients. However, such mechanisms for endophytic bacteria seem to be unlikely since these toxins cause irreversible damage to the host cell wall. A different mechanism that avoids any damage to the host cell plasma membrane was described for non-pathogenic bacteria, which induce partial degeneration of the host cell membrane and thus activate the K^+ efflux/H^+ influx exchange. Potassium is released into the intercellular space increasing the pH of intercellular fluid from 5.5 to 7.0–7.5. This pH change induces an efflux of sucrose, amino acids and inorganic ions which are utilized by internal colonizers. Similar processes might be initiated by endophytic bacteria and would explain the high numbers of bacteria found inter- and intracellularly in the plant.

Biological control

When discussing plant-health promotion it is necessary to differentiate between two strategies: (i) endophytic bacteria colonizing the internal tissue and occupying the ecological niche required by the pathogen and (ii) those bacteria primarily colonizing the cortex tissue where they produce metabolites that are suppressive to the pathogen or stimulate plant defence mechanisms. In the first strategy endophytic bacteria are randomly distributed throughout the plant, and present at the site of pathogen attack. This strategy was followed by Tomasino *et al.* (1995) when releasing transgenic *Clavibacter xyli* subsp. *cynodontis* expressing the δ-endotoxin from *Bacillus thuringiensis* subsp. *kurstaki* to control *Ostrinia nubilalis* on maize. The same strategy might be utilized by some endophytic colonizers of the vascular tissue to control vascular pathogens by pre-emptive colonization. However, this strategy bears the risk that bacterial enumeration might increase to such high numbers that they become deleterious to the plant by causing plugging of the vessels or excessive uptake of phytoassimilates. In contrast, the second strategy generally requires lower bacterial numbers for pathogen control and, due to their restricted colonization of the outer root tissue, competition for plant nutrients will be far less. If antimicrobial metabolites produced by the endophytic bacteria are the mechanism of pathogen control, increasing endophytic population densities should consequently result in enhanced control efficacy. In the case of induced resistance, total endophytic population numbers might be less important as long as they exceed a certain threshold level necessary to initiate the plant defence mechanisms. The following sections summarize some examples for endophytic bacteria controlling plant pathogens.

Bacterial pathogens

If pre-emptive colonization provides a satisfactory control mechanism the most obvious attempt at using endophytic bacteria to control bacterial pathogens would be to choose endophytic species colonizing the same ecological niche as the pathogen. Closely related species, for example *Agrobacterium radiobacter*, would then be suitable to control *A. tumefaciens*, *Pseudomonas fluorescens* to control *P. syringae* and *Enterobacter* spp. to control *Erwinia amylovora*. This strategy was successfully exploited within commercial development of the isolate *A. radiobacter* K84 to control *A. tumefaciens*. A completely different control mechanism is followed by some bacteria which induce plant defence mechanisms, such as for *Pseudomonas fluorescens* 89B-61 inducing systemic resistance against *P. syringae* pv. *lachrymans* (Liu *et al.*, 1995). An unclarified mode of action is represented by *Kluyvera ascorbata* and *Alcaligenes piechaudii* controlling *Xanthomonas campestris* pv. *campestris* (Assis *et al.*, 1998). Antagonistic activity has also been reported for endophytic bacteria isolated from potato tissue against *Clavibacter michiganensis* subsp. *sepedonicus*, the causal agent of bacterial ring rot of potato (van Buren *et al.*, 1993) and against *Erwinia carotovora* var. *atroseptica* causing potato soft rot (Sturz and Matheson, 1996). In the latter case the authors indicate that soft rot development was negatively correlated with the density of tuber populations of endophytic bacteria found able to inhibit *E. carotovora* var. *atroseptica* growth *in vitro*.

Fungal pathogens

Extensive work on endophytic bacteria controlling fungal pathogens has been conducted. Endophytic bacteria have been shown to control *Fusarium oxysporum* f. sp. *vasinfectum* on cotton (Chen *et al.*, 1995), *Fusarium oxysporum* f. sp. *pisi* on pea (Benhamou *et al.*, 1996), *Verticillium albo-atrum* and *Rhizoctonia solani* on potato (Nowak *et al.*, 1995), *Rhizoctonia solani* on potato (Pleban *et al.*, 1995) and rice (Krishnamurthy and Gnanamanickam, 1997), *Sclerotium rolfsii* on bean (Pleban *et al.*, 1995) and *Ceratocystis fagacearum* on oak (Brooks *et al.*, 1994). Antibiosis has been shown for endophytic bacteria isolated from potato against *Fusarium sambucinum*, *F. avenaceum*, *F. oxysporum*, *Rhizoctonia solani*, and from maize against *F. moniliforme* (Hinton and Bacon, 1995). Whereas these latter cases indicate that bacterial metabolites are involved in the control of fungal pathogens, Chen *et al.* (1995) assumed that enhanced host defence mechanisms were the preferred mode of action. However, if induced resistance is a potential defence mechanism mediated by endophytic bacteria one would also expect these endophytes to control foliar pathogens in a way similar to that reported for several rhizosphere bacteria (Wei *et al.*, 1996). Unfortunately, research work is limited to soil-borne pathogens. If and how endophytic bacteria control foliar pathogens is still unknown.

Plant-parasitic nematodes

As mentioned earlier in this chapter, plant-parasitic nematodes not only increase total endophytic population densities, but the bacteria also occur within galled tissue of nematode-infested plants. Does this make them superior biological control agents for nematodes? In greenhouse studies nematode control usually did not exceed 50% (Hallmann *et al.*, 1997c), and therefore did not differ from control efficacy described for rhizosphere bacteria (Becker *et al.*, 1988; Oostendorp and Sikora, 1990; Racke and Sikora, 1992). Bacterial isolates which gave consistent and reproducible control of *Meloidogyne incognita* on cotton included *Pseudomonas fluorescens* 89B-61 (Hallmann *et al.*, 1998), *Brevundimonas vesicularis* IN884 and *Serratia marcescens* 90–43 (Hallmann *et al.*, 1997c). Screening for new biocontrol strains is often based on time-consuming plant bioassays. In addition, there seems to be no superior screening source for detecting efficient biocontrol strains. Endophytic bacterial isolates of known biocontrol agent sources, bacteria isolated from cotton grown in nematode-suppressive soil, nor isolates with high chitinolytic activity gave superior results. Control of plant-parasitic nematodes in some cases seems to be more complex than control of fungal and bacterial pathogens. The nematode itself represents a robust organism covered by a thick cuticle and a strong stylet that allows inter- and intracellular movement through the plant tissue. However, the most damaging nematode species worldwide such as root-knot nematodes and cyst nematodes, become sedentary within the root tissue soon after infection and therefore would present an excellent target for endophytic bacteria. Root exudates that act as attractants and hatching factors might be metabolized by the bacteria and no longer stimulate nematode infestation in the way reported for plant health-promoting rhizosphere bacteria (Sikora, 1992).

Although the biocontrol potential of endophytic bacteria is well documented, a question that must be asked is whether biocontrol potential is restricted to bacteria only colonizing the outer root cortex? Comparative studies of endophytic bacteria with and without biocontrol potential demonstrated that *Bacillus pumilus* SE34 controlling Fusarium wilt on cotton and pea (Chen *et al.*, 1995; Benhamou *et al.*, 1996) and *Pseudomonas fluorescens* 89B-61 controlling various pathogens of cotton and cucumber (Chen *et al.*, 1995; Wei *et al.*, 1996) colonized the outer root cortex whereas the systemic colonizer *Enterobacter asburiae* JM22 had no biocontrol effect. In a different approach, Sturz *et al.* (1999) studied the potential of endophytic bacteria retrieved from different layers of the potato tuber periderm for antibiosis towards the soil-borne pathogens *Fusarium* spp. and *Phytophthora infestans*. Interestingly, biocontrol by the bacterial endophytes against these pathogens was significantly higher in isolates recovered from the outermost layer of tuber peel and decreased progressively towards the centre of the tuber. Even for bacteria recovered from all different layers, antibiosis to pathogens decreased with depth of recovery from the periderm to the inside of the tuber. The fact that endophytic bacteria with biocontrol potential seem to primarily colonize the outer root or tuber tissue might be a strategic development over time of the plant–endophytic bacteria association. This tissue will be attacked first by the pathogen and endophytic bacteria colonizing this tissue have evolved control mechanisms that lead to repression of the pathogen and thus protect their own habitat and the plant tissue simultaneously.

Mode of action

Compared to the large number of reports of endophytic bacteria controlling plant pathogens, considerably less work has been done examining the potential modes of action. In general, it is presumed that endophytic bacteria control plant pathogens by either: (i) pre-emptive colonization; (ii) direct antagonism by metabolites; or (iii) induced systemic resistance. Whereas the first strategy refers mainly to endophytic colonizers of the vascular tissue as potential antagonists of vascular-invading pathogens, such as *Verticillium*, *Fusarium* or *Ralstonia*, the second and third strategies are directed towards the pathogen at the site of attack such as the root cortex. Several recent papers have been published on the mode of action of plant growth- and plant health-promoting rhizosphere bacteria (PGPR/PHPR) (Sikora and Hoffmann-Hergarten, 1993; van Loon *et al.*, 1998). Considering the fact that endophytic bacteria are often closely related to PGPR/PHPR and several PGPR/PHPR have lately been found to colonize the root internally, we could assume that similar modes of action described for PGPR/PHPR also apply for endophytic bacteria.

The potential role of antimicrobial metabolites for biocontrol was demonstrated *in vitro* by Sturz *et al.* (1999). The authors studied the bacterial community of different layers of potato tubers for antibiosis against the fungal pathogens *Phytophthora infestans, Fusarium sambucinum, F. avenaceum* and *F. oxysporum*, and concluded that those endophytic bacteria isolated from the outermost layer showed significantly higher antibiosis against the four pathogens than bacteria from the deeper layers. However, very little is known about the actual antibiotic concentration *in situ*, and a dose–response association as known for rhizosphere bacteria (Raaijmakers *et al.*, 1995) has not yet been confirmed for endophytic bacteria.

The mode of action presently gaining the most interest is related to induction of plant defence mechanisms. In general, induced systemic resistance is promoted by several authors as a main biocontrol mechanism in many plant–bacteria interactions (van Loon *et al.*, 1998). Intensive work on systemic resistance induced by PGPR has pointed out the broad spectrum of pathogens being controlled by this mechanism including viruses, bacteria, fungi, nematodes and even insects (reviewed by van Loon *et al.*, 1998). Several of these PGPR were recently isolated from internal root tissue indicating that similar resistance mechanisms might be also applicable for endophytic bacteria (Hasky-Günther, 1996). Endophytic colonizers might even be better inducers of plant defence mechanisms since they establish a much closer relationship over an extended period of time with the plant as compared to rhizosphere bacteria which can be inhibited by competition with other microorganisms on the root surface. However, induction of plant defence mechanisms always requires some kind of plant–endophyte recognition. It has been postulated that gene products on the surface of the bacterial endophyte such as lipopolysaccharides (LPSs) function as elicitors to specifically bind to receptors on or near the plant cell surface. Plant receptors could be sugar-binding proteins called lectins. Reitz *et al.* (2000), studying the mechanisms of systemic resistance induced by *Rhizobium etli* G12 against the potato cyst nematode *Globodera pallida* in potato, identified purified LPS as causal inducer of the resistance mechanism. Similarly, LPS preparations from *Xanthomonas campestris, Pseudomonas solanacearum, P. fluorescens* and *P. putida* are also effective in inducing disease resistance (Leach *et al.*, 1983; Romeiro and Kimura, 1997; van Wees *et al.*, 1997). Whereas these authors agree on the overall importance of LPS as elicitor, there is still some controversy as to which fraction of the LPS causes the defence response. Whereas Duijff *et al.* (1977) refer to the O-antigenic side chain of the outer membrane LPSs of *Pseudomonas fluorescens* WCS417r involved in the induced resistance of tomato against the Fusarium wilt pathogen, Reitz (1999), using Tn5 mutants with truncated O-antigen, still received a defence response in the absence of the O-antigenic side chain and thus favoured the lipid A or core fraction of the LPS as inducive agent.

Most work on biological control of endophytic bacteria has been targeted at soil-borne pathogens. Almost nothing is known about how endophytic bacteria might induce plant defence mechanisms against foliar pathogens, and whether control efficacy could be enhanced by considering whole endophytic communities instead of single isolates. One of the first indications for plant defence mechanisms expressed in the leaves derived from observations made by Xia *et al.* (1997), who measured increased activities of enzyme activities associated with plant defence mechanisms such as peroxidase, esterase and superoxide dismutase in leaves and stems of cotton inoculated with endophytic bacteria and challenged with *Verticillium dahliae*. However, we are just beginning to look at the role of endophytic bacteria for biocontrol purposes and with the number of people interested in this research area constantly growing we will gain more insight into the modes of action from year to year.

Other beneficial effects

Besides their antagonistic activity towards plant pathogens, endophytic bacteria can provide additional benefits to the plant such as growth promotion or enhanced nodulation of leguminous plants. Whereas a single application of *Rhizobium leguminosarum* BV *trifolii* resulted in poor nodulation of red clover, coapplication with the endophytes *Bacillus insolitus*, *B. brevis* or *Agrobacterium rhizogenes* significantly increased the number of root nodules two- to fourfold (Sturz *et al.*, 1997). Similar effects are reported by Nishijima *et al.* (1988) who observed enhanced nodulation of soybean by *Bradyrhizobium japonicum* in the presence of *Pseudomonas fluorescens* SSJ2. Sturz *et al.* (1997) reported that total growth effects were greater when endophytic bacteria were coinoculated with *R. leguminosarum* BV *trifolii* than by sole application. These findings provide further evidence that a certain proportion of plant promotion effects are in fact the allelopathic side-effect of the competition between endophytes for the same ecological niche. This would also explain why tissue-culture plants perform much better following treatment with plant-associated bacteria. Endophytic bacteria have also been associated with the growth promotion of several crops as being summarized by Hallmann *et al.* (1997b). According to Sturz (1995), approximately 10% of the bacterial isolates recovered from within potato tubers were shown to promote plant growth. In general, plant growth promotion can be facilitated by several microbial processes such as nutrient solubilization, fixation of atmospheric N and production of plant growth hormones. A major effect on plant growth promotion is contributed to nitrogen fixation by free-living endophytic bacteria. Whereas the potential of N fixation has been described for a broad spectrum of bacterial genera such as *Acetobacter*, *Azoarcus*, *Azospirillum*, *Klebsiella* and *Enterobacter* (Döbereiner and

Pedrosa, 1987; Hurek *et al.*, 1994; Hecht-Buchholz, 1998), their contribution to the plant's total requirement for N is in most cases still speculative. However, good data on the N-fixing capacity by endophytic bacteria were provided for N-fixing symbionts of grass species, like *Azoarcus*, *Azospirillum*, etc. However, besides plant growth promotion effects based on microbial fixed N, we know very little about the presence of bacterial-derived growth-enhancing substances *in planta*.

Specificity

One of the most interesting questions concerning plant–endophyte associations relates to specificity. Do different plant species harbour similar endophytic bacteria and does one endophyte colonize several plant species? Specificity in general has been reviewed excellently by Smith and Goodman (1999). Within this context, specificity will be looked at in relation to endophytic bacteria. In the absence of specificity any potential endophyte present in the rhizosphere or phyllosphere should have an equal chance to colonize a host plant. However, as we know this is not the case, plants as well as potential endophytic bacteria must have developed certain traits which support plant–endophytic bacteria associations but not others.

Specificity related to plant species

In general, those endophytic bacterial species that contribute significantly to the total internal population of a given plant also represent major species of the rhizosphere environment. As a consequence, different plant species should reflect similar endophytic populations. However, when different host plants are grown side by side in the same field they usually differ in their endophytic community structure, although the initial rhizosphere population is still the same. These differences in endophytic bacterial community structure between different plant species can only be explained by plant species-specific mechanisms favouring some endophytic bacteria and excluding others. The specific composition of the plant root exudates, a major food source for potential endophytic bacteria, might play an important role in bacterial selectivity. As the composition of root exudates varies between plant species, they will directly affect the bacterial community structure in the rhizosphere and therefore the source for endophytic colonization.

Specificity related to plant genotype

In addition to plant species, genotypes can also significantly affect the endophytic community. Adams and Kloepper (1998), comparing the endophytic community of nine cotton cultivars, observed a significant genotype-specific effect for total numbers of endophytic bacteria and community structure. By using differential media, main effects were generally observed for Gram-negative bacteria including fluorescent pseudomonads. Similarly, Sturz *et al.* (1999), growing four potato cultivars in a randomized complete-block design, found differences in their bacterial community structure. Whereas *Curtobacterium flaccumfaciens* and *Pseudomonas cichorii* occurred in all four cultivars, certain bacterial species were only found in one cultivar, such as *Sphingomonas thalpophilum* in potato cv. 'Butte', *Curtobacterium luteum* in 'Kennebec' and *Curtobacterium albidum* in 'Russet Burbank'. Plant genotypes like plant species also vary in their physiological response and biochemical composition. As a consequence, the amount and composition of metabolites released into the apoplast and finally exudated into the rhizosphere represent the major food source for plant-associated bacteria, and will therefore directly affect the bacterial community. As mentioned before, any factor affecting the rhizosphere community will consequently affect the endophytic spectrum. However, this still does not answer the question of whether the rhizosphere is the only factor determining endophytic populations, or if the plant itself has mechanisms to specifically control endophytic colonization.

Specificity within the genotype might also change due to environmental conditions. When soil was treated with 1% chitin and planted with cotton 3 weeks later, the rhizosphere community at planting time consisted of 1.5% and 1.6% *Burkholderia cepacia* in the non-treated and treated soil, whereas *Phyllobacterium rubiacearum* was higher in the chitin-treated soil (5.5%) compared to the non-treated soil (0.7%) (Hallmann *et al.*, 1999). Assuming that endophytic bacteria colonize the host plant right after germination it would be expected that comparable population densities of *B. cepacia* and *P. rubiacearum* exist in the endorhiza. Interestingly, when the endorhiza was analysed 4 weeks after planting, *B. cepacia* dominated the endorhiza of cotton grown in chitin-treated soil (73%), whereas *P. rubiacearum* which originally occurred at higher densities in the rhizosphere of chitin-treated soil was found exclusively in cotton grown in non-treated soil where it accounted for about 61% of the total population. This endophytic community structure also did not match that of the rhizosphere population 4 weeks after treatment. *B. cepacia* accounted for 4.1% and 12.9% of the total population recovered from non-treated and treated soil compared to *P. rubiacearum* with 4.7% and 2.1%, respectively. These changes in endophytic community structure as a response to chitin treatment can only be explained by alterations of bacterial and/or host plant specificity. It is believed that the chitin amendment may have altered the physiological state of the plant which then led to support for a different spectrum of endophytic bacteria. In follow-up experiments it has been shown that chitin treatment induced systemic resistance in cotton against the root-knot nematode *Meloidogyne incognita* (Hallmann, unpublished). Systemic induced resistance was associated with increased chitinase and peroxidase activities in roots and leaves thus confirming changes in plant physiology.

As well as their effects on plant enzyme activity, the chitin treatment also enhanced soil chitinase, protease and esterase activity, and soil microbial activity. Unfortunately, the effect of the soil enzyme activity on bacterial specificity has not yet been studied.

Specificity related to bacteria species

The host spectrum of a single endophyte raises the question of whether they colonize primarily one plant species or a broad spectrum of different plant species. From the studies on endophytic communities of different host plants it becomes quite evident that certain species such as *Agrobacterium radiobacter* or *Pseudomonas fluorescens* are omnipresent whereas other species occur less frequently and are even unique for some plant species. When examining the often great physiological variability of bacterial isolates within one species the question as to whether a single endophytic isolate can also colonize a broad spectrum of host plants seems appropriate. Although this question has never been addressed directly, experiments performed independently with *Enterobacter asburiae* JM22 have shown that this species colonizes cotton, cucumber, bean, potato and *Arabidopsis* (Table 6.2). It is expected that when additional plant species are tested for colonization by JM22 the list of potential host plants will expand.

If the host spectrum of one bacterium is non-specific, are bacterial traits such as biocontrol potential also expressed across a broad spectrum of hosts? Recent work with *Rhizobium etli* G12, a biocontrol strain that can grow endophytically, has shown that this bacterium significantly suppresses the potato cyst nematodes *Globodera pallida* and *Globodera rostochiensis* as well as the root-knot nematode *Meloidogyne incognita* on potato and tomato. Similarly, *Bacillus pumilus* SE34 controls *Fusarium oxysporum* f. sp. *pisi* on pea (Benhamou *et al.*, 1996) and *Fusarium oxysporum* f. sp. *vasinfectum* on cotton (Chen *et al.*, 1995), whereas *Pseudomonas fluorescens* 89B-61 reduces disease

Table 6.2. Host spectrum of *Enterobacter asburiae* JM22.

Host plant	Colonized plant tissue	Reference
Cotton (*Gossypium hirsutum*)	Root, cotyledons, stem	McInroy and Kloepper, 1995a; Quadt-Hallmann and Kloepper, 1996
Cucumber (*Cucumis sativus*)	Root, cotyledons, stem	Quadt-Hallmann and Kloepper, 1996
Potato (*Solanum tuberosum*)	Root	Hallmann, unpublished
Bean (*Phaseolus vulgaris*)	Root, cotyledons, stem, leaf	Quadt-Hallmann and Kloepper, 1996
Arabidopsis thaliana	Root	Hallmann, unpublished

severity caused by *Meloidogyne incognita* (Hallmann *et al.*, unpublished) and *Fusarium oxysporum* f. sp. *vasinfectum* on cotton (Chen *et al.*, 1995), as well as *Pseudomonas syringae* pv. *lachrymans* on cucumber (Wei *et al.*, 1996). In general, it can be concluded that endophytic bacteria seem to have a broad spectrum of hosts, and express certain traits such as biocontrol potential in a range of host plants.

The second most often asked question concerns bacterial specificity as it relates to physiological differences of individuals of one bacterial species found internally in the plant versus those colonizing the rhizosphere – do endophytic bacteria differ from their rhizosphere counterparts? The work by Sturz *et al.* (1999) clearly showed that bacterial isolates of the same endophytic species collected from the outermost layer of a potato tuber expressed greater antibiosis activity against three *Fusarium* spp. and *Phytophthora infestans* than strains isolated from deeper tuber layers, indicating tissue type- and tissue site-specific bacterial adaptations. The authors refer to the structural differences between periderm cells and cortex cells immediately beneath the peel which have differences in cell biochemistry and host physiology that might explain the observed site-specific effects of bacterial isolates of the same species. It is assumed that bacterial adaptation occurs at certain tissue sites and among certain tissue types. Furthermore, the dynamic nature of bacterial phenotype expression, in this case antibiotic secretion, may be governed by a population density-dependent modulation of the bacterial phenotype (Sturz *et al.*, 1999). In a different approach, Duijff *et al.* (1997) studied the colonization of tomato roots by *Pseudomonas fluorescens* WCS417r with that of a mutant of this strain lacking the O-antigenic side chain of its outer membrane lipopolysaccharides. Whereas both strains colonized the rhizoplane at the same level, the wild-type strain of WCS417r colonized the root interior to a higher extent than the mutant strain, indicating the potential role of the O-antigenic side chain of the outer membrane lipopolysaccharides for endophytic colonization. This at least indicates that isolates of the same bacterial species differing slightly in their cell wall components do show different colonization behaviour. However, considering the enormous differences between internal and external habitats, and knowing how quickly endophytic bacteria adapt to internal habitats, it seems obvious that the process of endophytic colonization is accompanied by physiological modifications of the internal colonizers. Whereas physiological differences between internal and external colonizers of the same species have been confirmed, we know very little about the driving force of this process. Whether or not: (i) the plant is the major factor affecting bacterial behaviour; (ii) the bacteria change their physiology in response to plant tissue-specific conditons; or (iii) a combination of both occurs, is open to question. Does some form of communication exist between the plant and the endophytic bacteria? If the tissue type or tissue site-specific bacterial adaptations in potato tubers described previously by Sturz *et al.* (1999) occur *in situ* as proposed by the authors, it would require some form of communication between the plant and bacterial endophyte. Unfortunately, this process is still not adequately researched.

As indicated previously, bacterial specificity is not a constant factor and undergoes continuous modification as a result of biotic and abiotic conditions. How does this relate to endophytic bacteria? Metabolites of coexisting bacteria, for example, have been described that affect the host specificity of *Rhizobium* resulting in colonization of non-host legumes (Sturz *et al.*, 1997). When the authors isolated a broad spectrum of *Rhizobium* spp. from red clover not known to be symbionts of this legume, they concluded that microbial metabolites produced by either endophytic or rhizosphere bacteria facilitated non-host-specific rhizobia to colonize this legume. Similar results were achieved by Nishijima *et al.* (1988) who observed increased nodulation of soybean by *Bradyrhizobium japonicum* following preincubation with *Pseudomonas fluorescens* SSJ2. The authors assumed that *P. fluorescens* alters the *Rhizobium* and not the host plant specificity. Overall, it is still questionable whether these types of bacterial interactions can be used as models for all endophytic bacteria. It can be concluded that specificity as it relates to host plant or endophytic bacteria is a controversial field, but a fascinating one that will generate a great deal of interesting data in the future.

Practical Aspects and Conclusions

The potential areas for practical use of endophytic bacteria have not been fully explored. However, two major areas of potential use are biological control and plant growth promotion. Most of the potential uses being described for plant growth-promoting rhizobacteria might also be applicable for endophytic bacteria. The internal colonization strategy followed by endophytic bacteria, however, might provide additional benefits. The internal plant tissue is believed to provide a more uniform and protective environment for the bacteria than in the rhizosphere. The availability of water in the plant is more constant and excessive water supply such as irrigation or rainfall will not wash out endophytic bacteria as is the case for rhizosphere colonizers. Similarly, the supply of nutrients inside the plant tissue might be more consistent and of a more homogenous composition thus reducing stresses associated with bacterial nutrient requirement. With regards to endophytic colonizers the above-ground internal plant tissue will be less exposed to UV radiation than the leaf surface. The need to select bacterial isolates with high levels of rhizosphere competence which are often considered necessary for successful seed or root bacterization treatments would be eliminated. Finally, competition among organisms may be relatively low due to the presence of fewer types and numbers of organisms. All these factors might contribute to an overall increasing plant health promotion efficacy as well as control consistency when compared to rhizobacteria where inconsistent water and nutrient supply, and high levels of inter- and intraspecific competition, are believed to be major factors explaining the often inconsistent results observed under field conditions. However, endophytic bacteria also have to face challenges including low pH, exposure to plant defence mechanisms and genetic changes in cultivars, and therefore may require adaptations to survive in and exploit this environment that are distinct from those existing on the leaf or root surface (Beattie and Lindow, 1995).

Another important consideration for using endophytic bacteria is based on the fact that certain cultivars are known to respond more readily to endophytic bacteria than others (Conn *et al.*, 1997). Therefore, the possibility arises of designing or encouraging 'biased' communities of bacteria with various plant-enhancing qualities, both in the rhizosphere and within the plant (Sturz *et al.*, 1999). This approach was mentioned by Bird *et al.* (1980), who indicated that the observed resistance of multi-adversity resistance cotton lines against a broad spectrum of different pathogens cannot be explained by the genetic modifications alone but also had to be an effect of the associated endophytic community. Also soil amendments such as chitin can be used to stimulate shifts in the indigenous endophytic population towards enrichment of those bacterial species with antagonistic capacity against the targeted pathogens (Hallmann *et al.*, 1999).

Since the potential uses of endophytic bacteria in agriculture appear to be many, there is an urgent need for research to answer some of the questions raised within this chapter targeting: specificity, nutrient supply, modes of action and performance under field conditions. The answers to these questions will decide if and when endophytic bacteria will make the transition from being an exciting laboratory system to an integral part of modern agriculture.

Acknowledgements

The author wants to thank Richard Sikora and Andrea Quadt-Hallmann for their valuable comments on the manuscript and all colleagues supporting this work by providing laboratory facilities and moral support, especially Joseph W. Kloepper and Rodrigo Rodríguez-Kábana from Auburn University, Alabama, USA and Steven W. Lindow and William G. Miller from the University of California, Berkeley, California, USA. The author's work was funded by the Alexander von Humboldt Foundation and German Science Foundation.

References

Adams, P.D. and Kloepper, J.W. (1998) Effects of plant genotype on populations of indigenous bacterial endophytes on nine cotton cultivars grown under field conditions. *Phytopathology* 88, S2 (abstract).

Assis, S.M.P., da Silveira, E.B., de Mariano, L.R. and Menezes, D. (1998) Endophytic bacteria - method for isolation and antagonistic potential against cabbage black rot. *Summa Phytopathologica* 24, 216–220.

Bashan, Y. (1986) Migration of the rhizosphere bacteria *Azospirillum brasilense* and *Pseudomonas fluorescens* towards wheat roots in the soil. *Journal of General Microbiology* 132, 3407–3414.

Beattie, G.A. and Lindow, S.E. (1995) The secret life of foliar bacterial pathogens on leaves. *Annual Review of Phytopathology* 33, 145–172.

Becker, J.O., Zavaleta-Meija, Colbert, S.F., Schroth, M.N., Weinhold, A.R., Hancock, J.G. and Van Gundy, S.D. (1988) Effects of rhizobacteria on root knot nematodes and gall formation. *Phytopathology* 78, 1466–1469.

Bell, C.R., Dickie, G.A., Harvey, W.L.G. and Chan, J.W.Y.F. (1995) Endophytic bacteria in grapevine. *Canadian Journal of Microbiology* 41, 46–53.

Benhamou, N., Kloepper, J.W., Quadt-Hallmann, A. and Tuzun, S. (1996) Induction of defense-related ultrastructural modifications in pea root tissues inoculated with endophytic bacteria. *Plant Physiology* 112, 919–929.

Bird, L.S., Leverman, C., Thaxaton, P. and Percy, R.G. (1980) Evidence that microorganism in and on tissues have a role in a mechanism of multi-adversity resistance in cotton. In: Brown, J.M. (ed.) *Proceedings of Beltwide Cotton Production Research Conferences*. National Cotton Council, Memphis, Tennessee, pp. 283–285.

Brooks, D.S., Gonzales, C.F., Appel, D.N. and Filer, T.H. (1994) Evaluation of endophytic bacteria as potential biological control agents for oak wilt. *Biological Control* 4, 373–381.

Canny, M.J. (1995) Apoplastic water and solute movement: new rules for an old space. *Annual Review of Plant Physiology and Plant Molecular Biology* 46, 215–236.

Chanway, C.P. (1998) Bacterial endophytes: ecological and practical implications. *Sydowia* 50, 149–170.

Chen, C., Bauske, E.M., Musson, G., Rodríguez-Kábana, R. and Kloepper, J.W. (1995) Biological control of *Fusarium* wilt on cotton by use of endophytic bacteria. *Biological Control* 5, 83–91.

Conn, K.L., Nowak, J. and Lazorovita, G. (1997) A gnotobiotic bioassay for studying interactions between potatoes and plant growth-promoting rhizobacteria. *Canadian Journal of Microbiology* 43, 801–808.

De Boer, S.H. and Copeman, R.J. (1974) Endophytic bacterial flora in *Solanum tuberosum* and its significance in bacterial ring rot disease. *Canadian Journal of Plant Science* 54, 115–122.

Dong, Z., Canny, M.J., McCully, M.W., Roboredo, M.R., Cabadilla, C.F., Ortega, E. and Rodés, R. (1994) A nitrogen-fixing endophyte of sugarcane stems. *Plant Physiology* 105, 1139–1147.

Döbereiner, J. and Pedrosa, F.O. (1987) *Nitrogen-fixing Bacteria in Non-leguminous Crop Plants*. Science Tech Publishers, Madison, Wisconsin; Springer-Verlag, Berlin.

Duijff, B.J., Gianinazzi-Pearson, V. and Lemanceau, P. (1997) Involvement of the outer membrane lipopolysaccharides in the endophytic colonization of tomato roots by biocontrol *Pseudomonas fluorescens* strain WCS417r. *New Phytologist* 135, 325–334.

Fisher, P.J., Petrini, O. and Scott, H.M.L. (1992) The distribution of some fungal and bacterial endophytes in maize (*Zea mays* L.). *New Phytologist* 122, 299–305.

Gagné, S., Richard, C., Rousseau, H. and Antoun, H. (1987) Xylem-residing bacteria in alfalfa roots. *Canadian Journal of Microbiology* 33, 996–1000.

Gardner, J.M., Feldman, A.W. and Zablotowicz, R.M. (1982) Identity and behaviour of xylem-residing bacteria in rough lemon roots of Florida citrus trees. *Applied Environmental Microbiology* 43, 1335–1342.

Germida, J.J., Siciliano, S.D., de Freitas, J.R. and Seib, A.M. (1998) Diversity of root-associated bacteria associated with field-grown canola (*Brassica napus* L.) and wheat (*Triticum aestivum* L.). *FEMS Microbiology Ecology* 26, 43–50.

Hallmann, J., Kloepper, J.W. and Rodríguez-Kábana, R. (1997a) Application of the Scholander pressure bomb to studies on endophytic bacteria of plants. *Canadian Journal of Microbiology* 43, 411–416.

Hallmann, J., Quadt-Hallmann, A., Mahaffee, W.F. and Kloepper, J.W. (1997b) Bacterial endophytes in agricultural crops. *Canadian Journal of Microbiology* 43, 895–914.

Hallmann, J., Rodríguez-Kábana, R. and Kloepper, J.W. (1997c) Nematode interactions with endophytic bacteria. In: Ogoshi, A., Kobayashi, K., Homma, Y., Kodama, F., Kondo, N. and Akino, S. (eds) *Plant Growth-promoting Rhizobacteria - Present Status and Future Prospects*. Nakanishi Printing, Sapporo, pp. 243–245.

Hallmann, J., Quadt-Hallmann, A., Rodríguez-Kábana, R. and Kloepper, J.W. (1998) Interactions between *Meloidogyne incognita* and endophytic bacteria in cotton and cucumber. *Soil Biology and Biochemistry* 30, 925–937.

Hallmann, J., Rodríguez-Kábana, R. and Kloepper, J.W. (1999) Chitin-mediated changes in bacterial communities of the soil, rhizosphere and within roots of cotton in relation to nematode control. *Soil Biology and Biochemistry* 31, 551–560.

Hallmann, J., Quadt-Hallmann, A., Miller, W.G., Sikora, R.A. and Lindow, S.E. (2001) Endophytic colonization of plants by the biological control agent *Rhizobium etli* G12 in relation to *Meloidogyne incognita* infection. *Phytopathology* (in press).

Hasky-Günther, K. (1996) Untersuchungen zum Wirkungsmechanismus der antagonitischen Rhizosphärebakterien *Agrobacterium radiobacter* (Isolat G12) und *Bacillus sphaericus* (Isolat B43) gegenüber dem Kartoffelzystennematoden *Globodera pallida* an Kartoffel. PhD. thesis, University of Bonn.

Hecht-Buchholz, C. (1998) The apoplast-habitat of endophytic dinitrogen-fixing bacteria and their significance for the nitrogen nutrition of nonleguminous plants. *Zeitschrift für Pflanzenernährung und Bodenkunde* 161, 509–520.

Hinton, D.M. and Bacon C.W. (1995) *Enterobacter cloacae* is an endophytic symbiont of corn. *Mycopathologia* 129, 117–125.

Hollis, J.P. (1951) Bacteria in healthy potato tissue. *Phytopathology* 41, 350–367.

Hurek, T., Reinhold-Hurek, B., Van Montagu, M. and Kellenberger, E. (1994) Root colonization and systemic spreading of *Azoarcus* sp. Strain BH72 in grasses. *Journal of Bacteriology* 176, 1913–1923.

Jacobs, M.J., Bugbee, W.M. and Gabrielson, D.A. (1985) Enumeration, location, and characterization of endophytic bacteria within sugar beet roots. *Canadian Journal of Botany* 63, 1262–1265.

Krishnamurthy, K. and Gnanamanickam, S.S. (1997) Biological control of sheath blight of rice: induction of systemic resistance in rice by plant-associated *Pseudomonas* spp. *Current Science* 72, 331–334.

Lalande, R., Bissonnette, N., Coutlée, D. and Antoun, H. (1989) Identification of rhizobacteria from maize and determination of their plant-growth promoting potential. *Plant and Soil* 115, 7–11.

Lamb, T.G., Tonkyn, D.W. and Kluepfel, D.A. (1996) Movement of *Pseudomonas aureofaciens* from the rhizosphere to aerial plant tissue. *Canadian Journal of Microbiology* 42, 1112–1120.

Leach, J.E., Sherwood, J., Fulton, R.W. and Sequeira, L. (1983) Comparison of soluble proteins associated with disease resistance induced by bacterial lipopolysaccharide and by viral necrosis. *Physiological Plant Pathology* 23, 377–385.

Levanony, H. and Bashan, Y. (1989) Localization of specific antigens of *Azospirillum brasilense* Cd in its exopolysaccharide by immunogold staining. *Current Microbiology* 18, 145–149.

Liu, L., Kloepper, J.W. and Tuzun, S. (1995) Induction of systemic resistance in cucumber against bacterial angular leaf spot by plant growth-promoting rhizobacteria. *Phytopathology* 85, 843–847.

Mahaffee, W.F. and Kloepper, J.W. (1997) Temporal changes in the bacterial communities of soil, rhizosphere, and endorhiza associated with field-grown cucumber (*Cucumis sativus* L.). *Microbial Ecology* 34 210–223.

McInroy, J.A. and Kloepper J.W. (1995a) Survey of indigenous bacterial endophytes from cotton and sweet corn. *Plant and Soil* 173, 337–342.

McInroy, J.A. and Kloepper, J.W. (1995b) Population dynamics of endophytic bacteria in field-grown sweet corn and cotton. *Canadian Journal of Microbiology* 41, 895–901.

Mishagi, I.J. and Donndelinger, C.R. (1990) Endophytic bacteria in symptom-free cotton plants. *Phytopathology* 80, 808–811.

Mundt, J.O. and Hinkle, N.F. (1976) Bacteria within ovules and seeds. *Applied Environmental Microbiology* 32, 694–698.

Nishijima, F., Evans, W.R. and Vesper, S.J. (1988) Enhanced nodulation of soybean by *Bradyrhizobium* in the presence of *Pseudomonas fluorescens*. *Plant and Soil* 111, 149–150.

Nowak, J., Asiedu, S.K., Lazarovits, G., Pillay, V., Stewart, A., Smith, C. and Liu, Z. (1995) Enhancement of *in vitro* growth and transplant stress tolerance of potato and vegetable plantlets co-cultured with a plant growth promoting pseudomonad bacterium. In: Carre, F. and Chagvardieff, P. (eds) *Ecophysiology and Photosynthetic in Vitro Cultures*. Commissariat á l'energie atomique, France, pp. 173–179.

Oostendorp, M. and Sikora, R.A. (1990) *In-vitro* interrelationships between rhizosphere bacteria and *Heterodera schachtii*. *Revue Nematologie* 13, 269–274.

Pleban, S., Ingel, F. and Chet, I. (1995) Control of *Rhizoctonia solani* and *Sclerotium rolfsii* in the greenhouse using endophytic *Bacillus* spp. *European Journal of Plant Pathology* 101, 665–672.

Quadt-Hallmann, A. and Kloepper, J.W. (1996) Immunological detection and localization of the cotton endophyte *Enterobacter asburiae* JM22 in different plant species. *Canadian Journal of Microbiology* 42, 1144–1154.

Quadt-Hallmann, A., Benhamou, N. and Kloepper, J.W. (1997a) Bacterial endophytes in cotton: mechanisms of entering the plant. *Canadian Journal of Microbiology* 43, 577–582.

Quadt-Hallmann, A., Hallmann, J. and Kloepper, J.W. (1997b) Bacterial endophytes in cotton: location and interaction with other plant-associated bacteria. *Canadian Journal of Microbiology* 43, 254–259.

Raaijmakers, J.M., Leeman, M., van Oorschot, M.M.P., van der Sluis, I., Schippers, B. and Bakker, P.A.H.M. (1995) Dose–response relationships in biological control of *Fusarium* wilt of radish by *Pseudomonas* spp. *Phytopathology* 85, 1075–1081.

Racke, J. and Sikora, R.A. (1992) Isolation, formulation and antagonistic activity of rhizobacteria toward the potato cyst nematode *Globodera pallida*. *Soil Biology and Biochemistry* 24, 521–526.

Reinhold, B., Hurek, T. and Fendrik, I. (1987) Cross-reaction of predominant nitrogen-fixing bacteria with enveloped, round bodies in the root interior of Kaller grass. *Applied and Environmental Microbiology* 53, 889–891.

Reitz, M. (1999) Biochemische und molekularbiologische Untersuchungen zur bakterieninduzierten systemischen Resistenz in Kartoffel gegenüber dem Zystennematoden *Globodera pallida*. Ph.D. thesis, Bonn University.

Reitz, M., Hallmann, J. and Sikora, R.A. (2000) Lipopolysaccharides of *Rhizobium etli* Strain G12 act as inducing agents of systemic resistance in potato towards infection by the cyst nematode *Globodera pallida*. *Applied and Environmental Microbiology* 66, 3515–3518.

Romantchuk, M. (1992) Attachment of plant pathogenic bacteria to plant surfaces. *Annual Review of Phytopathology* 30, 225–243.

Romeiro, R.S. and Kimura, O. (1997) Induced resistance in pepper leaves infiltrated with purified bacterial elicitors from *Xanthomonas campestris* pv. *vesicatoria*. *Journal of Phytopathology* 145, 495–498.

Samish, Z., Etinger-Tulczynska, R. and Bick, M. (1961) Microflora within healthy tomatoes. *Applied Microbiology* 9, 20–25.

Sikora, R.A. (1992) Management of the antagonistic potential in agricultural ecosystems for the biological control of plant parasitic nematodes. *Annual Review of Phytopathology* 30, 245–270.

Sikora, R.A. and Carter, W.W. (1987) Nematode interactions with fungal and bacterial plant pathogens – fact or fantasy. In: Veech, J.A. and Dickson, D.W. (eds) *Vistas on Nematology*. Hyattsville Maryland, pp. 307–312.

Sikora, R.A. and Hoffmann-Hergarten, S. (1993) Biological control of plant parasitic nematodes with plant-health-promoting rhizobacteria. In: Lumsden, R.D. and Vaughn, J.L. (eds) *Pest Management: Biologically-based Technologies. Proceedings of Beltsville Symposium XVIII*, American Chemical Society, Washington, DC, pp. 166–172.

Smith, E.F. (1911) *Bacteria in Relation to Plant Diseases*, Vol. 2. Carnegie Institution of Washington, Washington, DC.

Smith, K.P. and Goodman, R.M. (1999) Host variation for interactions with beneficial plant associated microbes. *Annual Review of Phytopathology* 37, 473–491.

Sturz, A.V. (1995) The role of endophytic bacteria during seed piece decay and potato tuberization. *Plant and Soil* 175, 257–263.

Sturz, A.V. and Matheson, B.G. (1996) Populations of endophytic bacteria which influence host-resistance to *Erwinia*-induced bacterial soft rot in potato tubers. *Plant and Soil* 184, 265–271.

Sturz, A.V., Christie, B.R., Matheson, B.G. and Nowak, J. (1997) Biodiversity of endophytic bacteria which colonize red clover nodules, roots, stems and foliage and their influence on host growth. *Biology and Fertility of Soils* 25, 13–19.

Sturz, A.V., Christie, B.R., Matheson, B.G., Arsenault, W.J. and Buchanan, N.A. (1999) Endophytic bacterial communities in the periderm of potato tubers and their potential to improve resistance to soil-borne plant pathogens. *Plant Pathology* 48, 360–369.

Tomasino, S.F., Leister, R.T., Dimock, M.B., Beach, R.M. and Kelly, J.W. (1995) Field performance of *Clavibacter xyli* subsp. *cynodontis* expressing the insecticidal protein gene cryIA (c) of *Bacillus thuringiensis* against European corn borer in field corn. *Biological Control* 5, 442–448.

van Buren, A.M., Andre, C. and Ishimaru, C.A. (1993) Biological control of the bacterial ring rot pathogen by endophytic bacteria isolated from potato. *Phytopathology* 83, 1406 (abstract).

van Loon, L.C., Bakker, P.A.H.M. and Pieterse, C.M.J. (1998) Systemic resistance induced by rhizosphere bacteria. *Annual Review of Phytopathology* 36, 453–483.

Van Wees, S.C.M., Pieterse, C.M.J., Trijssenaar, A., van't Westende, Y.A.M, Hartog, F. and van Loon, L.C. (1997) Differential induction of systemic resistance in *Arabidopsis* by biocontrol bacteria. *Molecular Plant–Microbe Interactions* 10, 716–724.

Vance, C.P. (1983) *Rhizobium* infection and nodulation: a beneficial plant disease? *Annual Review of Microbiology* 37, 399–424.

Wei, G., Kloepper, J.W. and Tuzun, S. (1996) Induced systemic resistance to cucumber diseases and increased plant growth by plant growth-promoting rhizobacteria under field conditions. *Phytopathology* 86, 221–224.

Xia, Z.J., Gu, B.K., Wu, A.M. and Fu, Z.Q. (1997) Isoenzymatic activities involved in induced resistance of cotton to *Verticillium dahliae* by endophytic bacteria. *Jiangsu Journal of Agricultural Sciences* 13, 99–101.

You, C. and Zhou, F. (1989) Non-nodular endorhizospheric nitrogen fixation in wetland rice. *Canadian Journal of Microbiology* 35, 403–408.

Are Chitinolytic Rhizosphere Bacteria Really Beneficial to Plants?

7

W. de Boer and J.A. van Veen

Netherlands Institute of Ecology, Centre for Terrestrial Ecology, Department of Plant–Microorganism Interactions, PO Box 40, 6666 ZG Heteren, The Netherlands

Background

In the early 1960s, Mitchell and Alexander published several papers about mycolytic activities of chitinolytic soil bacilli and pseudomonads (Mitchell and Alexander, 1961, 1963). They postulated that fungal cell walls, which contain chitin as a structural component, could be degraded by the action of bacterial chitinases. More recent research using purified chitinases, chitinase-negative mutants and chitinase-positive transformants have clearly demonstrated the involvement of chitinases in mycolysis (Sundheim *et al.*, 1988; Chet *et al.*, 1990; Lim *et al.*, 1991; Chernin *et al.*, 1997). It is now evident that especially the hyphal apex, which contains nascent chitin fibrils, is sensitive to bacterial chitinases (Ordentlich *et al.*, 1988).

The actual role of bacterial chitinases in the mycolytic process is, however, not clear. It has already been noted by Mitchell and Alexander that, in addition to chitinases, bacteria required other factors to lyse fungal hyphae (Mitchell and Alexander, 1963). *In vitro* tests have shown that the extension of fungal hyphae is inhibited by some chitinase-producing bacterial colonies, while growth goes on unimpeded in the chitinase-containing zones of other colonies (De Boer *et al.*, 1998; Frändberg and Schnürer, 1998). Thus, it appears that mycolytic properties may differ among chitinolytic soil bacteria. De Boer *et al.* (1998) suggested that this might be attributed to the involvement of antibiotics in mycolysis. They hypothesized that antibiotics differ between chitinolytic strains and that the susceptibility of a particular fungus to lysis by a bacterial strain was more a factor of its sensitivity to the specific antibiotic rather than to the bacterial chitinases. Nielsen *et al.* (1998) reported that the composition of

 Biotic Interactions in Plant–Pathogen Associations (eds M.J. Jeger and N.J. Spence)

antibiotics produced by even closely related bacterial strains, e.g. within the same species, could differ widely. Differences between antifungal activity of chitinases themselves may also explain the selectivity. Roberts and Selitrennikoff (1988) observed that endo-chitinases, which cleave chitin randomly at internal sites, had a stronger effect on hyphal extension than exo-chitinases. Bacteria produce several endo-chitinases and exo-chitinases. Hence, the composition of the chitinase pool may also help to determine antifungal effects.

The discovery of mycolysis by chitinolytic bacteria has stimulated research on possible application of these bacteria for biocontrol purposes. Such research has focused on rhizosphere bacteria, as they should be best adapted to the environment where plant-pathogenic fungi infect roots. In several cases, strains with proven *in vitro* antifungal effects reduced disease symptoms significantly under controlled greenhouse conditions (Ordentlich *et al.*, 1988; Inbar and Chet, 1991; Kobayashi *et al.*, 1995). However, application of such strains under field conditions has been far less successful (Maloy, 1993). In fact, consistent mycolytic activities under field conditions have yet to be demonstrated. In order to resolve this issue, additional information is needed about the ecological function of the chitinase-producing bacteria, and the roles that their mycolytic activities play under natural environmental conditions.

Functional Aspects of Bacterial Chitinases

Chitin-degrading abilities

The most obvious advantage for the possession of chitinase genes by certain bacterial strains is their use in chitin consumption, with its yield of carbon, nitrogen and energy (Cohen-Kupiec and Chet, 1998). Indeed, this role for chitinases has been well established for aquatic bacteria (Gooday, 1990). In soils, however, filamentous fungi and actinomycetes appear to be the dominant chitin degraders (Gooday, 1990; De Boer *et al.*, 1999). The dominance of filamentous microorganisms over non-filamentous ones in terrestrial soils can most likely be explained by their ability to bridge air-filled gaps in the soil matrix and to penetrate chitinous substrates (De Boer *et al.*, 1996). Only non-filamentous bacteria that are strongly motile, e.g. *Cytophagae*, are able to compete with the rapid and efficient chitin degradation by filamentous microorganisms (Gooday, 1990). However, most chitinolytic strains isolated from the rhizosphere are non-filamentous cells, e.g. pseudomonads and bacilli, that are restricted in their motility in a solid matrix and in their ability to degrade crystalline chitin (Gould *et al.*, 1981; De Boer *et al.*, 1999).

Mycoparasitism

It has been suggested that chitinases may enable non-filamentous soil bacteria to grow mycoparasitically, i.e. at the expense of living hyphae (De Boer *et al.*, 1998). Chitinolytic strains have been shown to proliferate in liquid medium containing living fungal hyphae as the only growth substrate (Podile and Prakash, 1996). However, it is not absolutely clear that such growth was really the result of lytic activity of the bacteria since autolysis of starving fungal hyphae is common (Sahai and Manocha, 1993). It is also possible that exudates of living hyphae may have stimulated bacterial growth. Recently, we compared the growth of chitinolytic and non-chitinolytic pseudomonads in soil microcosms in response to the colonization by mycelium of the saprophytic fungus *Mucor hiemalis*. It appeared that the growth of chitinolytic pseudomonads was stimulated by the growing hyphae whereas that of non-chitinolytic ones was not (Fig. 7.1). Similar observations were made for the fungi *Chaetomium globosum* and *Fusarium culmorum* (not shown). The lack of stimulation observed for non-chitinolytic strains indicates that growth of the chitinolytic strains was not due to fungal exudates. Microscopic inspection showed that many hyphae were covered by chitinolytic bacteria (Fig. 7.2) and that several of these hyphae were irregularly shaped. These results suggest that chitinolytic pseudomonads enhanced the release of substrates from hyphae. Mycoparasitic growth of chitinolytic rhizosphere bacteria could be an important mechanism to control plant-pathogenic fungi. It should be realized, however, that chitinase production in pseudomonads as well as in many other

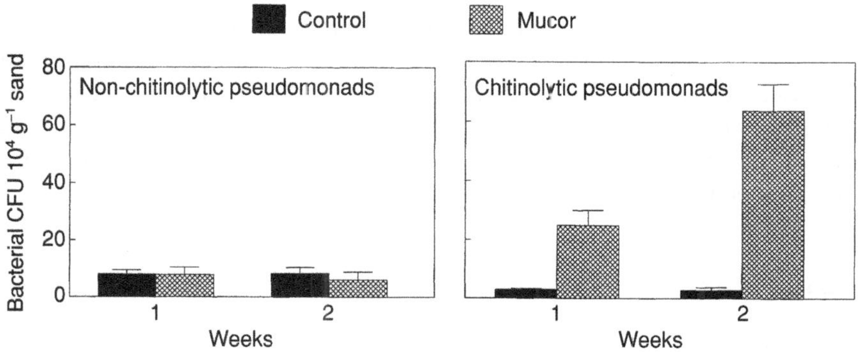

Fig. 7.1. Dynamics of colony-forming units (CFU) of chitinolytic and non-chitinolytic pseudomonads in sterile sand in the mycelial zone of the fungus *Mucor hiemalis* and in the comparable zone without fungal inoculum. At the start of the experiment the sand was inoculated with 10^4 cells g^{-1} sand of a mixture of ten bacterial strains. Extension of mycelium into the sand was from a potassium dextrose agar disc on top of a glass slide. The extension stopped between 1 and 2 weeks of incubation. In control treatments, potassium dextrose agar discs without fungus were placed on the glass slide. Data represent mean and standard deviation of two replicates that were harvested at the indicated time interval.

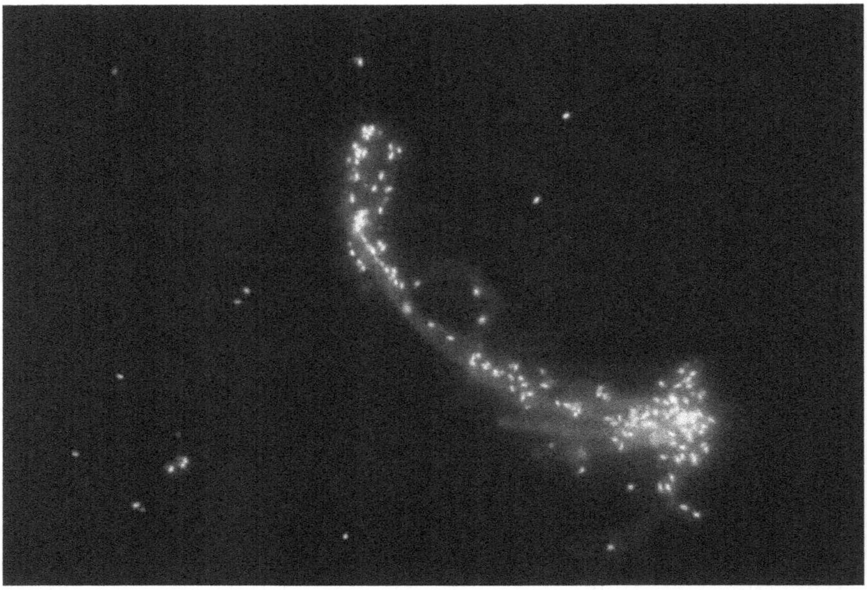

Fig. 7.2. Chitinolytic pseudomonads covering a hyphal fragment of the fungus *Chaetomium globosum*. Hyphal fragments were obtained from the experiment described in Fig. 7.1 and were examined microscopically for the presence of bacteria after staining with DAPI. Magnification factor: 1000×.

soil bacteria, including potential biocontrol strains, is repressed by small organic substrates like sugars and amino acids (De Boer *et al.*, 1998). Thus, mycolytic activity may only occur when no other growth substrates are available, i.e. under conditions of carbon starvation. The release of organic compounds by roots could repress the mycolytic activity of many chitinolytic rhizosphere bacteria. This would be analogous to the negative effect of organic substrates on chitinase production and mycoparasitic activity of saprotrophic fungi, e.g. *Trichoderma harzianum*, and streptomycetes (Gooday, 1990; Sahai and Manocha, 1993).

Defence

Under low nutrient conditions hyphae of certain fungi are reported to be strongly attracted to microcolonies of soil bacteria (Barron, 1988). The bacteria in the colonies are lysed by the fungi and the contents apparently absorbed as a nutrient source. Chitinolytic bacteria living in aggregates may protect themselves against attacking fungi by lysing invading hyphae. In this way, bacterial chitinases would have a function that is comparable with that of plant chitinases (Graham and Sticklen, 1994).

Recently, it was shown that chitinase production in the soil bacterium *Chromobacterium violaceum* is regulated by a density-dependent mechanism (quorum sensing) using *N*-acyl-homoserine lactone (Chernin *et al.*, 1998). Hence, for this bacterium a high cell density appears to be a prerequisite for chitinase production. Interestingly, other physiological traits of *C. violaceum*, such as the production of antibiotics, cyanide and exoproteases are also controlled via quorum sensing. Similarly, production of both the antibiotic phenazine and of chitinases by *Pseudomonas aureofaciens* apparently depends on the concentration of a homoserine lactone (Winson *et al.*, 1995). Density-dependent regulation of mycolytic compounds would agree well with the proposed defence function of chitinases since aggregates of bacteria are more likely to be exposed to fungal attack than dispersed cells. It may well be that a critical amount of chitinase production can only be achieved by a certain minimum of cells.

Cross-communication

For both mycoparasitic and defence strategies, the production of mycolytic compounds, including chitinases, could effectively be regulated by the use of fungal signal molecules. So far, the existence of such signal molecules and, consequently, cross-communication between fungi and bacteria, has received little attention. However, two recent studies demonstrate the existence and potential significance of such cross-communication. Fedi *et al.* (1997) demonstrated that the plant-pathogenic fungus *Pythium ultimum* produced diffusible compounds that altered gene expression in *Pseudomonas fluorescens*. Patterson and Bolis (1997) observed that cell wall components of aquatic fungi selectively elicited the accumulation of an antifungal tolytoxin in the cyanobacterium *Scytonema ocellatum*. The authors proposed an ecological role for tolytoxin as an inducible chemical defence agent capable of protecting *S. ocellatum* against fungal invasion. Interestingly, in the latter study, chitin oligomers were most effective in eliciting tolytoxin production. This shows a strong similarity with the production of phytoalexins in plants for which several compounds, including chitin oligomers, are elicitors (Graham and Sticklen, 1994).

Chitobiose and chitotriose, as well as the monomer acetylglucosamine, are known to stimulate chitinase production by the marine bacterium *Vibrio harveyi* (Montgomery and Kirchman, 1994). These oligomeric sugars are the end products of hydrolysis of chitin by chitinases. In the case of *Vibrio harveyi*, a low level of constitutive chitinase production enables the bacteria to explore the environment for the presence of chitin via the release of the oligomeric sugars. Several soil bacteria also showed a low level of chitinase activity under conditions of carbon starvation (De Boer *et al.*, 1998). Hence, this could allow for the possibility that approaching hyphae are sensed by the release of chito-sugars, leading to the induction of mycolytic activity. It remains to be seen if these compounds affect the expression of antifungal genes in soil bacteria. If so, then chitinases can contribute at two different levels to mycolysis, namely by direct degradation of the cell wall and by promoting the production of mycolytic compounds.

Implications and Perspectives

Strategies to select for biocontrol strains

A general scheme for chitinase gene regulation in pseudomonads is given in Fig. 7.3. While this is no doubt an oversimplification, this scheme may well encompass many of the key factors regulating chitinase production across a wide range of potentially mycolytic bacterial taxa. If so, chitinase-mediated mycolytic activity would not be expected to occur in those parts of the rhizosphere where sugars and amino acids are released in relatively large quantities, such as the root tips and the young roots (Jaeger *et al.*, 1999). Unfortunately, these are precisely the parts of the root system that are

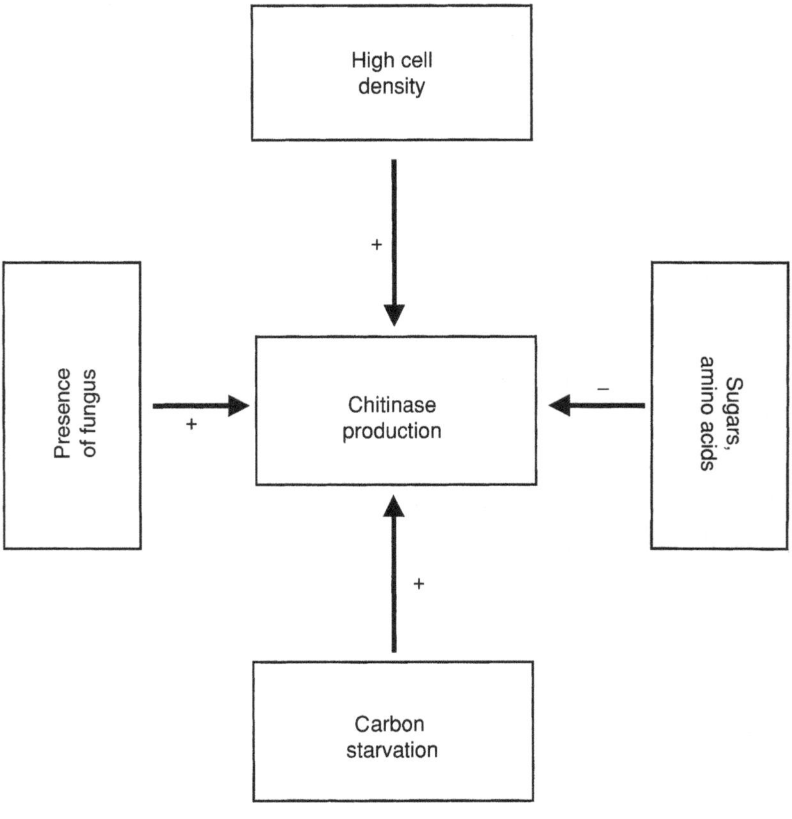

Fig. 7.3. Factors affecting expression of *Pseudomonas* chitinase genes.

most vulnerable to infection by plant-pathogenic fungi. In addition, density-dependent regulation of chitinase expression may require unrealistic densities of biocontrol strains along the roots. Before such negative feedbacks can be overcome in successful biocontrol programmes, the mechanisms underlying *in situ* chitinase gene expression must be better understood.

For biocontrol purposes, mycolytic bacteria which produce chitinases in the presence of root exudates appear to be of most interest. Indeed, some rhizosphere bacteria seem to deviate from the general regulation mechanism of bacterial chitinase gene expression since their chitinase production is not repressed by sugars and amino acids (De Boer *et al.*, 1998). Hence, such bacteria should be screened for biocontrol properties.

Chitinase genes of potential biocontrol strains have been successfully cloned into bacteria with good root-colonizing properties (Sundheim *et al.*, 1988; Koby *et al.*, 1994). However, the success of application of these modified strains will strongly depend on their fitness and the stability of the chitinase inserts under field conditions. So far, the successful field application of such strains has yet to be reported. In addition, public concern about the use of genetically modified organisms in food production will probably hamper the application of chitinase-positive transformants. An alternative to creating desired traits in rhizosphere bacteria by genetic modification is to screen for mutant strains with desired phenotypes. The recent description by Sato *et al.* (1988) of spontaneous mutants of a streptomycete in which chitinase production was not repressed by glucose indicates that such a strategy has potential. Furthermore, strains that respond strongly to the presence of fungal hyphae may be screened as putative biocontrol agents.

Interactions with beneficial microorganisms

A potential negative effect of using chitinolytic bacteria as biocontrol agents may be exerted upon mycorrhizal development. As mentioned above, the absence of easily degradable organic substrates in the bulk soil, into which mycorrhizal hyphae are extending from the root, may promote mycolytic activities. In some soils, lysis of external mycorrhizal hyphae has been observed, although colonization of roots was not inhibited (Linderman and Paulitz, 1990). This could well be the result of mycolysis by chitinolytic bacteria outside the zone of repressing action of root exudates. Therefore, study of interactions between chitinolytic strains and mycorrhizal fungi should be part of biocontrol studies.

Chitinolytic bacteria may also have a negative effect on nodule formation of legume roots. Nod factors, which induce root nodule formation, are lipochito-oligosaccharides that are excreted by rhizobia. It has been shown that Nod factors can be hydrolysed by a chitinase from the biocontrol strain *Serratia marscescens* (Krishnan *et al.*, 1999). Hence, production of bacterial chitinases in the rhizosphere may reduce nodule formation.

Concluding Remarks

The use of chitinolytic microorganisms as biocontrol agents has not proved to be the panacea against fungal plant pathogens that was once hoped, but our failure to utilize the chitinase activities may lie in the ignorance of the ecological functions and mechanisms of expression. A better understanding of *in situ* expression of bacterial chitinases seems to be a prerequisite to explore fully the biocontrol potential of chitinolytic rhizosphere bacteria. In addition, a better evaluation of risks of possible negative effects of chitinolytic rhizosphere bacteria on beneficial microorganisms is required.

Acknowledgements

We thank George Kowalchuk for critical reading of the manuscript.

References

Barron, G.L. (1988) Microcolonies of bacteria as a nutrient source for lignicolous and other fungi. *Canadian Journal of Botany* 66, 2505–2510.

Chernin, L.S., De La Fuente, L., Sobolev, V., Haran, S., Vorgias, C.E., Oppenheim, A.B. and Chet, I. (1997) Molecular cloning, structural analysis, and expression in *Escherichia coli* of a chitinase gene from *Enterobacter agglomerans*. *Applied and Environmental Microbiology* 63, 834–839.

Chernin, L.S., Winson, M.K., Thompson, J.M., Haran, S., Bycroft, B.W., Chet, I., Williams, P. and Stewart, G.S.A.B. (1998) Chitinolytic activity in *Chromobacterium violaceum*: substrate analysis and regulation by quorum sensing. *Applied and Environmental Microbiology* 180, 4435–4441.

Chet, I., Ordentlich, A., Shapira, R. and Oppenheim, A. (1990) Mechanisms of biocontrol of soil-borne plant pathogens by rhizobacteria. *Plant and Soil* 129, 85–92.

Cohen-Kupiec, R. and Chet, I. (1998) The molecular biology of chitin digestion. *Current Opinions in Biotechnology* 9, 270–277.

De Boer, W., Klein Gunnewiek, P.J.A. and Parkinson, D. (1996) Variability of N mineralization and nitrification in a simple, simulated microbial forest soil community. *Soil Biology and Biochemistry* 28, 203–211.

De Boer, W., Klein Gunnewiek, P.J.A., Lafeber, P., Janse, J.D., Spit, B.E. and Woldendorp, J.W. (1998) Anti-fungal properties of chitinolytic dune soil bacteria. *Soil Biology and Biochemistry* 30, 193–203.

De Boer, W., Gerards, S., Klein Gunnewiek, P.J.A. and Modderman, R. (1999) Response of the chitinolytic community to chitin amendments of dune soils. *Biology and Fertility of Soils* 29, 170–177.

Fedi, S., Tola, E., Moënne-Loccoz, Y., Dowling, D.N., Smith, L.M. and O'Gara, F. (1997) Evidence for signaling between the phytopathogenic fungus *Pythium ultimum* and *Pseudomonas fluorescens* F113: *P. ultimum* represses the expression of genes in *P. fluorescens* F113, resulting in altered ecological fitness. *Applied and Environmental Microbiology* 63, 4261–4266.

Frändberg, E. and Schnürer, J. (1998) Antifungal activity of chitinolytic bacteria isolated from airtight stored cereal grain. *Canadian Journal of Microbiology* 44, 121–127.

Gooday, G.W. (1990) The ecology of chitin degradation. In: Marshall, K.C. (ed.) *Advances in Microbial Ecology*, Vol. 11. Plenum Press, New York, pp. 387–430.

Gould, W.D., Bryant, R.J., Trofymow, J.A., Anderson, R.V., Elliott, E.T. and Coleman, D.C. (1981) Chitin decomposition in a model soil system. *Soil Biology and Biochemistry* 13, 487–492.

Graham, L.S. and Sticklen, M.B. (1994) Plant chitinases. *Canadian Journal of Botany* 72, 1057–1083.

Inbar, J. and Chet, I. (1991) Evidence that chitinase produced by *Aeromonas caviae* is involved in the biological control of soil-borne plant pathogens by this bacterium. *Soil Biology and Biochemistry* 23, 973–978.

Jaeger, C.H., Lindow, S.E., Miller, W., Clark, E. and Firestone, M.K. (1999) Mapping of sugar and amino acid availability in soil around roots with bacterial sensors of sucrose and tryptophan. *Applied and Environmental Microbiology* 65, 2685–2689.

Kobayashi, D.Y., Guglielmoni, M. and Clarke, B.B. (1995) Isolation of the chitinolytic bacteria *Xanthomonas maltophilia* and *Serratia marcescens* as biological control agents for summer patch disease of turfgrass. *Soil Biology and Biochemistry* 27, 1479–1487.

Koby, S., Schickler, H., Chet, I. and Oppenheim, A.B. (1994) The chitinase encoding Tn7-based chiA gene endows *Pseudomonas fluorescens* with the capacity to control plant pathogens in soil. *Gene* 147, 81–83

Krishnan, H.B., Kim, K.Y. and Krishnan, A.H. (1999) Expression of a *Serratia marcescens* chitinase gene in *Sinorhizobium fredii* USDA191 and *Sinorhizobium meliloti* RCR2011 impedes soybean and alfalfa nodulation. *Molecular Plant–Microbe Interactions* 12, 748–751.

Lim, H.-S., Kim, Y.-S. and Kim, S.-D. (1991) *Pseudomonas stutzeri* YPL-1 genetic transformation and antifungal mechanism against *Fusarium solani*, an agent of plant root rot. *Applied and Environmental Microbiology* 57, 510–516.

Linderman, R.G. and Paulitz, T.C. (1990) Mycorrhizal–rhizobacterial interactions. In: Cook, R.J., Hennis, Y., Ko, W.H., Rovira, A.D., Schippers, B. and Scott, P.R. (eds) *Biological Control of Soil-borne Plant Pathogens*. CAB International, Wallingford, UK, pp. 261–283.

Maloy, O.C. (1993) *Plant Disease Control: Principles and Practice*. John Wiley & Sons, Chichester.

Mitchell, R. and Alexander, M. (1961) The mycolytic phenomenon and biological control of *Fusarium* in soil. *Nature (London)* 190, 109–110.

Mitchell, R. and Alexander, M. (1963) Lysis of soil fungi by bacteria. *Canadian Journal of Microbiology* 9, 169–177.

Montgomery, M.T. and Kirchman, D.L. (1994) Induction of chitin-binding proteins during the specific attachment of the marine bacterium *Vibrio harveyi* to chitin. *Applied and Environmental Microbiology* 60, 4282–4288.

Nielsen, M.N., Sørensen, J., Fels, J. and Pedersen, H.F. (1998) Secondary metabolite- and endo-chitinase dependent antagonism toward plant-pathogenic microfungi of *Pseudomonas fluorescens* isolates from sugar beet rhizosphere. *Applied and Environmental Microbiology* 64, 3563–3569.

Ordentlich, A., Elad, Y. and Chet, I. (1988) The role of chitinase of *Serratia marcescens* in biocontrol of *Sclerotium rolfsii*. *Phytopathology* 78, 84–88.

Patterson, G.M. and Bolis, C.M. (1997) Fungal cell-wall polysaccharides elicit an antifungal secondary metabolite (phytoalexin) in the cyanobacterium *Scytonema ocellatum*. *Journal of Phycology* 33, 54–60.

Podile, A.R. and Prakash, A.P. (1996) Lysis and biological control of *Aspergillus niger* by *Bacillus subtilis* AF1. *Canadian Journal of Microbiology* 42, 533–538.

Roberts, W.K. and Selitrennikoff, C.P. (1988) Plant and bacterial chitinases differ in antifungal activity. *Journal of General Microbiology* 134, 169–176.

Sahai, A.S. and Manocha, M.S. (1993) Chitinases of fungi and plants: their involvement in morphogenesis and host-parasite interaction. *FEMS Microbiological Reviews* 11, 317–338.

Saito, A., Fuji, T., Yoneyama, T. and Miyashita, K. (1988) glKA is involved in glucose repression of chitinase production in *Streptomyces lividans*. *Journal of Bacteriology* 180, 2911–2914.

Sundheim, L., Poplawsky, A.R. and Ellingboe, A.H. (1988) Molecular cloning of two chitinase genes from *Serratia marcescens* and their expression in *Pseudomonas* species. *Physiological and Molecular Plant Pathology* 33, 484–491.

Winson, M.K., Camara, M., Lafiti, A., Foglino, M., Chabra, S.R., Daykin, M., Bally, M., Chapon, V., Salmond, G.P.C., Bycroft, B.W., Lazdunski, A., Stewart, G.S.A.B. and Williams, P. (1995) Multiple N-acyl-L-homoserine lactone signal molecules regulate production of virulence determinants and secondary metabolites in *Pseudomonas aeruginosa*. *Proceedings of the National Academy of Sciences USA* 92, 9427–9431.

8

Diversity and Interactions Among Strains of *Fusarium oxysporum*: Application to Biological Control

C. Alabouvette, V. Edel, P. Lemanceau, C. Olivain, G. Recorbet and C. Steinberg

CMSE-INRA, U.M.R. Biochimie, Biologie Cellulaire et Ecologie des Interactions Plantes Microorganismes, F 21034 Dijon Cedex, France

Introduction

The species *Fusarium oxysporum* is well represented among the communities of soil-borne fungi, in every type of soil, all over the world (Burgess, 1981). This species is also considered as a normal constituent of the fungal communities in the rhizosphere of plants (Gordon and Martyn, 1997). All strains of *F. oxysporum* are successful saprophytes and are able to grow and survive for long periods on organic matter in soil and in the rhizosphere of many plant species (Garrett, 1970). Moreover, some strains of *F. oxysporum* are pathogenic to different plant species; they penetrate into the roots and provoke either root-rots or tracheomycosis when they invade the vascular system. The wilt-inducing strains of *F. oxysporum* are responsible for severe damage on many plant species of economic importance. These pathogens show a high level of host specificity and, based on the plant species and plant cultivars they are able to infect, are classified into more than 120 special forms (or *formae speciales*) and races (Armstrong and Armstrong, 1981).

Soils also harbour very large populations of non-pathogenic strains of *F. oxysporum*, which play an important role in soil microbial ecology, especially in soils suppressive to fusarium wilt. Indeed it has been well established that suppressiveness of several soils is, at least partly, due to the abundance of *F. oxysporum* (Smith and Snyder, 1971; Louvet *et al.*, 1976; Rouxel *et al.*, 1979). Since these early studies showing the involvement of non-pathogenic strains of *F. oxysporum* in the mechanisms of soil suppressiveness, many papers have reported the capacity of non-pathogenic strains to control fusarium wilt, at least, under experimental conditions (Ogawa and

Komada, 1984; Paulitz *et al.*, 1987; Mandeel and Baker, 1991; Postma and Rattink, 1992; Alabouvette *et al.*, 1993; Larkin *et al.*, 1996). Today, plant pathologists are not only interested in the pathogenic strains of *F. oxysporum* but also the non-pathogenic ones, especially when studying the interactions between pathogenic and non-pathogenic strains.

The aim of this chapter is to present methods available for the characterization of *F. oxysporum*, whether pathogenic or non-pathogenic, to describe the different types of interactions among populations of *F. oxysporum*, and to show how this knowledge can be applied to biological control of fusarium diseases.

Characterization of the Diversity that Affects Populations of *Fusarium oxysporum*

The genus *Fusarium* is a heterogeneous genus including very different species with respect to their morphological and physiological characters. Several taxonomic systems have been proposed that recognize from nine (Snyder and Hansen, 1940) to 78 species (Nelson *et al.*, 1983).

These classical taxonomic systems are based on morphological characters such as the size and the shape of macroconidia, the presence or absence of microconidia and chlamydospores, the coloration of the colonies, and the structure of the conidiophores (Windels, 1992). Based on these criteria the differentiation of some species is difficult, since it relies only on minor differences. This is the case for *F. oxysporum* and several species corresponding to the sections Elegans and Liseola. Alternative methods based on molecular tools have been developed for the identification of *Fusarium* species. For example, Edel *et al.* (1997a) have proposed a PCR–RFLP (restriction fragment length polymorphism) method to identify *Fusarium* species. The method targets a fragment of ribosomal DNA (rDNA) including the internal transcribed spacer (ITS1 + 5,8S RNA gene + ITS2) and the 5′ end of the 28S RNA genes. The combination of only four restriction enzymes enables the differentiation of 14 species and two groups of two or three species (Fig. 8.1). The initial polymorphism found in the species *F. oxysporum* was due to the misidentification of strains of *F. redolens* (Edel *et al.*, 1997a). There is in fact no polymorphism for the species *F. oxysporum* in the studied sequences, and this method is currently used to differentiate *F. oxysporum* from other species (Edel *et al.*, 2000).

The same rDNA region was further used to develop an oligonucleotide probe and a PCR assay specific for *F. oxysporum* (Edel *et al.*, 2000). These tools will facilitate the isolation of *F. oxysporum* populations from field samples.

Once achieved, the identification of the species *F. oxysporum* does not provide all the information needed by the plant pathologist who needs to identify the *formae speciales* and races of the pathogenic strains or the non-pathogenic ability of strains isolated from plants or soil.

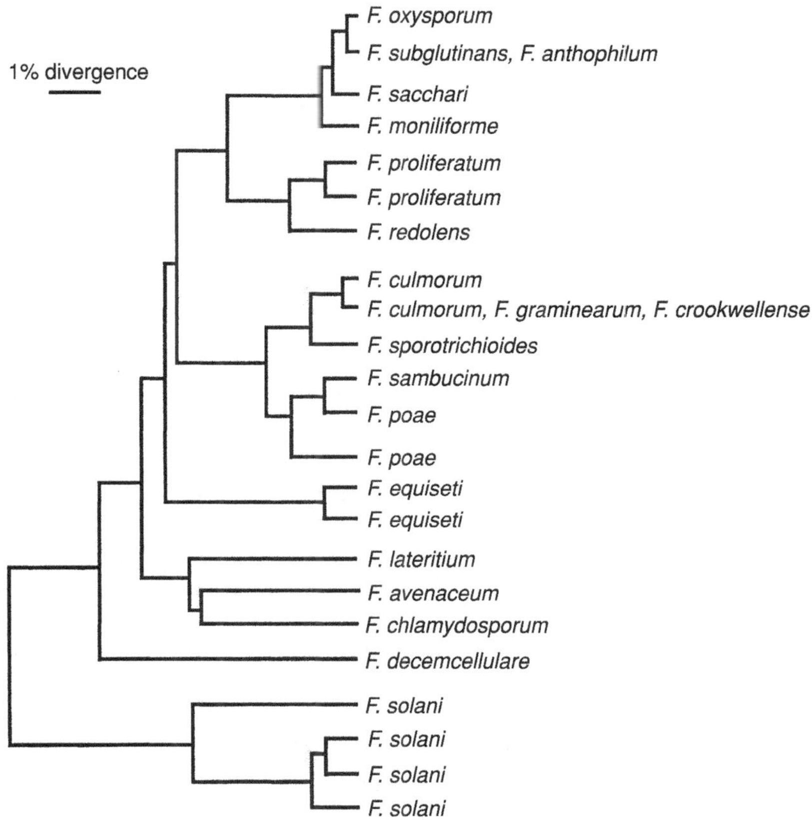

Fig. 8.1. Dendogram (UPGMA) showing the relationships among the species of *Fusarium* characterized by the PCR–RFLP method targeting a fragment of the rDNA including the ITS.

Most of the methods of species characterization based on phenotypic and physiological traits have also been used to try to differentiate *formae speciales* or races of *F. oxysporum*. These methods take into account the capacity of the strains to utilize specific nutrients, produce different types of secondary metabolites such as enzymes or toxins, resist antibiotics or other toxic compounds. All methods reveal a great diversity in *Fusarium* both at the inter- and intraspecific levels (Wasfy *et al.*, 1987; Thrane and Hansen, 1995). Seifert and Brisset (1998) have proposed an identification key for species of *Fusarium*, based on the Biolog system. When applied to *F. oxysporum*, this method showed a great diversity among strains, but profiles could not be correlated with *formae speciales* or races. Similarly, characterization of enzymatic polymorphisms has been proposed to identify *Fusarium* species (Rataj-Guranowska and Wolko, 1991; Huss *et al.*, 1996) and also to study the diversity among strains of *F. oxysporum* (Elias and Schneider, 1992; Skovgaard and Rosendahl, 1998). The observed polymorphism cannot be satisfactorily correlated with *formae speciales* or races of *F. oxysporum*. Finally, serological methods have also been utilized for the characterization of given *formae speciales* or races of *F. oxysporum* (Iannelli *et al.*, 1983; Del Sorbo *et al.*, 1993). With few exceptions, cross-reactions do not enable an easy identification of *formae speciales*.

Puhalla (1985) proposed a system of classification of strains of *F. oxysporum*, based on their vegetative compatibility and described a method based on pairing nitrate non-utilizing mutants to determine the vegetative compatibility group (VCG) of each strain. Some *formae speciales* correspond to a single VCG, whereas others include several VCGs. In a recent review, Katan (1999) recognized 59 VCGs among 38 *formae speciales*. With non-pathogenic populations, many isolates are single member VCG and some are even self-incompatible (Gordon *et al.*, 1992; Kondo *et al.*, 1997). Therefore, although useful, the determination of VCG cannot be used as a universal tool to identify *formae speciales* or non-pathogenic isolates.

At the present time, the only valid method to identify *formae speciales* or races is through a bioassay in which putative host plants are confronted by the fungus. Although useful to verify the pathogenicity of a few strains isolated from a given plant species, this method cannot be used to characterize non-pathogenic strains. Indeed, as more than 120 *formae speciales* and races have been described, it would be necessary to inoculate the unidentified strains to many different plant species. This is obviously not possible, especially when a large collection of soil isolates has to be characterized. Therefore it must be stressed that the so-called 'non-pathogenic' strains of *F. oxysporum* are strains that failed to provoke a disease on a limited number of plant species to which they have been inoculated. This negative definition of the non-pathogenic strains of *F. oxysporum* does not help plant pathologists willing to use these non-pathogenic strains for biological control.

In order to characterize the diversity that affects the natural population of *F. oxysporum* and to study genetic relationships between strains of *F. oxysporum*, different molecular methods have been proposed. In our group, Edel *et al.* (1995) have described three molecular methods showing different levels of discrimination.

The first method is based on the PCR–RFLP analysis of the intergenic spacer (IGS) of the rDNA unit. This non-transcribed region shows a lesser degree of conservation than the ITS region used to identify the species, and presents polymorphism among strains of *F. oxysporum*. The second method utilizes prokaryotic motif primers derived from ERIC and REP sequences. It is a sort of RAPD (random amplified polymorphic DNA) fingerprinting utilizing ERIC and REP primers. The last method, the most discriminatory, is based on RFLP analysis of total DNA after hybridization with a random DNA probe.

In studies on the diversity within a small collection of 60 strains isolated from soil and roots, Edel *et al.* (1995) characterized 11 IGS types, 27 PCR-fingerprinting groups and 40 RFLP types. The results obtained with each method are consistent, since there is no overlap between groups defined with the three methods at varying degrees of discrimination. Thus, depending on the type of study and the level of accuracy needed, we can use one or another of the methods. However, to characterize the diversity among large populations of *F. oxysporum* the IGS typing method is preferred.

These molecular methods demonstrate diversity among strains of *F. oxysporum*, but do not provide any knowledge on the significance of this diversity in relation to the physiological or ecological behaviour of these strains.

Steinberg *et al.* (1997a) characterized the same 60 strains used in molecular studies for vegetative compatibility, growth kinetics on four different carbon sources and pathogenicity on three hosts plants: flax, tomato and melon. Principal component analysis of growth parameter estimates (*K*, asymptotic population density; *r*, growth rate; *P*, time of inflexion point) gave a good discrimination of the isolates and revealed a high level of diversity within this population (Fig. 8.2). The 60 strains were assigned to 40 VCG also indicating a great diversity. Strains grouped in the same VCG had similar growth patterns. It was not possible to show any clear relationship between the genotypic diversity previously described for these strains and their physiological diversity. In fact, trophic characterization and vegetative compatibility grouping were more discriminating than the genotypic characterization. This study indicated that the VCG could be considered the population unit among natural populations of *F. oxysporum* in soil.

The comparison of the different methods available to characterize the diversity within populations of *F. oxysporum* showed that the phenotypic methods are too laborious and probably too discriminatory for ecological studies involving a great number of strains. Similarly, the PCR fingerprinting methods resulted in complex patterns difficult to compare among a large number of strains. Therefore, the IGS typing was the method chosen to characterize the diversity among strains of *F. oxysporum* in an experiment aimed at assessing the possible impact of cultivation of several plant species on the indigenous soil populations of *F. oxysporum* (Edel *et al.*, 1997b). Four plant species: flax, melon, tomato and wheat were cultivated in separate samples of the same soil. Forty soil-borne isolates of *F. oxysporum* and 40 root-colonizing isolates of each plant species were sampled during the first and the fourth crop. Sixteen IGS types were defined among the 400 isolates analysed. The distributions of the soil isolates among IGS types were similar at both sampling dates. The structure of *F. oxysporum* populations associated with roots of flax or melon did not differ from the structure of soil-borne populations. In contrast, the structure of *F. oxysporum* populations associated with roots of wheat or tomato differed from the structure of soil-borne populations and also differed from each other, indicating that the plant species exerted selection on the soil-borne populations of *F. oxysporum* (Fig. 8.3). These results show that strains of non-pathogenic *F. oxysporum* might be more or less adapted to the colonization of the roots of different plant species, explaining why some strains are better biocontrol agents in one crop than in another.

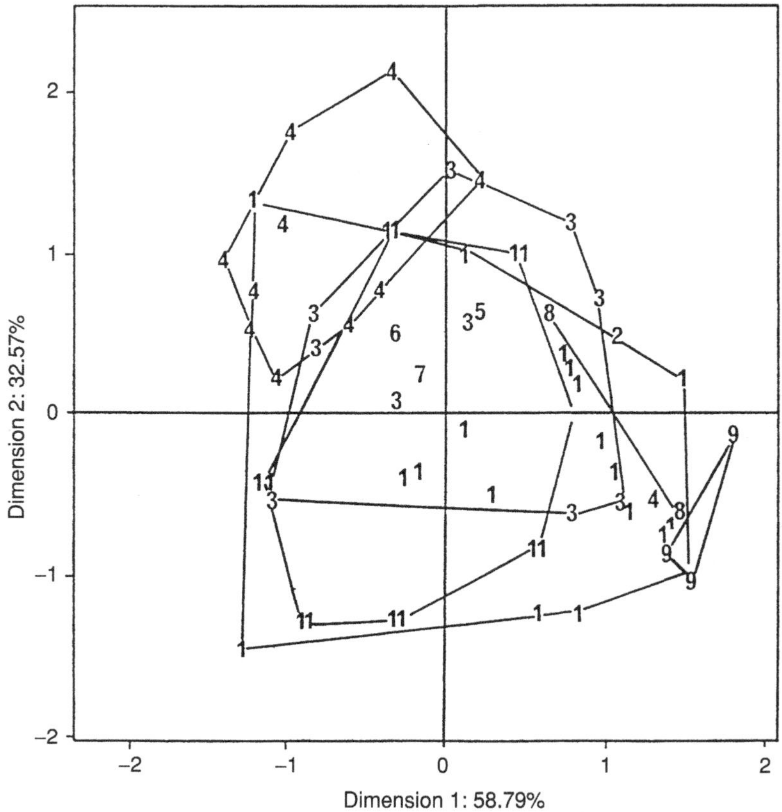

Fig. 8.2. Principal component analysis (PCA) performed on the growth parameters of each replicate for each strain of *F. oxysporum* on glucose. PCA was performed using the five replicates for each parameter but only the average was plotted on the figure. The strains are represented by the number of the IGS type to which they were assigned. The peripheral points of each cloud of strains belonging to the same IGS type are joined. Strains on the right on the figure had higher *r* and *K* values than strains on the left. Above the first component axis are strains with a longer lag phase than strains situated along this axis.

Together the results demonstrate that the soil-borne populations of *F. oxysporum* present a great diversity, whatever the method used to reveal this diversity. Soil-borne strains of *F. oxysporum* differ at the molecular level but also in many phenotypic and physiological traits. They show different growth rates on the same substrates (Steinberg *et al.*, 1999a,b) and in soil (Couteaudier and Alabouvette, 1990), produce different quantities of secondary metabolites such as fusaric acid (Kern, 1972; Landa del Castillo, 1999), and have different abilities to colonize the root surface of different plant species (Nagao *et al.*, 1990). Therefore these strains, having different capacities but sharing the same ecological niches in the soil, probably interact strongly in different soils and in the rhizosphere of plants.

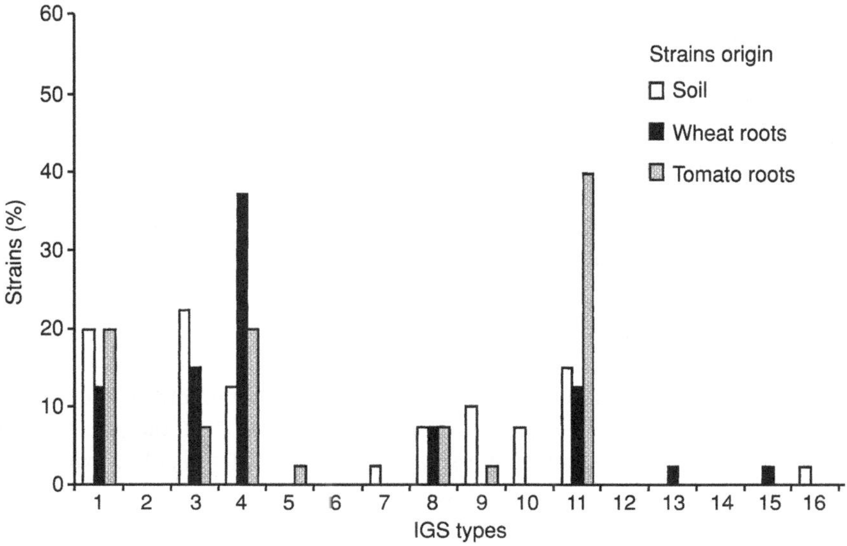

Fig. 8.3. Distribution among 16 IGS types of *F. oxysporum* isolated from soil, wheat roots and tomato roots.

Interactions Among Strains of *F. oxysporum*

Studies devoted to the mechanisms of soil suppressiveness to fusarium wilt have drawn attention to the role of non-pathogenic *F. oxysporum* interacting with the pathogen, resulting in a decrease in disease incidence.

The interaction can take place in the soil, the rhizosphere or even within the plant itself. But these interactions do not exist only between pathogenic and non-pathogenic strains of *F. oxysporum*. As stated above, there exists a very large diversity among natural populations of *F. oxysporum* in soils, and this diversity affects both the pathogenic and non-pathogenic strains. Therefore one must consider that all strains of *F. oxysporum* able to share the same ecological niche potentially interact independently of their capacity to induce disease or not. Most of the studies reported below have been conducted with non-pathogenic strains interacting with a given pathogen, but these non-pathogenic strains included strains pathogenic on plant species other than those utilized in the experiments (Couteaudier and Alabouvette, 1990; Eparvier and Alabouvette, 1994).

Direct interactions in soil and the rhizosphere

Twenty-five years ago, the observation that the suppressive soil from Châteaurenard harboured high populations of *F. oxysporum* and *F. solani* led to the hypothesis that these natural populations of non-pathogenic *Fusarium* spp.

were involved in the mechanisms of suppression. Indeed suppressiveness was destroyed by heat treatment at 55°C, which eliminated the thermo-sensitive microflora including the *Fusarium* spp., and was restored by artificial introduction of strains of non-pathogenic *F. oxysporum* or *F. solani* in the heat-treated soil (Rouxel *et al.*, 1977, 1979) (Fig. 8.4). In the absence of any evidence of antibiosis between non-pathogenic and pathogenic strains of *F. oxysporum*, the hypothesis of trophic interactions was proposed. Indeed, at that time, there was a common belief that pathogenic strains were less competitive during their saprophytic growth phase than non-pathogenic strains (Baker and Cook, 1974). More precisely, the hypothesis of competition for carbon was proposed based on the fact that a single addition of glucose to the suppressive soil was sufficient to make it conducive (Louvet *et al.*, 1976).

To demonstrate the validity of the hypothesis of competition for carbon between strains of *F. oxysporum*, Couteaudier and Alabouvette (1990) compared the growth kinetics of a small collection of strains of *F. oxysporum* introduced into a sterilized soil amended with glucose. By modelling of the growth curve, Couteaudier and Steinberg (1990) calculated the growth rate and the yield coefficient (i.e. the number of propagules formed per unit of glucose consumed) for each strain. Results showed a great diversity among the seven strains compared, with the yield coefficient varying from 1×10^6 to 8×10^6 propagules formed per mg of glucose consumed (Fig. 8.5). Six of these strains were then confronted with the seventh strain, a pathogenic strain *F. o.* f. sp. *lini* (Foln35) resistant to benomyl. Each strain was introduced into the sterilized soil in combination with the pathogenic strain Foln35 at five different inoculum ratios. Following the kinetics of growth of each strain in mixture it

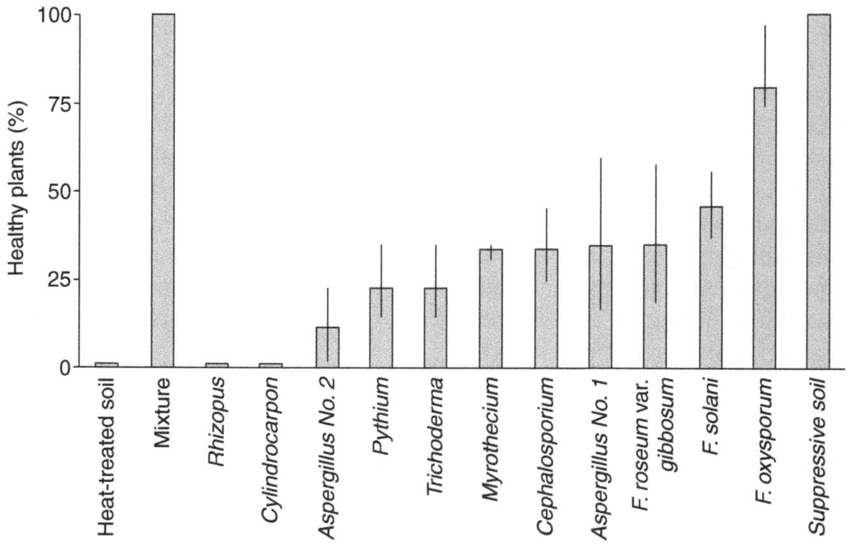

Fig. 8.4. Ability of the main fungi native to the suppressive soil to re-establish suppressiveness in the steamed soil.

was possible to calculate a 'competitiveness index' for each strain. Indeed there exists a good relationship between the ratio of inoculum densities of the two challenging strains at the asymptotes and the ratio of inoculum densities of the strains introduced into the sterilized soil (Fig. 8.6a). The slope of the regression line varies according to the relative competitive ability of the six strains when confronted with the same pathogenic strain. Therefore the slope can be regarded as a 'competitiveness index'. It corresponds to the ratio of the number of propagules formed by the non-pathogen to the number of propagules formed by the pathogen, when the two strains are co-inoculated into the sterilized soil in equal quantities. These coefficients ranged from 1.3 to 3.5, indicating a large diversity in the ability of these six strains to compete in soil with the pathogenic strain *F. o.* f. sp. *lini* (Fig. 8.6b). Subsequently, Lemanceau *et al.* (1993) have confirmed, *in vitro*, that carbon was the major nutrient for which a pathogenic strain of *F. o.* f. sp. *dianthi* was competing in soilless culture with the biocontrol agent Fo47.

These results enable us to conclude that different strains of *F. oxysporum* having different abilities to utilize carbon sources efficiently are competing for this nutrient in soil and the rhizosphere. These interactions among strains of *F. oxysporum* are taking place in the soil environment and are influenced by both abiotic and biotic soil factors. It is obvious that nutrient availability for *F. oxysporum* depends not only on the concentration of carbon substrates but also on the activity of the microbial biomass consuming these nutrients (Alabouvette *et al.*, 1985). Moreover, the capacity of a given population to utilize one nutrient efficiently is also dependent on the activity of other microbial populations. Lemanceau *et al.* (1993) have clearly demonstrated that the

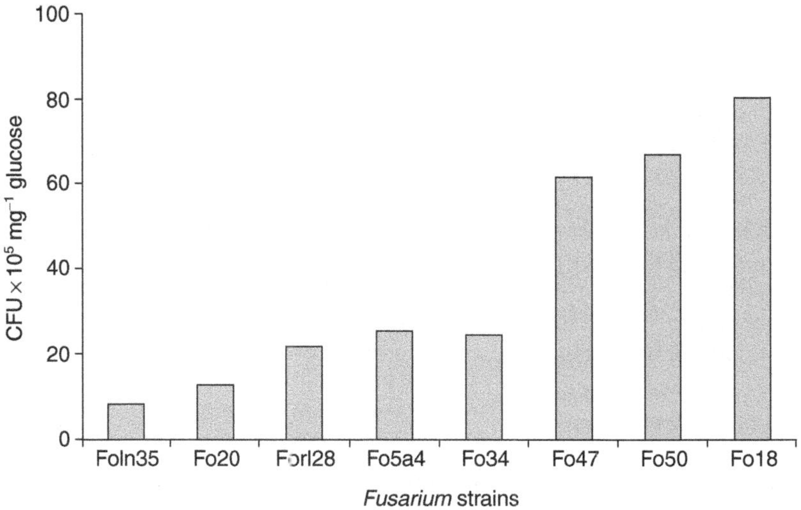

Fig. 8.5. Yield coefficient (expressed as number of CFU mg^{-1} glucose consumed) of different strains of *F. oxysporum* introduced at 2×10^3 CFU g^{-1} soil into glucose-amended soil.

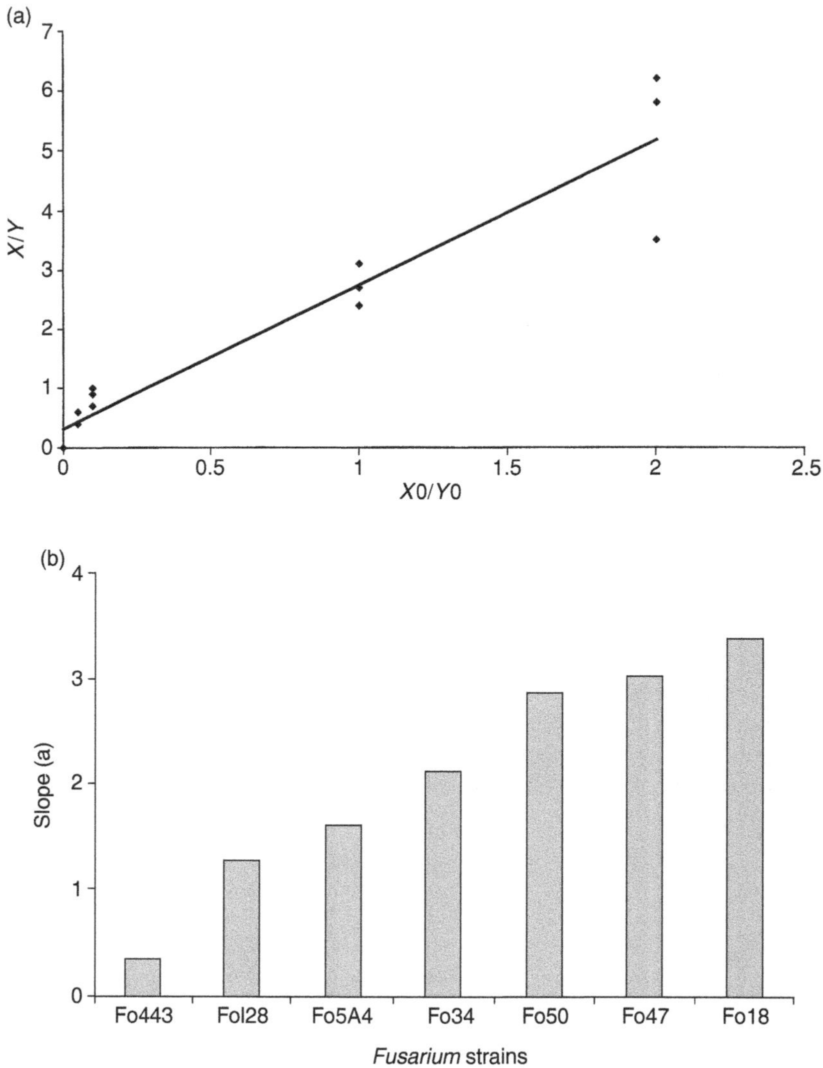

Fig. 8.6. (a) Relationship between the ratio of inoculum densities at the plateau *X/Y* and the ratio of inoculum densities incorporated into the sterilized soil *X0/Y0* (*X* = Fo47, *Y* = Foln35). (b) Competitiveness index of strains of non-pathogenic *F. oxysporum* confronted to the same pathogenic strain (Foln35). The competitiveness index corresponds to the slope of the regression line presented in (a).

yield coefficient of *F. oxysporum* utilizing glucose depends on the availability of iron, which is dependent on the activity of siderophore-producing *Pseudomonas* spp. These results confirmed that both non-pathogenic *F. oxysporum* and *P. fluorescens* were involved in soil suppressiveness to fusarium wilt.

Direct interactions on the root surface and in the plant

The second place where competition can occur is the root surface. Mandeel and Baker (1991) stated that the root surface presented a finite number of infection sites that could be protected by increasing the inoculum density of the non-pathogenic strain. In fact, good control of the disease requires not only a high inoculum density of the biocontrol agent but also a high ratio of the non-pathogenic versus the pathogenic strain (Alabouvette *et al.*, 1993; Lemanceau *et al.*, 1993; Larkin and Fravel, 1999). Recently, Olivain and Alabouvette (1997, 1998) clearly showed, using GUS-transformed strains, that both pathogenic strains and a non-pathogenic strain were able to colonize actively the surface of the tomato root. Both were able to penetrate the epidermal cells and to colonize the cortex to some extent. The plant reacted to this fungal invasion by expressing defence reactions, such as wall thickening and intracellular plugging, that were more intense in the case of the non-pathogen. As a result these defence reactions always prevented the non-pathogen from reaching the stele, whereas the pathogenic strain grew quickly towards and invaded the vessels. These observations are concordant with the hypothesis of competition between strains of *F. oxysporum* at the root surface and during root colonization. Indeed, both strains colonized the same sites at the root surface and showed great similarities in the colonization process. These observations are also in agreement with results obtained previously by Eparvier and Alabouvette (1994). A GUS-transformed strain of a pathogenic *F. oxysporum* was confronted with the wild-type non-pathogenic strain Fo47 in the presence of the plant root. Using an ELISA method the total biomass of fungi having colonized the plant root was assessed, and the glucuronidase activity in root tissues provided a measure of the metabolic activity of the pathogen. The results clearly demonstrated that the co-inoculation of the pathogen with the non-pathogen resulted in a total fungal biomass comparable to that measured when the pathogen or the non-pathogen were inoculated alone, but resulted in a significant decrease of the GUS activity of the pathogen in comparison to the single inoculation of the pathogen to the plant (Fig. 8.7). The same experiment was conducted with a small collection of non-pathogenic strains of *F. oxysporum*. The results showed that different strains have different abilities to compete with the same pathogen at the surface and in the root tissues of the host plant.

Finally, Postma and Luttikholt (1996) considered the hypothesis of a direct competition between two strains of *F. oxysporum* in the vessels of the host plant. They compared the growth of a pathogenic strain of *F. oxysporum* f. sp. *dianthi* and of several non-pathogenic strains in the stele of carnation after artificial inoculation of these strains, alone or in combination, into the vessels of the plant. They showed that some non-pathogenic strains were able to reduce the stem colonization by the pathogen resulting in a decrease in disease severity. Locally induced resistance or direct competition between strains within the vessels could cause this disease-suppressive effect after

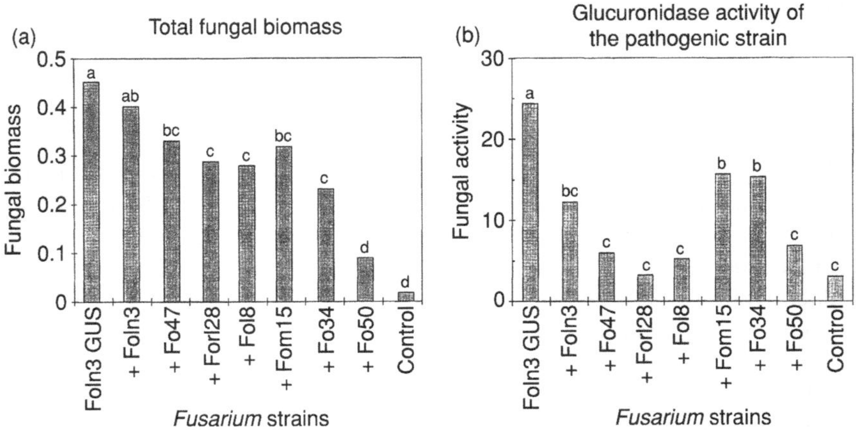

Fig. 8.7. Ability of six different strains of non-pathogenic *F. oxysporum* to inhibit (a) the total root colonization of flax by *F. oxysporum* f. sp. *lini* (Foln3 GUS) assessed by ACP–ELISA; and (b) the activity of the pathogenic Foln3 GUS determined by measuring the glucuronidase activity.

mixed inoculation into the stem. These observations are in agreement with the results of Ogawa and Komada (1984) who selected a non-pathogenic strain of *F. oxysporum* able to control fusarium wilt of sweet potato when introduced into the stem of the plant.

Indirect interactions through the plant

It has been known for many years that pre-inoculation of a plant with an incompatible strain of *F. oxysporum* (either a non-pathogenic strain or a pathogenic strain belonging to another *forma specialis*) results in the mitigation of symptoms when the plant is subsequently inoculated with a compatible strain (Matta, 1989). This phenomenon was described as cross-protection or premunition. Today this phenomenon is considered as an expression of induced systemic resistance, a general mechanism of plant response to microbial infection or stresses derived from various origins. Induced systemic resistance (ISR) has been extensively studied since it could explain the disease control provided by non-pathogenic strains of *F. oxysporum*. Biles and Martyn (1983) were the first authors to attribute to ISR the control of fusarium wilt of watermelon achieved by several strains of non-pathogenic *F. oxysporum*. Many papers reported experiments where a non-pathogenic strain applied to some roots of a host plant can delay symptom expression induced by the specific pathogen applied separately to other roots (Fig. 8.8) or directly into the stem of the plant (Biles and Martyn, 1983; Mandeel and Baker, 1991; Kroon *et al.*, 1992; Olivain *et al.*, 1995; Fuchs *et al.*, 1997; Larkin and Fravel, 1999). Since this split-root method prevented any direct interaction between

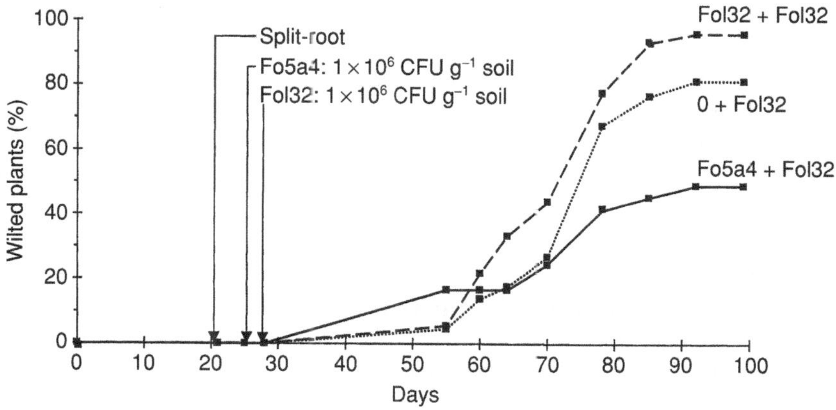

Fig. 8.8. Ability of the non-pathogenic strains Fo5a4 to protect tomato against the pathogenic strain Fol32 through induced systemic resistance.

the two microorganisms, the protection has to be attributed to increased plant defence reactions in response to root colonization by the non-pathogenic strain. In contrast to other plant–microorganism interactions where ISR has been clearly correlated with an increased production of plant enzymes involved in plant resistance, there is little evidence for the involvement of these mechanisms in the case of plants inoculated with non-pathogenic strains of *F. oxysporum*. Tamietti *et al.* (1993) were the first to show an increased activity of several plant enzymes related to plant defence reactions in tomato plants transplanted in sterilized soil infested with a strain of non-pathogenic *F. oxysporum*. Fuchs *et al.* (1997) attributed the biocontrol activity of the non-pathogenic strain Fo47 to induced resistance in tomato, which was associated with an increased activity of chitinase, β-1,3-glucanase and β-1,4-glucosidase. Duijff *et al.* (1998) showed that the non-pathogenic strain Fo47 was less effective in inducing systemic resistance in tomato than a strain of *Pseudomonas fluorescens*, although inoculation of the non-pathogenic *Fusarium* resulted in an increase of pathogenesis-related (PR) proteins whereas the inoculation of the bacteria did not. On the contrary, Recorbet *et al.* (1998) showed an overall increased activity of constitutive glycosidase isoforms in response to infection by *F. o.* f. sp. *lycopersici* that did not occur in roots colonized with non-pathogenic strains. The pathogen induced the expression of at least one additional β-1,3-glucanase, leading the authors to conclude that the post-infectional accumulation of chitinases and β-1,3-glucanases did not play a significant role in restricting the root colonization by the pathogenic strain of *F. oxysporum*. These contrasting results obtained with the same strain of non-pathogenic *F. oxysporum* applied to tomato show that the biochemical response of the plant is poorly understood and has to be accurately described before resistance induced in tomato by non-pathogenic *F. oxysporum* can be compared with other plant–pathogen models where the cascade of host defence responses is better known.

In conclusion, we can say that many different types of interactions take place between strains of *F. oxysporum* whether pathogenic or not. Studies of soil suppressive to fusarium wilt led to the hypothesis that interactions between pathogenic and non-pathogenic strains can result in the control of the disease. Therefore it is important now to summarize data related to biological control of fusarium wilt achieved with non-pathogenic strains of *F. oxysporum*.

Application to Biological Control

Clear demonstration of the involvement of populations of non-pathogenic *F. oxysporum* in soil suppressiveness to fusarium wilt and indication of the possible role of induction of systemic resistance in plants inoculated with non-pathogenic strains, has stimulated research towards the use of non-pathogenic *F. oxysporum* for biological control. Knowledge of the interactions between pathogenic and non-pathogenic strains of *F. oxysporum* is very useful in characterizing the modes of action of the biocontrol agents but it is not sufficient to define the conditions under which biological control can be applied and will be effective. There is a need for applied research dealing specifically with the development of biological control methods.

Screening for effective strains of non-pathogenic F. oxysporum

The first step in developing a biocontrol method is obviously the screening of an effective strain. Studies dealing with non-pathogenic *F. oxysporum* prove that not all non-pathogenic strains are effective in controlling fusarium wilt. Studies on diversity among strains of *F. oxysporum* failed to establish any clear relationship between a molecular or easily characterized phenotypic trait and either the pathogenicity or the antagonistic activity of the strains studied. Therefore, the only available and reliable method to screen for efficient strains is a bioassay. Indeed although competition for nutrients is one of the main modes of action of the non-pathogenic strains, the studies presented above showed that determination of the competitiveness index was too time consuming to be used as a screening method. Similarly, there is no easy biochemical method available to detect rapidly strains inducing resistance in the plant. Therefore, we still prefer a standardized bioassay in which the potential biocontrol agents are confronted in disinfested soil with the pathogen in the presence of the host plant. In practice, flax is cultivated in heat-treated soil infested with a mixture of the non-pathogenic strain and the pathogen *F. o.* f. sp. *lini* in three population ratios: 10, 100 and 1000 times more inoculum of the non-pathogenic strain. The first symptoms are observed within 23–25 days after sowing. Statistical analysis of data is made using a calculated 'mean survival time' (MST) of a plant and then by comparing the MST in the protected treatment to that in the control. The advantage of such a bioassay is that it takes into account most of the possible interactions among microorganisms, and between microorganisms and the plant, that can lead to biological control. Results show a large diversity among strains of non-pathogenic *F. oxysporum* for their ability to control fusarium wilt of flax (Fig. 8.9).

Modes of action of the biocontrol strains

All the interactions among strains of *F. oxysporum* described above are possible modes of action of the non-pathogenic strains used to control fusarium wilt. Relevant examples illustrating each mode of action can be found in the literature. Competition for nutrients has been shown to be involved in the mode of action of the strains Fo47 (Couteaudier and Alabouvette, 1990; Lemanceau *et al.*, 1993), 618.12 (Postma and Rattink, 1992), and C5 and C14 (Mandeel and Baker, 1991). Competition for infection sites or root colonization has been proposed as a mode of action for the strains Fo47 (Eparvier and Alabouvette, 1994; Olivain and Alabouvette, 1998) and C14 (Mandeel and Baker, 1991). Induced systemic resistance is a mechanism through which several strains protect the plant; for Fo47 (Fuchs *et al.*, 1997; Olivain *et al.*, 1995), C5 and C14 (Mandeel and Baker, 1991), several non-pathogenic strains (Biles and Martyn, 1983), and the *F. solani* strain CS1 and *F. oxysporum* strain CS20 (Larkin and Fravel, 1999). Other mechanisms by which antagonistic microorganisms can control diseases such as antibiosis and hyperparasitism have not yet been demonstrated to be possible modes of action of non-pathogenic *F. oxysporum*.

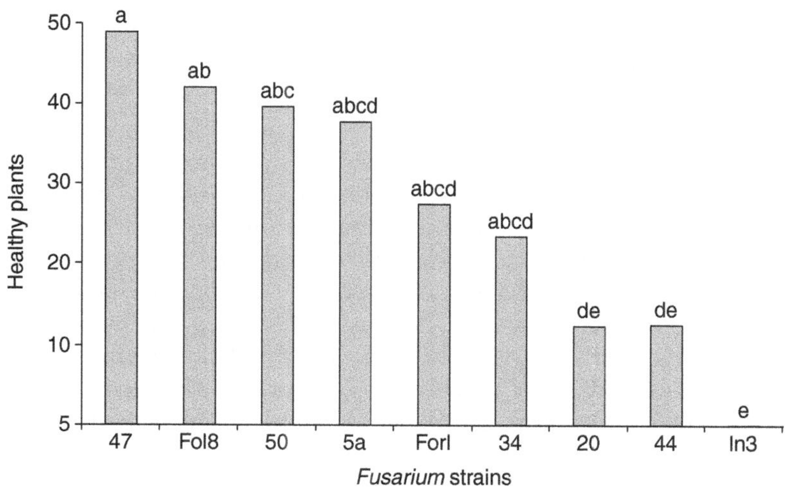

Fig. 8.9. Ability of different strains of non-pathogenic *Fusarium oxysporum* to control fusarium wilt of flax. Sterilized soil precolonized by the non-pathogenic strain was infested with 1×10^4 CFU ml^{-1} of *F. oxysporum* f. sp. *lini*. Disease incidence was scored after 5 weeks of culture.

It must be emphasized that these different modes of action do not exclude each other. On the contrary, the same non-pathogenic strain can express several modes of action. This is the case for the well-studied strain Fo47 for which several teams have reported the involvement of competition for nutrients in soil, competition during root colonization and induced systemic resistance. This is also the case for the non-pathogenic C5 and C14 isolated by Mandeel and Baker (1991) and 618–12 (Postma and Rattink, 1992). Mandeel and Baker (1991) summarized the attributes associated with an efficient biological control strain: 'ability to penetrate and colonize host-tissue without pathogenicity, induction of resistance reactions in living cells capable of providing such reactions, capacity for repeated penetration of moving infection courts to provide continuous induction of resistance, and efficient suppressive activity in environments that are conducive for high inoculum potentials of the pathogen'. One might expect that a strain expressing several modes of action would be more efficient and provide more consistent control than a strain having a single mode of action. However, there is one notable exception, the strain CS20 that works only through induced resistance, and seems to be very effective (Larkin and Fravel, 1999).

Other strain characteristics might be very important in determining the biocontrol activity of a non-pathogenic strain of *F. oxysporum*. For example, the capacity of a strain to grow saprophytically in soil is probably one of the main components of its competitive ability. This competitive ability partly determines the capacity of a strain to establish in soil and in the plant rhizosphere and is probably involved in the capacity of a strain to colonize the root surface. Nagao *et al.* (1990) have shown that different strains have different capacities to colonize a heat-treated soil, the population density at the asymptote being significantly different (Fig. 8.10). Moreover, when flax was grown in soil fully colonized by the non-pathogenic strains, root colonization was also drastically different (Fig. 8.11), and there was no correlation between the population density in the soil and colonization rate of the roots. Steinberg *et al.* (1999a,b) have compared the ability of a small collection of strains of *F. oxysporum* to utilize substrates present in cell walls and to grow actively in close vicinity to a tomato root. These results led to the conclusion that growth habits related to carbohydrate utilization are unique to each strain of *F. oxysporum* and that these traits are not related to pathogenicity or antagonistic ability. Similarly, Recorbet and Alabouvette (1997) showed that adhesion of *F. oxysporum* conidia to tomato roots corresponded to a binding process with specific sites at the root surface. Differences in the ability of strains to bind at the root surface were not related to pathogenic or non-pathogenic ability.

Many different properties are probably involved in determining of the antagonistic activity of non-pathogenic strains of *F. oxysporum*. Thus far it has proved very difficult to assess the relative importance of each mode of action. To accurately study the importance of these diverse modes of action an approach based on the selection of mutants with respect to one or several of these traits would be necessary. Such an approach is being developed in our

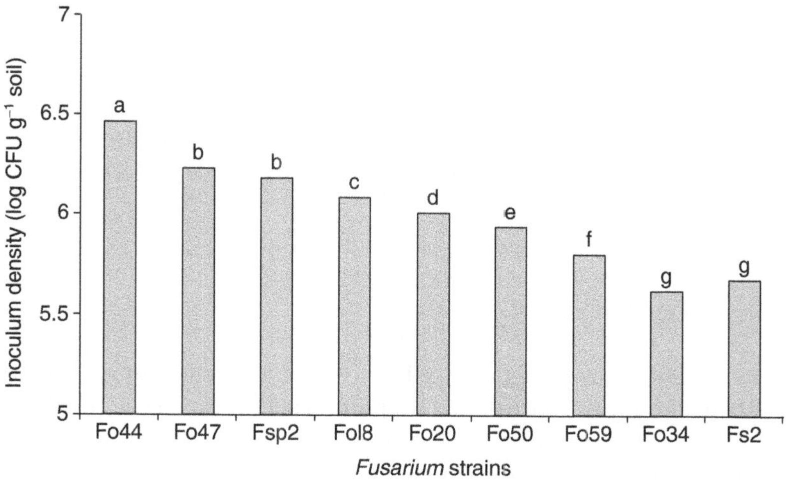

Fig. 8.10. Inoculum density at the plateau (2 weeks after infestation) of nine strains of *Fusarium* spp. added to sterilized soil at the initial density of 1×10^3 CFU g^{-1} soil.

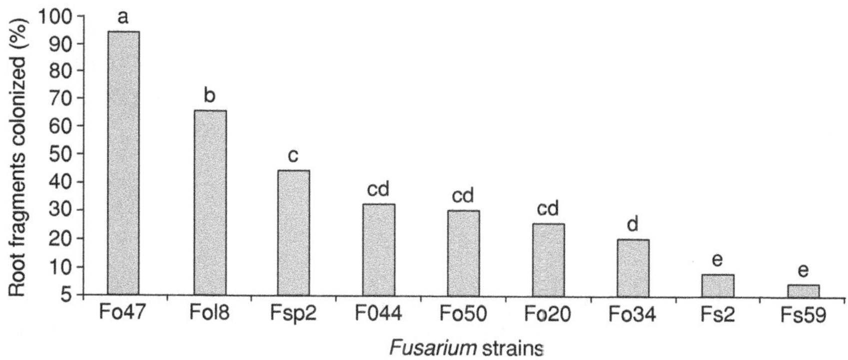

Fig. 8.11. Ability of different strains of non-pathogenic *F. oxysporum* to colonize flax root. Percentage of flax root fragments colonized after 3 weeks of culture in soil previously colonized by the non-pathogenic strains (cf. Fig. 8.10).

group using transposon mutagenesis. We have already selected mutants of Fo47 showing either an increased or a decreased biocontrol capacity (Trouvelot *et al.*, 2000).

Mass production and formulation

Having selected an efficient strain, it is necessary to mass produce and to formulate it in such a way that it can be applied easily. It is well established that the practical efficacy of a biological control method depends greatly on the quality of the inoculant, itself a function of the production and formulation processes (Lewis *et al.*, 1991; Lumdsen *et al.*, 1995; Whipps, 1997).

Strains of non-pathogenic *F. oxysporum* are easy to grow both in submerged fermentation and solid-state fermentation. Both processes are being used to produce the strain Fo47 for large-scale experiment and for commercialization.

Submerged fermentation

The strain Fo47 has been produced in a 400 l fermentor, in an appropriate growth medium with control of O_2 supply and pH (Durand *et al.*, 1989). At the end of the growth phase (after 3–5 days of culture), the growth medium is removed by filtration and the propagules, mainly 'bud cells', are mixed with talcum powder used as an inert carrier. The talcum containing the propagules is dried for 48 h at 18–20°C. During the drying process the propagules tend to develop a thick wall and have been called micro-chlamydospores. With residual humidity below 5%, the inoculant can be stored for several weeks at room temperature and several months at 4°C. Indeed, after more than 1 year of storage at 4°C, the propagule concentration in talcum was only decreased by 23% and by 35% after 2 years (Alabouvette and Couteaudier, 1992). This type of inoculant is easy to apply by mixing it with nutrient solution delivered to plants in soilless cultures or by mixing with potting mixtures.

Solid-state fermentation

The strain Fo47 can also be produced by solid-state fermentation either in sterilized peat or in calcinated clay enriched with an appropriate nutrient solution. For example, in peat, whatever the initial concentration, the strain Fo47 will reach, at the asymptote, a concentration higher than 1×10^7 propagules g^{-1}. In both cases the peat or the calcinated clay provides the carrier for the inoculum, there is no need for further formulation. However, this type of formulation does not enable application of the inoculant as a suspension in water and is adapted for horticultural usage where it is mixed with potting mixtures. This inoculant can be stored at 4°C or even at room temperature without loss of density, or activity (Olivain *et al.*, 1999). Indeed, inoculum density is only one of the parameters needed to reach efficacy. Another factor, much more difficult to assess, is the 'quality' of the propagules, and a method for quality control of the inoculant is one of the most important tasks for manufacturers. The molecular characterization of the strain enables easy identification of the strain and will allow the detection of any contaminant or variant. However, the efficacy of the inoculant has to be tested through a bioassay, which is time and space consuming. Therefore it would be very useful to develop a biochemical assay for use in controlling the viability of the propagules and their 'activity'. For example, different enzymatic activities have been correlated with the capacity of propagules to germinate rapidly. A method based on the measurement of dehydrogenase activity is under development in our laboratory and will allow control of the quality of each batch of production before sale.

Formulation

As stated above, the substratum used for solid-state fermentation provides the carrier for the inoculum, it is therefore not necessary to develop a formulation process. In some cases, however, it would be of interest to improve the efficacy of an inoculant, and different approaches can be followed. Adding to the inoculant a specific food base that will favour the growth of the biocontrol agent after release has been proposed. Steinberg *et al.* (1997) compared the population kinetics and the biological efficacy of several formulations of Fo47. A formulation made of microgranules enriched with a food base provided a better survival and a better biocontrol efficacy than the traditional talcum formulation used in the laboratory (Fig. 8.12). Minuto *et al.* (1997) have compared different strains of non-pathogenic *F. oxysporum* and different commercial formulations of these strains used to control fusarium wilt of basil. They concluded that the formulation is a key point for success of biological control since the efficacy of a given strain greatly depends on the formulation.

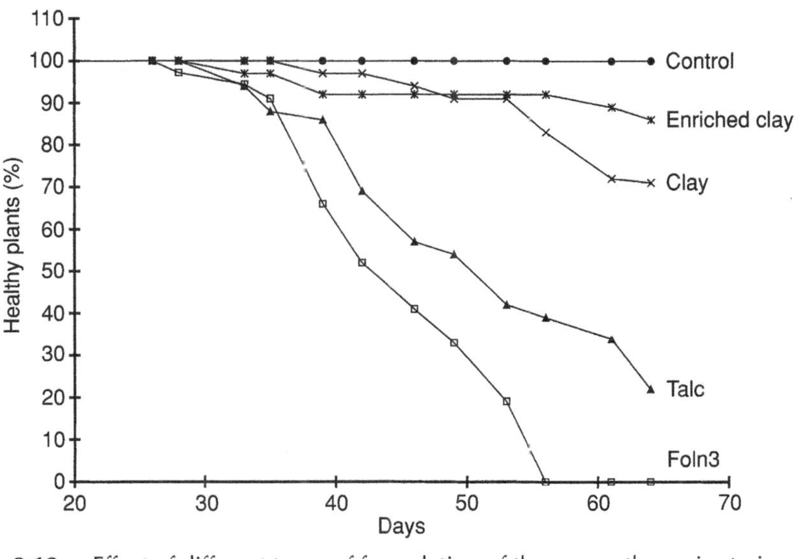

Fig. 8.12. Effect of different types of formulation of the non-pathogenic strain Fo47 on its efficacy in controlling fusarium wilt of flax.

For application to soilless cultures where the easiest way is to apply the inoculant through the drip irrigation system, a liquid formulation would be needed. Until now, it has not been possible to develop a liquid formulation enabling the survival of the inoculum for a sufficient time compatible with the commercial practices. To our knowledge, all the inoculants made of *Fusarium* spp. are formulated as wettable powder or granules and there is a clear need for more research in the field of production and formulation processes (Hebbar *et al.*, 1996).

Delivery: doses and conditions for application

Only large-scale field experiments will provide enough information to determine the best conditions (doses, time, location etc.) for application of a biocontrol agent. But before going to the field it is necessary to study the dose–response relationships leading to effective control. In a recent paper Larkin and Fravel (1999), studying the dose–response relationships between three *formae speciales* of pathogenic *F. oxysporum* and three non-pathogenic strains, proposed several mathematical models for analysing these relationships and calculating an 'effective biocontrol dose'. Depending on the mode of action of the non-pathogenic strains the relationships between inoculum density of the pathogens and inoculum density of the non-pathogens varied greatly. For example, the strain CS20 was able to control the disease at an inoculum density as low as 100–500 propagules g^{-1} soil, even in the presence of a high inoculum density of the pathogen (1×10^5); whereas efficacy of Fo47 required population densities substantially greater than that of the pathogen. These results are in agreement with previous ones showing that, under experimental conditions, the maximum efficacy of Fo47 will be reached with a concentration 100 times higher than that of the pathogen (Alabouvette *et al.*, 1993). In practice, the pathogenic strain is introduced at 5×10^2 or 1×10^3 CFU (colony-forming unit) ml^{-1} soil and the strain Fo47 at 5×10^4 or 1×10^5 CFU ml^{-1} soil. Similarly, Katzube *et al.* (1994) demonstrated that the non-pathogenic strain must be introduced at a concentration 10 times higher than the inoculum density of the pathogen to achieve biological control of fusarium wilt of spinach. However disease suppression was not significantly affected when the inoculum density of the pathogen was greater than 10^4 g^{-1} soil. These high inoculum densities of the non-pathogen necessary to achieve biological control would be difficult to reach in the field; for this reason, the use of non-pathogenic *F. oxysporum* has been mostly limited to crops cultivated in potting mixtures or soilless systems.

When the efficacy of a biocontrol strain is dependent on the inoculum density of the pathogen, it would be useful (if difficult) to determine the pathogen density in the soil. The temptation is always to introduce high inoculum densities of the biocontrol strain. However, increasing the dose of the antagonist more than 100-fold might not be the best option. Indeed Raaijmackers *et al.* (1995), studying dose–response in relation to biological control by *Pseudomonas* spp., showed that the maximum efficacy was not reached with the highest dose of the antagonist but with an intermediate dose. In the control of fusarium wilt in horticultural crops such as carnation, cyclamen or basil, the antagonist will be introduced in the cropping medium before planting. These potting mixtures or soilless substrates are pathogen-free, and the pathogen is usually introduced with irrigation water or with contaminated seeds or plants. In that case, it will be sufficient to introduce the non-pathogenic strain at a concentration close to the carrying capacity of the substratum, which is easy to determine (Alabouvette and Steinberg, 1995). If the mixture can be prepared some time in advance, it would be possible to introduce a lower concentration since the antagonist will usually grow by itself to reach the carrying capacity of the substratum. Additional inoculum would be wasted.

Finally, an important topic for research has been neglected thus far, concerning the relationships between the soil abiotic characteristics and establishment of biological control agents. Most of the soils suppressive to fusarium wilt have distinct physico-chemical properties in common; for example, they are usually clay soils with a pH higher than 7 (Stover, 1962; Höper and Alabouvette, 1996). The nature of the clay minerals plays an important role in relation to the spread of fusarium wilt, and influences the microbial composition and activities in the soil (Stotzky and Martin, 1963; Stotzky and Rem, 1966). For example, Höper *et al.* (1995) showed that addition of different types of clay minerals to a sandy, acidic conducive soil can make it suppressive. Illite was more effective than montmorillonite and kaolinite due to it promoting an increase in the population density of non-pathogenic *F. oxysporum*. Abiotic conditions necessary to establish populations of non-pathogenic *F. oxysporum* in soil over a sufficient period of time to control the disease still tend to be ignored. Unfortunately, it seems that only a very few research groups in the world are addressing this problem.

Discussion and Conclusion

Although soils suppressive to fusarium wilt were identified more than 100 years ago and despite progress made in the basic understanding of the interactions involved in disease suppression, examples of practical use of this knowledge are rare. Two approaches can be followed for biocontrol of fusarium diseases: either increase the level of natural suppressiveness that exists in every soil or isolate microorganisms responsible for suppressiveness and develop these as biological control agents. This second approach has been preferred for the last 20 years. Many teams have published papers related to the efficacy of non-pathogenic strains of *F. oxysporum* to control most of the important wilts: banana (Gerlach *et al.*, 1999), basil (Minuto *et al.*, 1994), carnation (Tramier *et al.*, 1983; Garibaldi *et al.*, 1986, 1990; Postma and Rattink, 1992; Minuto *et al.*, 1994), cucumber (Mandeel and Baker, 1991), cyclamen (Mattusch, 1990; Minuto *et al.*, 1995), flax (Alabouvette *et al.*, 1993), gladiolus (Magie, 1980), melon (Rouxel *et al.*, 1979; Tamietti and Alabouvette, 1986; Alabouvette, 1995), tomato (Lemanceau and Alabouvette, 1991; Fuchs *et al.*, 1999), spinach (Katzube *et al.*, 1994) and strawberry (Tezuka and Makino,1991). But among these papers only a few present data obtained under commercial conditions, and most practical applications of biological agents to control fusarium diseases remain confidential.

It is interesting to consider the causes of this apparent failure. One is linked to the fact that scientists from public institutions are encouraged to study the basic mechanisms of microbial–plant interactions rather than studying conditions necessary to apply biological control agents successfully. At the same time, the agrochemical industry is mostly not interested in biological control of plant pathogens, which is based on totally different mechanisms to chemical control and presents a high level of specificity not compatible with large markets at the world scale. A second reason is linked to the lack of consistency of biological control based on the application of a single biological agent. Indeed, studies of suppressive soils have shown that suppressiveness is always a complex phenomenon involving several microorganisms and several mechanisms. Therefore, it is time to consider application of biological control agents only as one factor contributing to integrated control involving other components. Only a few research groups have followed this approach where non-pathogenic strains of *F. oxysporum* are used in association with cultural practices aimed at decreasing the conduciveness of the crop to the disease. For example, Hervas *et al.* (1997) have studied the interactions between the chickpea genotype, inoculum concentration of the pathogen, application of a *Bacillus* sp. and biological control provided by a strain of non-pathogenic *F. oxysporum*. It is clear from this work that many different considerations apply to obtain efficacy of biological control (Landa *et al.*, 1999). Reist *et al.* (1996) observed increased yields in susceptible tomato grafted on a resistant rootstock in the presence of the non-pathogenic strain Fo47. These results invite us to consider the influence of the host plant on the efficacy of biological control and that the selection of plant cultivars 'susceptible' to the biocontrol agents should seriously be considered.

The progress made in characterizing the diversity that affects populations of *F. oxysporum* has provided the tools for studying the ecology of this species, and for assessing the effects of the abiotic soil characteristics, cultural practices, and plant species and cultivars on the establishment and activity of different strains of non-pathogenic *F. oxysporum* in soil and the rhizosphere of cultivated plants. With this knowledge it might become possible to modify microbial communities, through environment-friendly techniques, to control fusarium diseases of cultivated plants.

References

Alabouvette, C. (1995) Biological control of fusarium wilts. In: Manka, M. (ed.) *Environmental Biotic Factors in Integrated Plant Disease Control*. The Polish Phytopathological Society, Poznan, pp. 61–68.

Alabouvette, C. and Couteaudier, Y. (1992) Biological control of fusarium wilts with nonpathogenic *Fusaria*. In: Tjamos, E.C., Cook, R.J. and Papavizas, G.C. (eds) *Biological Control of Plant Diseases*. Plenum Press, New York, pp. 415–426.

Alabouvette, C. and Steinberg, C. (1995) Suppressiveness of soils to invading micro-organisms. In: Hokkanen, H.T.M. and Lynch, J.M. (eds) *Biological Control: Benefits and Risks*. Cambridge University Press, Cambridge, pp. 3–12.

Alabouvette, C., Couteaudier, Y. and Louvet, J. (1985) Soils suppressive to Fusarium wilts: mechanisms and management of suppressiveness. In: Parker, C.A., Rovira, A.D., Moore, K.J., Wong, P.T.W. and Kollmorgen, J.F. (eds) *Ecology and Management of Soilborne Plant Pathogens*. The American Phytopathological Society, St Paul, Minnesota, pp. 101–106.

Alabouvette, C., Lemanceau, P. and Steinberg, C. (1993) Recent advances in biological control of fusarium wilts. *Pesticide Science* 37, 365–373.

Armstrong, G.M. and Armstrong, J.K. (1981) *Formae speciales* and races of *Fusarium oxysporum* causing wilt diseases. In: Nelson, P.E., Toussoun, T.A. and Cook, R.J. (eds) *Fusarium: Diseases, Biology, and Taxonomy*. Pennsylvania State University Press, University Park and London, pp. 391–399.

Baker, K.F. and Cook, R.J. (1974) *Biological Control of Plant Pathogens*. The American Phytopathological Society, St Paul, Minnesota.

Biles, C.J. and Martyn, R.D. (1983) Local and systemic resistance induced in watermelon by *formae speciales* of *Fusarium oxysporum*. *Phytopathology* 79, 856–860.

Burgess, L.W. (1981) General ecology of the Fusaria. In: Toussoun, T.A., Nelson, P.E. and Cook, R.J. (eds) *Fusarium: Diseases, Biology, and Taxonomy*. Pennsylvania State University Press, University Park and London, pp. 225–235.

Couteaudier, Y. and Alabouvette, C. (1990) Quantitative comparison of *Fusarium oxysporum* competitiveness in relation with carbon utilization. *FEMS Microbiology Ecology* 74, 261–268.

Couteaudier, Y. and Steinberg, C. (1990) Biological and mathematical description of growth pattern of *Fusarium oxysporum* in sterilized soil. *FEMS Microbiology Ecology* 74, 253–260.

Del Sorbo, G.S., Capparelli, R., Iannelli, D. and Noviello, C. (1993) Differentiation between two *formae speciales* of *Fusarium oxysporum* by antisera produced in mice immunologically tolerized at birth through lactation. *Phytopathology* 84, 534–540.

Duijff, B.J., Pouhair, D., Olivain, C., Alabouvette, C. and Lemanceau, P. (1998) Implication of systemic induced resistance in the suppression of fusarium wilt of tomato by *Pseudomonas fluorescens* WCS417r and nonpathogenic *Fusarium oxysporum* Fo47. *European Journal of Plant Pathology* 104, 903–910.

Durand, A., Vergoignan, C. and Almanza, S. (1989) Studies of the survival of *Fusarium oxysporum* conidia produced by submerged culture. *Biotechnology Letters* 11, 503–508.

Edel, V., Steinberg, C., Avelange, I., Laguerre, G. and Alabouvette, C. (1995) Comparison of three molecular methods for the characterization of *Fusarium oxysporum* strains. *Phytopathology* 85, 579–585.

Edel, V., Steinberg, C., Gautheron, N. and Alabouvette, C. (1997a) Evaluation of restriction analysis of polymerase chain reaction (PCR)-amplified ribosomal DNA for the identification of *Fusarium* species. *Mycological Research* 101, 179–187.

Edel, V., Steinberg, C., Gautheron, N. and Alabouvette, C. (1997b) Populations of nonpathogenic *Fusarium oxysporum* associated with roots of four plant species compared to soilborne populations. *Phytopathology* 87, 693–697.

Edel, V., Steinberg, C., Gautheron, N. and Alabouvette, C. (2000) Ribosomal DNA-targeted oligonucleotide probe and PCR assay specific for *Fusarium oxysporum*. *Mycological Research* 104, 518–526.

Elias, K.S. and Schneider, R.W. (1992) Genetic diversity within and among races and vegetative compatibility groups of *Fusarium oxysporum* f. sp. *lycopersici* as determined by isozyme analysis. *Phytopathology* 82, 1421–1427.

Eparvier, A. and Alabouvette, C. (1994) Use of ELISA and GUS-transformed strains to study competition between pathogenic and non-pathogenic *Fusarium oxysporum* for root colonization. *Biocontrol Science and Technology* 4, 35–47.

Fuchs, J.-G., Moënne-Loccoz, Y. and Défago, G. (1997) Nonpathogenic *Fusarium oxysporum* strain Fo47 induces resistance to Fusarium wilt in tomato. *Plant Disease* 81, 492–496.

Fuchs, J.G., Moënne-Loccoz, Y. and Defago, G. (1999) Ability of nonpathogenic *Fusarium oxysporum* Fo47 to protect tomato against Fusarium wilt. *Biological Control* 14, 105–110.

Garibaldi, A., Brunatti, F. and Gullino, M. L. (1986) Suppression of Fusarium wilt of carnation by competitive non pathogenic strains of Fusaria. *Mededelingen van de Faculteit Landbouwwetenschappen Rijksuniversiteit Gent* 51, 633–638.

Garibaldi, A., Gullino, M.L. and Aloi, C. (1990) Biological control of Fusarium wilt of carnation. *Proceedings of the Brighton Crop Protection Conference*. Brighton, 3 A-1, pp. 89–95.

Garrett, S.D. (1970) *Pathogenic Root-infecting Fungi*. Cambridge University Press, London.

Gerlach, K.S., Bentley, S., Moore, N.Y., Aitken, E.A.B. and Pegg, K.G. (1999) Investigation of non-pathogenic strains of *Fusarium oxysporum* for suppression of Fusarium wilt of banana in Australia. In: Alabouvette, C. (ed.) *Second International Fusarium Biocontrol Workshop*. INRA, Dijon.

Gordon, T.R. and Martyn, R.D. (1997) The evolutionary biology of *Fusarium oxysporum*. *Annual Review of Phytopathology* 35, 111–128.

Gordon, T.R., Okamoto, D. and Milgroom, M.G. (1992) The structure and interrelationship of fungal populations in native and cultivated soils. *Molecular Ecology* 1, 241–249.

Hebbar, K.P., Lewis, J.A., Poch, S.M. and Lumsden, R.D. (1996) Agricultural by-products as substrates for growth, conidiation and chlamydospore formation by a potential mycoherbicide, *Fusarium oxysporum* strain EN4. *Biocontrol Science and Technology* 6, 263–275.

Hervas, A., Landa, B. and Jimenez-Diaz, R. (1997) Influence of chickpea genotype and *Bacillus* sp. on protection from Fusarium wilt by seed treatment with nonpathogenic *Fusarium oxysporum*. *European Journal of Plant Pathology* 103, 631–642.

Höper, H. and Alabouvette, C. (1996) Importance of physical and chemical soil properties in the suppressiveness of soils to plant diseases. *European Journal of Soil Biology* 32, 41–58.

Höper, H., Steinberg, C. and Alabouvette, C. (1995) Involvement of clay type and pH in the mechanisms of soil suppressiveness to fusarium wilt of flax. *Soil Biology and Biochemistry* 27, 955–967.

Huss, M.J., Campbell, C.L., Jennings, D.B. and Leslie, J.F. (1996) Isozyme variation among biological species in the *Gibberella fujikuroi* species complex (*Fusarium* section *Liseola*). *Applied and Environmental Microbiology* 62, 3750–3756.

Iannelli, D., Capparelli, R., Marziano, F., Scala, F. and Noviello, C. (1983) Production of hybridomas secreting monoclonal antibodies to the genus *Fusarium*. *Mycotaxon* 17, 523–532.

Katan, T. (1999) Current status of vegetative compatibility groups in *Fusarium oxysporum*. *Phytoparasitica* 27, 51–64.

Katzube, K., Akasaka, Y. and Nakatani, F. (1994) Biocontrol of Fusarium wilt of spinach by using nonpathogenic *Fusarium oxysporum*. 2. Investigation of inoculation methods. *Annual Report of Plant Protection North Japan* 45, 72–75.

Kern, H. (1972) Phytotoxins produced by fusaria. In: Wood, R.K.S. and Graniti, A. (eds) *Phytotoxin in Plant Diseases*. Academic Press, New York, pp. 35–44.

Kondo, N., Kodama, F. and Ogoshi, A. (1997) Vegetative compatibility groups of *Fusarium oxysporum* f. sp. *adzukicola* and nonpathogenic *Fusarium oxysporum* on adzuki bean isolated from adzuki bean fields in Hokkaido. *Annals of the Phytopathological Society of Japan* 63, 8–12.

Kroon, B.A.M., Scheffer, R.J. and Elgersma, D.M. (1992) Induced resistance in tomato plants against fusarium wilt involved by *Fusarium oxysporum* f. sp. *dianthi*. *Netherlands Journal of Plant Pathology* 97, 401–408.

Landa, B.B., Navas-Cortes, J.A., Hervas, A. and Jimenez-Diaz, R.M. (1999) Use of sowing date, host resistance and biocontrol treatments for the integrated control of Fusarium wilt of chickpea. In: Alabouvette, C. (ed.) *Second International Biocontrol Workshop*. INRA, Dijon, p. 40.

Landa del Castillo, B. (1999) Control biologica de la fusariosis vascular del garbanzo mediante antagonistas microbianos. PhD thesis, Universidad de Cordoba, Cordoba.

Larkin, R.P. and Fravel, D.R. (1999) Mechanisms of action and dose-response relationships governing biological control of Fusarium wilt of tomato by nonpathogenic *Fusarium* spp. *Phytopathology* 89, 1152–1161.

Larkin, R.P., Hopkins, D.L. and Martin, F.N. (1996) Suppression of fusarium wilt of watermelon by nonpathogenic *Fusarium oxysporum* and other microorganisms recovered from a disease-suppressive soil. *Phytopathology* 86, 812–819.

Lemanceau, P. and Alabouvette, C. (1991) Biological control of fusarium diseases by fluorescent *Pseudomonas* and non-pathogenic *Fusarium*. *Crop Protection* 10, 279–286.

Lemanceau, P., Bakker, P.A.H.M., De Kogel, W.J., Alabouvette, C. and Schippers, B. (1993) Antagonistic effect on nonpathogenic *Fusarium oxysporum* strain Fo47 and pseudobactin 358 upon pathogenic *Fusarium oxysporum* f. s.p. *dianthi*. *Applied Environmental Microbiology* 59, 74–82.

Lewis, J.A., Papavizas, G.C. and Lumsden, R.D. (1991) A new formulation system for the application of biocontrol fungi to soil. *Biocontrol Science and Technology* 1, 59–69.

Louvet, J., Rouxel, F. and Alabouvette, C. (1976) Recherches sur la résistance des sols aux maladies I. Mise en évidence de la nature microbiologique de la résistance d'un sol au développement de la fusariose vasculaire du melon. *Annales de Phytopathologie* 8, 425–436.

Lumsden, R.D., Lewis, J.A. and Fravel, D.R. (1995) Formulation and delivery of biocontrol agents for use against soilborne plant pathogens. In: Hall, F.R. and Barry, J.W. (eds) *Biorational Pest Control Agents – Formulation and Delivery*. American Chemical Society, Washington, DC, pp. 166–182.

Magie, R.O. (1980) Fusarium disease of gladioli controlled by inoculation of corms with non-pathogenic Fusaria. *Proceedings of the Florida State Horticultural Society* 93, 172–175.

Mandeel, Q. and Baker, R. (1991) Mechanisms involved in biological control of fusarium wilt of cucumber with strains of nonpathogenic *Fusarium oxysporum*. *Phytopathology* 81, 462–469.

Matta, A. (1989) Induced resistance to fusarium wilt diseases. In: Tjamos, E.C. and Beckman, C.H. (eds) *Vascular Wilt Diseases of Plants – Basic Studies and Control*. Springer Verlag, NATO ASI Series, Berlin, pp. 175–196.

Mattusch, P. (1990) Biologische Bekämpfung von Fusarium oxysporum an einigen gärtnerischen Kulturpflanzen. *Nachrichtenbl Deut Planzenschutzd* 42, 148–150.

Minuto, A., Garibaldi, A. and Gullino, M.L. (1994) Biological control of Fusarium wilt of basil (*Ocimum basilicum* L.). *Proceedings of the Brighton Crop Protection Conference*, Brighton, 6D 18, pp. 811–816.

Minuto, A., Migheli, Q. and Garibaldi, A. (1995) Evaluation of antagonistic strains of *Fusarium* in the biological and integrated control of Fusarium wilt of cyclamen. *Crop Protection* 14, 221–226.

Minuto, A., Minuto, J., Migheli, Q., Mocioni, M. and Gullino, M.L. (1997) Effect of antagonistic *Fusarium* spp. and of different commercial biofungicide formulations on Fusarium wilt of basil (*Ocimum basilicum* L.). *Crop Protection* 17, 765–769.

Nagao, H., Couteaudier, Y. and Alabouvette, C. (1990) Colonization of sterilized soil and flax roots by strains of *Fusarium oxysporum* and *Fusarium solani*. *Symbiosis* 9, 343–354.

Nelson, P.E., Toussoun, T.A. and Marasas, W.F.O. (1983) *Fusarium Species. An Illustrated Manual for Identification*. The Pennsylvania State University Press, University Park and London.

Ogawa, K. and Komada, H. (1984) Biological control of Fusarium wilt of sweet potato by non-pathogenic *Fusarium oxysporum*. *Annals of the Phytopathological Society of Japan* 50, 1–9.

Olivain, C. and Alabouvette, C. (1997) Colonization of tomato root by a nonpathogenic strain of *Fusarium oxysporum*. *The New Phytologist* 137, 481–494.

Olivain, C. and Alabouvette, C. (1998) Process of tomato root colonization by a pathogenic strain of *Fusarium oxysporum* f. sp. *lycopersici* discussed in comparison to a non-pathogenic strain. *The New Phytologist* 141, 497–510.

Olivain, C., Steinberg, C. and Alabouvette, C. (1995) Evidence of induced resistance in tomato inoculated by nonpathogenic strains of *Fusarium oxysporum*. In: Manka, M. (ed.) *Environmental Biotic Factors in Integrated Plant Disease Control*. The Polish Phytopathological Society, Poznan, pp. 427–430.

Olivain, C., Steinberg, C., Dreumont, C., Stawiecki, K., Pouhair, D., Wadoux, P. and Alabouvette, C. (1999) Production of a mixed inoculum of *Fusarium oxysporum* Fo47 and *Pseudomonas fluorescens* C7 in peat. In: Alabouvette, C. (ed.) *Second International Fusarium Biocontrol Workshop*. INRA, Dijon.

Paulitz, T.C., Park, C.S. and Baker, R. (1987) Biological control of Fusarium wilt of cucumber with nonpathogenic isolates of *Fusarium oxysporum*. *Canadian Journal of Microbiology* 33, 349–353.

Postma, J. and Luttikholt, A.J.G. (1996) Colonization of carnation stems by a nonpathogenic isolate of *Fusarium oxysporum* and its effect on wilt caused by *Fusarium oxysporum* f. sp. *dianthi*. *Canadian Journal of Botany* 74, 1841–1851.

Postma, J. and Rattink, H. (1992) Biological control of fusarium wilt of carnation with a nonpathogenic isolate of *Fusarium oxysporum*. *Canadian Journal of Botany* 70, 1199–1205.

Puhalla, J.E. (1985) Classification of strains of *Fusarium oxysporum* on the basis of vegetative compatibility. *Canadian Journal of Botany* 63, 179–183.

Raaijmakers, J.M., Leeman, M., van Oorschot, M.M.P., van der Sluis, I., Schippers, B. and Bakker, P.A.H.M. (1995) Dose–response relationships in biological control of Fusarium wilt of radish by *Pseudomonas* spp. *Phytopathology* 85, 1075–1081.

Rataj-Guranowska, M. and Wolko, B. (1991) Comparison of *Fusarium oxysporum* and *Fusarium oxysporum* var. *redolens* by analyzing the isozyme and serological patterns. *Journal of Phytopathology* 132, 287–293.

Recorbet, G. and Alabouvette, C. (1997) Adhesion of *Fusarium oxysporum* conidia to tomato roots. *Letters in Applied Microbiology* 25, 375–379.

Recorbet, G., Bestel, G., Gaudot-Dumas, E., Gianinazzi, S. and Alabouvette, C. (1998) Differential accumulation of β-1,3-glucanase and chitinase isoforms in tomato roots in response to colonization by either pathogenic or non-pathogenic strains of *Fusarium oxysporum*. *Letters in Applied Microbiology* 153, 257–263.

Reist, A., Gillioz, J.M. and Corbaz, R. (1996) Comparaison de plants de tomate, greffés ou non, en culture sur substrat et en présence de pathogènes. *Revue Suisse Viticulture Arboriculture Horticulture* 28, 327–331.

Rouxel, F., Alabouvette, C. and Louvet, J. (1977) Recherches sur la résistance des sols aux maladies II – Incidence de traitements thermiques sur la résistance microbiologique d'un sol à la Fusariose vasculaire du melon. *Annales de Phytopathologie* 9, 183–192.

Rouxel, F., Alabouvette, C. and Louvet, J. (1979) Recherches sur la résistance des sols aux maladies IV – Mise en évidence du rôle des *Fusarium* autochtones dans la résistance d'un sol à la Fusariose vasculaire du Melon. *Annales de Phytopathologie* 11, 199–207.

Seifert, K. and Brissett, J. (1998) Substrate utilization profiles as identification aids in *Fusarium*. In: Brayford, D. (ed.) *8th International Fusarium Workshop*. Egham, UK, p. 90.

Skovgaard, K. and Rosendahl, S. (1998) Comparison of intra- and extracellular isozyme banding patterns of *Fusarium oxysporum*. *Mycological Research* 102, 1077–1084.

Smith, S.N. and Snyder, W.C. (1971) Relationship of inoculum density and soil types to severity of fusarium wilt of sweet potato. *Phytopathology* 61, 1049–1051.

Snyder, W.C. and Hansen, H.N. (1940) The species concept in *Fusarium*. *American Journal of Botany* 27, 64–67.

Steinberg, C., Edel, V. and Alabouvette, C. (1997a) Rôle du mode de formulation sur la survie et l'activité antagoniste d'agents de lutte biologique contre les fusarioses de plantes cultivées *Cryptogamie, Mycologie* 18, 139–143.

Steinberg, C., Edel, V., Gautheron, N., Abadie, C., Vallaeys, T. and Alabouvette, C. (1997b) Phenotypic characterization of natural populations of *Fusarium oxysporum* in relation to genotypic characterization. *FEMS Microbiology Ecology* 24, 73–85.

Steinberg, C., Whipps, J.M., Wood, D.A., Fenlon, J. and Alabouvette, C. (1999a) Effect of nutritional sources on growth of one non-pathogenic strain and four strains of *Fusarium oxysporum* pathogenic on tomato. *Mycological Research* 103, 1210–1216.

Steinberg, C., Whipps, J.M., Wood, D.A., Fenlon, J. and Alabouvette, C. (1999b) Mycelial development of *Fusarium oxysporum* in the vicinity of tomato roots. *Mycological Research* 103, 769–778.

Stotzky, G. and Martin, T. (1963) Soil mineralogy in relation to the spread of Fusarium wilt of banana in Central America. *Plant Soil* 18, 317–337.

Stotzky, G. and Rem, L.T. (1966) Influence of clay minerals on microorganisms I. Montmorillonite and kaolinite on bacteria. *Canadian Journal of Microbiology* 12, 547–563.

Stover, R.H. (1962) Fusarial wilt (Panama disease) of bananas and other *Musa* species. *CMI, Phytopathological Papers* 4, 117pp.

Tamietti, G. and Alabouvette, C. (1986) Résistance des sols aux maladies: XIII – rôle des *Fusarium oxysporum* non pathogènes dans les mécanismes de résistance d'un sol de Noirmoutier aux fusarioses vasculaires. *Agronomie* 6, 541–548.

Tamietti, G., Ferraris, L., Matta, A. and Abbattista Gentile, I. (1993) Physiological responses of tomato plants grown in Fusarium suppressive soil. *Journal of Phytopathology* 138, 66–76.

Tezuka, N. and Makino, T. (1991) Biological control of Fusarium wilt of strawberry by nonpathogenic *Fusarium oxysporum* isolated from strawberry. *Annals of the Phytopathological Society of Japan* 57, 506–511.

Thrane, U. and Hansen, U. (1995) Chemical and physiological characterization of taxa in the *Fusarium sambucinum* complex. *Mycopathologia* 129, 183–190.

Tramier, R., Pionnat, J.C. and Tebibel, N. (1983) Role of the fungi in the induction of suppressiveness into substrates to Fusarium wilt carnation. *Acta Horticulturae* 141, 55–59.

Trouvelot, S., Edel, V., Recorbet, G., Olivain, C., Daboussi, M.J. and Alabouvette, C. (2000) Recovery of strains impaired in their antagonistic activity after transposition of the *Fot1* element in *Fusarium oxysporum*. In: *Fifth European Conference on Fungal Genetics*, Arcachon, France.

Wasfy, E.H., Bridge, P.D. and Brayford, D. (1987) Preliminary studies on the use of biochemical and physiological tests for the characterization of Fusarium isolates. *Mycopathologia* 99, 9–13.

Whipps, J.M. (1997) Developments in the biological control of soil-borne plant pathogens. *Advances in Botanical Research* 26, 1–134.

Windels, C.E. (1992) *Fusarium*. In: Singleton, L.L. and Rush, C.M. (eds) *Methods for Research on Soilborne Phytopathogenic Fungi*. American Phytopathological Society Press, St Paul, Minnesota, pp. 115–128.

The Use of Avirulent Mutants of *Ralstonia solanacearum* to Control Bacterial Wilt Disease

9

J.J. Smith and G.S. Saddler

CABI Bioscience UK Centre (Egham), Bakeham Lane, Egham, Surrey TW20 9TY, UK

Introduction

Bacterial wilt disease is endemic in tropical, subtropical and warm temperate regions, where it represents a major constraint on the production of numerous agricultural crops. It remains one of the most intractable bacterial plant diseases. The causative organism, *Ralstonia solanacearum*, is highly diverse and, as a consequence, a number of attempts have been made to circumscribe infraspecific variation, with varying degrees of success (see Hayward, 1991). Our current, incomplete understanding of infraspecific diversity, coupled with gaps in knowledge on dissemination, infection and disease development have hampered the development of effective disease management strategies for many crops (Sequeira, 1994). Ecology of the species overall is highly diverse, though certain infraspecific groups exhibit a more restricted distribution and may be found in association with only a small number of plants. As a consequence, no universal control measures exist which are effective across the wide host range on which this pathogen occurs.

In general, breeding for host resistance remains one of the most effective avenues for controlling bacterial plant disease. However, this approach has met with only limited success in the case of bacterial wilt disease. Resistance breeding has been effective with crops such as tobacco and groundnut, but success with solanaceous hosts, potato, tomato, etc., appears to be regional or linked to climatic conditions (Hayward, 1991; French, 1994; Prior *et al.*, 1994). Generally with solanaceous crops, disease tolerance is the best that can be achieved, and even this is frequently unstable. Production of somatic hybrids using wild potatoes such as *Solanum commersonii*, as the fusion parent,

 Biotic Interactions in Plant–Pathogen Associations (eds M.J. Jeger and N.J. Spence)

with *S. tuberosum* has provided some encouraging results (Laferriere *et al.*, 1998). But as yet the performance of these hybrids under varying field conditions and temperature regimes remains unknown.

Cultural control methods are also of value, growth of susceptible tobacco cultivars with French marigolds (*Tagetes patula*) not only reduced the incidence of bacterial wilt disease but also the numbers of pathogens found in soil (Terblanche and de Villiers, 1998). Control was thought to have been achieved through the presence of thiopenes, natural broad-spectrum biocides secreted from the roots of the marigold. Similarly, disease control has been achieved by intercropping potatoes with beans (Autrique and Potts, 1987). Soil amendments have been used successfully to control bacterial wilt in naturally infested fields of tobacco and tomato (Yao *et al.*, 1994). It is interesting to note that control was reduced if the amendment mixture was heat treated prior to application, suggesting that control derived from an element that was biological in its origin.

Biological Control of Bacterial Plant Disease

Biological control is based on microbial antagonism which can be direct (parasitism, competition, antibiosis) or indirect (induced resistance of the host). The definitions and general mechanisms involved in biological control of phytopathogenic microorganisms have been reviewed (Leong, 1986; Baker, 1987; Fravel, 1988; Weller, 1988; Campbell, 1989). Many biological control strategies developed in the laboratory often fail under natural conditions (Fravel, 1988) as they are based on the assumption that the biocontrol agent is able to compete under conditions that favour the pathogen. In the soil/rhizosphere microenvironment, for example, the biocontrol agent must contend with complex biological and physical factors, including soil composition and structure, moisture and pH. All of which can influence the structure of the microbial community (Nesmith and Jenkins, 1985; Weller, 1988). Equally, parameters affecting direct antagonism on agar media are often not precisely known and the conditions required to maximize these effects *in vivo* can frequently not be reproduced. It has also been suggested that direct antagonism may not be significant *in planta* in controlling some plant disease and that other mechanisms, e.g. induced resistance, may have great significance (Sequeira and Hill, 1974; Sequeira *et al.*, 1977; Wakimoto, 1987; Trigalet and Trigalet-Demery, 1990).

A number of bacteria which are antagonistic towards *R. solanacearum* have been isolated from various sources, e.g. suppressive soils and host plant rhizospheres, and evaluated as control agents (Table 9.1). Inoculation of potatoes with *Pseudomonas fluorescens* brought about a significant reduction in the severity of bacterial wilt (Ciampi-Panno *et al.*, 1989; Gallardo *et al.*, 1989). The control agent was applied as a seed dressing and appeared to be well adapted to saprophytic survival in naturally infested soils. *P. fluorescens*

Table 9.1. Saprophytic bacteria that have been studied as biological control agents against bacterial wilt disease.

Bacteria	Host	Reference
Pseudomonas aeruginosa ATCC 7700	Tomato	Furuya *et al.*, 1997
Pseudomonas fluorescens PfG32	Tomato	Mulya *et al.*, 1996
Burkholderia (*Pseudomonas*) *glumae*	Tomato	Furuya *et al.*, 1991a; Wakimoto, 1987
Bacillus spp.	Tomato	Da Silveira *et al.*, 1995
Bacillus subtilis NB22	Tomato	Phae *et al.*, 1992
Streptomyces mutabilis	Tomato	El Shanshoury *et al.*, 1996
Bacillus polymyxa FU6, *Pseudomonas fluorescens*	Tomato/potato	Aspiras and de la Cruz, 1985
Pseudomonas fluorescens W163	Potato	Kempe and Sequeira, 1983
Pseudomonas fluorescens BC8	Potato	Ciampi *et al.*, 1997
Pseudomonas fluorescens, Bacillus subtilis, Bacillus sp., actinomycetes	Potato	Shekhawat *et al.*, 1993
P. fluorescens B33 and B36 and *Bacillus* spp.	Banana/aubergine/tomato	Anuratha and Gnanamanickam, 1990
Bacillus spp.	Tobacco	Liu *et al.*, 1999

was able to penetrate the plant via its root system, however, protection was not stable nor totally efficient as latent infection of daughter tubers occurred.

To bypass the impact of environmental factors or incompatible plant–microbe interactions restricting the actions and survival of the control agent, a number of studies have focused on antagonists which are closely related to or derived from the wild-type pathogen. In general, these microorganisms are adapted to survive in the same plant microenvironment in which the pathogen operates. The use of such antagonists has several advantages, principally that the antagonist will be able to exploit conditions which also favour the pathogen and, that once established, may be able to persist and thereby provide continuous protection. In essence, the ideal antagonist must be able to colonize and survive asymptomatically on the host under all the varying conditions in which the host is grown.

The most successful implementation of this approach has been achieved through the biological control of crown gall. Certainly, this approach has spawned a number of attempts to mimic this success in the control of other bacterial plant diseases. *Pseudomonas savastanoi* pv. *savastanoi* deficient in the production of indoleacetic acid (Surico *et al.*, 1984) has shown some promise as a biocontrol agent against olive knot disease (Varvaro and Martella, 1993). Similar effects have been seen with *Xanthomonas oryzae* pv. *oryzae* (Liu *et al.*, 1998), *X. transluscens* pv. *graminis* (Schmidt, 1988), *Burkholderia glumae* (Furuya *et al.*, 1991b), *Pseudomonas syzygii* (Hartati *et al.*, 1991) and *Erwinia amylovora* (Tharaud *et al.*, 1997).

The Crown Gall/*Agrobacterium* Story

Plant pathogenicity in agrobacteria manifests itself as oncogenic (tumorigenic or rhizogenic) activity (see Ryder and Jones, 1990). The host range of tumorigenic *Agrobacterium* is wide: 640 plant species belonging to 331 genera in 93 families of dicotyledonous and gymnospermous plants (De Cleene and De Ley, 1976). The disease, commonly known as crown gall, seldom kills the host, but growth is impaired and economic loss occurs (De Cleene, 1979).

Tumorigenic activity is primarily mediated by plasmid-borne genes on the *Ti* plasmid. Genes on the *Ti* plasmid comprise T-DNA, virulence (*vir*) and opine synthesis genes. The bacterium enters the plants through wounds, activating the *vir* genes which facilitate the transfer of the T-DNA component into the plant nucleus (Chyi *et al.*, 1986). T-DNA contains the genes for tumour induction, in particular auxin and kinin synthesis. In addition, opines, unusual amino acid derivatives, are synthesized. These compounds are not utilized by plants nor by most microorganisms, but serve as nutrients restricted to the tumorigenic strain (Schell *et al.*, 1979).

Biological control agents for crown gall have been in commercial production for the last 25 years (Ryder and Jones, 1990). The method is reliant on the use of an avirulent strain of *Agrobacterium*, frequently referred to in the literature as *A. radiobacter* K84 (New and Kerr, 1972; Kerr and Htay, 1974). Seeds, roots or wounded plant surfaces are dipped in suspensions of K84, which produces a bacteriocin, agrocin 84. The bacteriocin selectively inhibits the pathogen (Moore and Warren, 1979; Kerr, 1980, 1991) and in particular, is highly effective against pathogenic *Agrobacterium* strains harbouring a nopaline/agrocinopine A-type *Ti* plasmid. These strains are largely responsible for causing damage in orchards and nurseries (Kerr and Tate, 1984). For the bacteriocin to be effective it must be taken up by the pathogen. In this regard, agrocin 84 is a partial mimic of agrocinopine A and is selectively taken up via agrocinopine permease, found in pathogenic strains harbouring the nopaline/agrocinopine A-type *Ti* plasmid (Ellis and Murphy, 1981). The mode of action of the bacteriocin has not been fully elucidated, however, as an adenine nucleotide it is likely that it inhibits DNA synthesis (Murphy and Roberts, 1979). Effective control can only be achieved if the control agent is able to colonize the host. Previous studies have shown that other bacteriocin-producing agrobacteria effective *in vitro* are unable to control crown gall in the field as a result of their poor colonization potential (Macrae *et al.*, 1988).

The bacteriocin is encoded on the pAgK84 plasmid which can be transmitted by conjugation to other agrobacteria. Resultant transconjugants can produce the bacteriocin and immunity is also conferred. Plasmid transfer to pathogenic strains in the field has been reported and the resultant strains are no longer susceptible to the biological control agent (Panagopoulous *et al.*, 1979). To counteract this threat a transfer-deficient (Tra^-) agrocin 84 plasmid has been constructed, deleting a 5.9 kb fragment from the plasmid (Jones *et al.*, 1988). Strain K1026, harbouring the Tra^- plasmid, retains the ability to produce agrocin 84 but lacks the ability to transfer its plasmid. K1026 is commercially available (NOGALL™, Bio-Care Technology, NSW, Australia) and represents the first commercial release of a genetically modified biological control agent (Kerr, 1991).

The *Ralstonia solanacearum* Story

Spontaneous avirulent mutants of *R. solanacearum*, deficient in exopolysaccharide (EPS) production, which are able to colonize the host albeit in a limited capacity, have been known for some time (Kelman, 1954; Averre and Kelman, 1964; Kelman and Hiruschka, 1973). Using avirulent *R. solanacearum* strains, some of which also produced bacteriocins, protection has been achieved with tobacco (Chen *et al.*, 1982; Luo and Wang, 1983; Chen and Echandi, 1984; Tanaka *et al.*, 1990), tomato (Luo and Wang, 1983; Tsai *et al.*, 1985; Ren *et al.*, 1988) and potato (Kempe and Sequeira, 1983; McLaughlin and Sequeira, 1988; Quezado-Soares and Lopes, 1994). In general, mutants that did not produce bacteriocins were only partially effective at halting the development of disease symptoms and protection was heavily dependent on environmental conditions. In some cases, however, additional benefits were provided by the control agent as potatoes inoculated with an avirulent *R. solanacearum* strain showed reduced infection by root-knot nematodes (McLaughlin *et al.*, 1990). The ability of these strains, however, to control bacterial wilt disease in the longer term was limited (Tsai *et al.*, 1985; Grimault and Prior, 1994). In general, systemic spread was restricted and numbers declined as host defence mechanisms took effect (see Sequeira, 1982). In this regard, a strong correlation between EPS production, virulence and root invasiveness has been demonstrated (Cook and Sequeira, 1991; Denny and Baek, 1991; Trigalet-Demery *et al.* 1993; Kao and Sequeira, 1994). Certainly, weakly virulent mutants of *R. solanacearum* have been found to induce better protection than strictly non-virulent mutants (Hara and Ono, 1991). The reasons for this phenomenon are unclear but may be related to the fact that weakly virulent strains are able to penetrate and persist in the vascular tissue, thus inducing the host response, whereas EPS mutants are not.

Although promising under controlled conditions, none of the early attempts to develop a biocontrol agent against bacterial wilt have proven universally effective in the natural environment. In most cases field experiments were too limited and protection was not sufficient to warrant commercial development (Anuratha and Gnanamanickam, 1990; Tanaka *et al.*, 1990). Equally, in some cases protection failed because root colonization of the biocontrol agent was poor (Chen and Echandi, 1984) or was highly dependent on environmental greenhouse conditions (McLaughlin and Sequeira, 1988).

As with most phytopathogenic Gram-negative bacteria, *R. solanacearum* possesses a *hrp* gene cluster (25 kb) governing the hypersensitive response on incompatible hosts and production of disease symptoms in susceptible plants

(Boucher *et al.*, 1987, 1992; Marenda *et al.*, 1996). Mutants obtained by random insertion of the Tn5-transposon within the *hrp* gene cluster (Boucher *et al.*, 1985) lack the ability to cause the hypersensitive response and disease, but retain wild-type EPS production. These mutants retain the ability to colonize the host (Trigalet and Demery, 1986) and have been tested as putative control agents (Trigalet and Trigalet-Demery, 1990). Initially, the *hrp*⁻ mutants produced were not suitable for field use as the mutations were potentially unstable and transposition of Tn5 to other bacteria inhabiting the soil and rhizosphere remained a distinct possibility. To counteract these potentially deleterious effects, characterization of GMI1353, the most promising Tn5 mutant, located the mutation in the *hrp*O gene. This gene codes for a hydrophobic protein with potential membrane-spanning domains that produces an efflux of plant electrolytes when in association with tomato and tobacco leaves (Gough *et al.*, 1993). This gene had been previously found to be widely distributed amongst strains of *R. solanacearum* (Boucher *et al.*, 1988) and as a consequence new *hrp*O mutants were produced using a non-transposable omega intersposon encoding for kanamycin resistance (Ω-Km; Frey *et al.*, 1994). These mutants are non-reversible and the genetic element cannot be transferred to any other bacteria (Fellay *et al.*, 1987). Colonization and biological control assays were conducted using *hrp*O⁻ on tomato using challenge inoculation tests. In comparison to the pathogen, mutants were able to colonize taproot and collar tissues, but did not reach the fruits and fruit production was unaffected. Inoculation with avirulent mutants was correlated with a significant reduction in disease severity when the plants were subsequently challenged with a pathogenic strain. Despite the reduction in disease severity, virulent bacteria were still able to multiply in protected tissues, attaining population levels five to six orders of magnitude greater than those attained by the avirulent strain (Trigalet *et al.*, 1994).

These data indicate that protection was unlikely to have resulted from the general exclusion of the virulent strain. Initially, there were concerns that the avirulent strains may themselves be susceptible to antagonistic effects from pathogenic strains (Trigalet *et al.*, 1998). For example, out of the 24 virulent strains isolated on Martinique and on Guadeloupe (Prior and Steva, 1990), 13 isolates produced growth inhibition zones on agar medium against indicator strains. However, the impact of bacteriocin production on control assays appears minimal as the use of avirulent strains possessing broad-spectrum resistance to *R. solanacearum* bacteriocins provided no appreciable difference in protection when compared to sensitive strains (Trigalet *et al.*, 1998). Whilst it is possible to detect bacteriocin production *in planta* (Frey *et al.*, 1996), it would appear to play at best a minor role in the efficacy, or lack of it, achieved using avirulent mutants.

Microscopic studies of the colonization of the vascular tissues of pathogenic strains and *hrp*O⁻ mutants have been conducted in isolation or in challenge testing (Vasse *et al.*, 1995; Etchebar *et al.*, 1998). By using two different marker genes (β-galactosidase and β-glucuronidase), it was possible to monitor both wild-type and control agent simultaneously. In roots, both strains occurred together in infected root tips and in lateral root emergence sites (lateral root cracks), but these bacterial strains subsequently invaded separate xylem vessels in the root system. At the hypocotyl level, the competitors could coexist or exclusively colonize separate xylem vessels. The population of the pathogen was reduced in the presence of the avirulent strain, whilst in contrast the avirulent strain numbers increased in the presence of the pathogen, when compared to controls. It seems likely that control is achieved through competition for space or nutrients, as no evidence of direct antagonism was found. Certainly, these data confirm that prior root inoculation with *hrpO*⁻ mutants limits disease development upon subsequent inoculation with virulent *R. solanacearum*, even under conditions most favourable for disease.

Use of Avirulent *hrpO*⁻ Mutants to Control Bacterial Wilt in Potato

In our laboratory, research into the use of *hrpO*⁻ mutants has been underway for several years. The series of projects has mainly focused on potato production in the Kenyan Highlands (between 1200 and 2800 m) in which between 75,000 and 100,000 ha are given over to potato production. Socioeconomic pressures ensure near continuous potato cultivation, a practice favourable to the build-up of disease and pests within potato systems. In this regard losses due to bacterial wilt have been serious in recent years (Ajanga, 1993). Potato seed tubers harbouring latent infection of *R. solanacearum* are considered the main source of disease outbreaks, ensuring carryover of the disease into the subsequent growing season and to new regions (Nyangeri *et al.*, 1984).

At the outset, it was considered essential to have a detailed knowledge of the indigenous pathogenic population, in order to facilitate the selection of control agents and to predict the stability of any future control strategy. In the case of potato bacterial wilt, particularly in cooler environments, the disease is generally caused by the highly homogeneous infraspecific grouping, race 3/biovar 2a (Cook *et al.*, 1989; Smith *et al.*, 1995b). A total of 116 *R. solanacearum* isolates were obtained from diseased potato plants in a single sampling trip of the most significant potato-growing regions in Kenya. Subsequent characterization of these strains using established methods (Hayward, 1964; Lozano and Sequeira, 1970) showed that the vast majority of isolates belong to race 3/biovar 2a. Diversity within this relatively homogeneous group was further defined using a series of biochemical and molecular techniques (Smith *et al.*, 1995a,b, 1998a,b). In particular, the data obtained by pulsed gel electrophoresis analysis of macro-restriction digests enabled the circumscription of ten groups within the indigenous population and allowed the rationale selection of isolates for development as non-pathogenic biocontrol agents. All ten selected isolates were transformed using the Ω-Km intersposon described by Frey *et al.* (1994). Pathogenicity assessment of the resultant *hrpO*⁻ mutants confirmed their avirulence on potato. In keeping with the findings of colonization on tomato (Vasse *et al.*, 1995; Etchebar *et al.*, 1998), avirulent mutants were able to colonize potato, albeit in a limited capacity when compared with wild-type strains (Smith *et al.*, 1998a).

Experimental protocols were based on challenge inoculation of the biocontrol agent and wild-type organism (Smith *et al.*, 1988a). Preliminary screening of all 10 *hrpO*⁻ mutants showed no significant variation in effect. Accordingly, a detailed assessment was performed on *hrpO*⁻ mutant IMI370233 with its wild-type parent IMI360280; the representative of the dominant indigenous group. Three-week-old (5–15 cm height) potato plants were typically inoculated using a soil drench of liquid culture of the biocontrol agent, 20 ml broth culture (10^8 CFU ml^{-1}), and subsequently challenged 3 days later with a pathogenic isolate, 20 ml broth culture (10^6 or 10^7 CFU ml^{-1}). Disease data were recorded from the first day on which bacterial wilt symptoms were evident.

The assessment and the results, presented graphically, are shown in Figs 9.1 and 9.2. From the control data, it was evident that all plants inoculated with a higher concentration of virulent strain (10^7 CFU ml^{-1}) wilted after 25 days, whereas only 80% wilting was achieved after 35 days when the *hrpO*⁻ mutant was applied. At the lower concentration (10^6 CFU ml^{-1}), only 30% of plants wilted after 30 days and 10% when the *hrpO*⁻ mutant was applied. The data from the challenge experiments show that for both concentrations of

Fig. 9.1. Greenhouse experiments into the use of avirulent *R. solanacearum* as biological control agents. Healthy rows of plants are those inoculated with the agent or water controls, whilst dead and dying plants are infected with the pathogen alone.

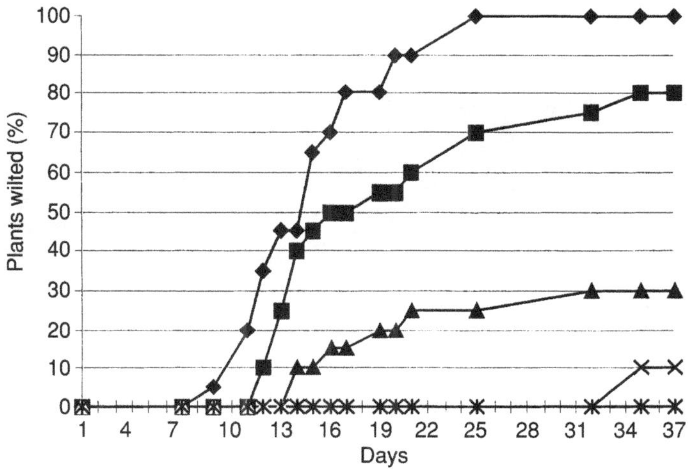

Fig. 9.2. Biocontrol assessment with *hrpO*⁻ mutant IMI370233 against wild-type parent, IMI360280. Key: (◆) virulent inoculant (IMI360280; 10^7 cells ml^{-1}); (■) virulent inoculant (IMI360280; 10^7 cells ml^{-1}) plus biocontrol (IMI370233; 10^8 cells ml^{-1}); (▲) virulent inoculant (10^6 cells ml^{-1}); (**X**) virulent inoculant (10^6 cells ml^{-1}) plus biocontrol (10^8 cells ml^{-1}); (*) control (no inoculation).

challenge inocula the biological control agent was able to provide protection in some plants. In cases where the control measures were ultimately overcome, the onset of disease was considerably delayed, when compared with the positive control.

In subsequently repeated assays, performed under controlled conditions, the application of avirulent strains has consistently decreased the severity of bacterial wilt disease on potato (data not shown). Additional assessments have also centred on developing formulations for the application of the *hrpO*⁻ mutant on to tubers prior to planting. Field experiments are currently planned in South Africa and Kenya using naturally infested soils.

Additionally, a separate line of research has focused on the biosafety implications of using a genetically modified control agent. Results presented in Fig. 9.3 demonstrate that the survival of the biological agent is similar to that of the wild-type pathogen, in that numbers begin to fall over time. This decline is undiminished in the presence of the host plant. Equally no evidence of exchange to the wild-type population, or any other bacteria, has been recorded to date (data not shown).

Conclusions and Future Prospects

In the last 20 years numerous studies have described the development of control strategies against bacterial wilt disease and its causative organism *R. solanacearum*. The potential of avirulent *hrpO*⁻ mutants of *R. solanacearum* as

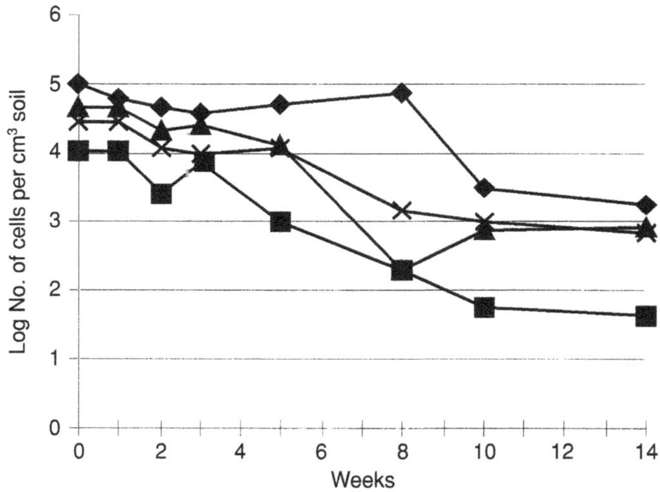

Fig. 9.3. Persistence of the *hrpO*⁻ mutant IMI370233 and a wild-type *R. solanacearum,* in the presence and absence of potato. Key: (◆) wild type and potato; (■) wild type – fallow; (▲) *hrpO*⁻ mutant and potato; (**X**) *hrpO*⁻ mutant – fallow.

biocontrol agents has been established previously on tomato (Trigalet and Trigalet-Demery, 1990; Frey *et al.*, 1994; Trigalet *et al.*, 1994, 1998). Comparable results have been achieved through our own research into the potato pathosystem. Taken together, it is clear that the use of *hrpO*⁻ mutants confers greater protection against bacterial wilt than previous studies involving the use of spontaneous avirulent mutants. Further, in both cases protection by *hrpO*⁻ mutants was effective when their numbers were a factor of ten greater than the pathogen in challenge tests. Under natural field conditions the resident virulent inoculum will probably be less concentrated and protection may be even more efficient. Certainly, results from the tomato research, which is more advanced than that on potato, suggest that this approach can succeed under field conditions (Frey *et al.*, 1994).

The mechanism of control remains unclear as yet, but induction of host defence mechanisms seems likely. Certainly, bacteriocin production *in planta* seems to be of marginal importance. As colonization of the avirulent control agent is generally limited to the root system and lower parts of the stem, direct antagonism against the pathogen seems unlikely. Similar results have been obtained from research into biological control strategies for grape crown gall, caused by a separate group of tumorogenic agrobacteria to the disease on stone fruits and other crops. This research demonstrated that the loss of bacteriocin production by the avirulent control strain did not affect disease control (Burr *et al.*, 1997; Burr and Otten, 1999). This does not rule out the use of avirulent bacteria capable of producing bacteriocins in the control of bacterial wilt. The construction of *hrpO*⁻ mutants able to produce broad-spectrum bacteriocins, thereby combining direct and indirect antagonisms, could offer considerable advantages. However, from our relatively limited knowledge of *R. solanacearum* bacteriocins it is evident that they are high molecular mass compounds (Cuppels *et al.*, 1978; Xie *et al.*, 1989), quite different from the small nucleotide bacteriocins produced by avirulent agrobacteria.

The modifications to the overall approach established by Trigalet and colleagues for the development of a biological control agent for potatoes in Kenya could be adapted to any crop/region. This approach offers the opportunity to develop local control agents thus negating the need to introduce non-indigenous organisms to control disease. Equally, as there exists strong homology among *hrp* gene clusters within Gram-negative bacterial phytopathogens (Boucher *et al.*, 1987, 1992; Arlat *et al.*, 1991; Marenda *et al.*, 1996), the possibility exists to extend this approach to include the biocontrol of other bacterial diseases of economic importance.

Acknowledgements

This chapter contains details of outputs from research projects funded by the UK Department for International Development (DFID) for the benefit of developing countries. The views expressed are not necessarily those of DFID. The authors gratefully acknowledge support from the DFID RNRKS Crop Protection Programme (Project Code R6629) and the work was carried out under licence within our laboratories and controlled growth rooms (MAFF PHL 36A/2774 and MAFF PHL 36A/2854). The authors would like to express their sincere thanks to Dr André Trigalet (INRA, Toulouse, France) for an exceptionally fruitful collaboration throughout the last 5 years. Equally, support from colleagues at NARL, KARI, Nairobi, Kenya, in particular Dr Gilbert Kibata and Mr Kinyua Murimi, is also gratefully acknowledged. Thanks are also expressed for the support received from Mr Nico Mienie and Ms Reinette Gouws, ARC, South Africa. Finally, a special thank you to Mrs Lisa Offord for all her efforts throughout the life of these projects.

References

Ajanga, S. (1993) Status of bacterial wilt of potato in Kenya. In: Hartman, G.L. and Hayward, A.C. (eds) *Bacterial Wilt*. Proceedings of an International Conference held in Kaohsiung, Taiwan, 28–30 October 1992 (ACIAR Proceedings No. 45). Australian Centre for International Agricultural Research, Canberra, pp. 338–340.

Anuratha, C. and Gnanamanickam, S. (1990) Biological control of bacterial wilt caused by *Pseudomonas solanacearum* in India with antagonistic bacteria. *Plant and Soil* 124, 109–116.

Arlat, M., Gough, C., Barber, C., Boucher, C. and Daniels, M. (1991) *Xanthomonas campestris* contains a cluster of *hrp* genes related to the larger *hrp* cluster of *Pseudomonas solanacearum*. *Molecular and Plant–Microbe Interactions* 4, 593–601.

Aspiras, R.B. and de la Cruz, A.R. (1985) Potential biological control of bacterial wilt in tomato and potato with *Bacillus polymyxa* FU6 and *Pseudomonas fluorescens*. In: Persley, G.J. (ed.) *Bacterial Wilt Disease in Asia and the South Pacific* (ACIAR Proceedings No. 13). Australian Centre for International Agricultural Research, Canberra, pp. 89–92.

Autrique, A. and Potts, M.J. (1987) The influence of mixed cropping on the control of potato bacterial wilt. *Annals of Applied Biology* 111, 125–133.

Averre, C. and Kelman, A. (1964) Severity of bacterial wilt as influenced by ratio of avirulent cells of *Pseudomonas solanacearum* in inoculum. *Phytopathology* 54, 779–783.

Baker, K. (1987) Evolving concepts of biological control of plant pathogens. *Annual Review of Phytopathology* 25, 67–85.

Boucher, C., Barberis, P., Trigalet, A. and Demery, D. (1985) Transposon mutagenesis of *Pseudomonas solanacearum*: isolation of Tn5-induced avirulent mutants. *Journal of General Microbiology* 131, 2449–2457.

Boucher, C., Van Gijsegern, F., Arlat, M. and Zischek, C. (1987) *Pseudomonas solanacearum* genes controlling both pathogenicity and hypersensitivity on tobacco are clustered. *Journal of Bacteriology* 169, 5626–5632.

Boucher, C., Barberis, P. and Arlat, M. (1988) Acridine Orange selects for deletion of *hrp* genes in all races of *Pseudomonas solanacearum*. *Molecular Plant–Microbe Interactions* 1, 282–288.

Boucher, C., Gough, C. and Arlat, M. (1992) Molecular genetics of pathogenicity determinants of *Pseudomonas solanacearum* with special emphasis on *hrp* genes. *Annual Review of Phytopathology* 30, 443–461.

Burr, T.J. and Otten, L. (1999) Crown gall of grape: biology and disease management. *Annual Review of Phytopathology* 37, 53–80.

Burr, T.J., Reid, C.L., Tagliati, E., Bazzi, C. and Süle, S. (1997) Biological control of grape crown gall by strain F2/5 is not associated with agrocin production or competition from attachment sites on grape cells. *Phytopathology* 87, 706–711.

Campbell, R. (1989) *Biological Control of Microbial Plant Pathogens*. Cambridge University Press, Cambridge.

Chen, W. and Echandi, E. (1984) Effects of avirulent bacteriocin-producing strains of *Pseudomonas solanacearum* on the control of bacterial wilt of tobacco. *Plant Pathology* 33, 245–253.

Chen, W., Echandi, E. and Spurr, H.W. (1982) Protection of tobacco plants from bacterial wilt with avirulent bacteriocin-producing strains of *Pseudomonas solanacearum*. In: Lozano, J.C. (ed.) *Proceedings of the 5th International Conference on Plant Pathogenic Bacteria, August 16–23 1981*. Centro Internacional de Agricultura Tropical, Cali, Colombia, pp. 482–492.

Chyi, Y.S., Jorgenson, R.A., Goldstein, D., Tanksley, S.D. and Loaiza-Figueroa, F. (1986) Locations and stability of *Agrobacterium*-mediated T-DNA insertions in the *Lycopersicon* genome. *Molecular and General Genetics* 204, 64–69.

Ciampi, L., Burzio, L.O. and Burzio, L.A. (1997) Carriers for *Pseudomonas fluorescens*, antagonistic to *Pseudomonas* (*Ralstonia*) *solanacearum*, causal agent of bacterial wilt. *Fitopatologia* 32, 64–70.

Ciampi-Panno, L., Fernandez, C., Bustamante, P., Andrade, N., Ojeda, S. and Contreras, A. (1989) Biological control of bacterial wilt of potatoes caused by *Pseudomonas solanacearum*. *American Potato Journal* 66, 315–332.

Cook, D. and Sequeira, L. (1991) Genetic and biochemical characterization of a *Pseudomonas* cluster required for extracellular polysaccharide production and virulence. *Journal of Bacteriology* 173, 1654–1662.

Cook, D., Elizabeth, B. and Sequeira, L. (1989) Genetic diversity of *Pseudomonas solanacearum*: detection of restriction fragment length polymorphisms with DNA probes that specify virulence and hypersensitive response. *Molecular Plant–Microbe Interactions* 2, 113 –121.

Cuppels, A., Hanson, R. and Kelman, A. (1978) Isolation and characterization of a bacteriocin produced by *P. solanacearum*. *Journal of General Microbiology* 109, 295–303.

Da Silveira, E.B., Mariano, R.L.R., Michereff, S.J. and Menezes, M. (1995) Antagonism of *Bacillus* spp. against *Pseudomonas solanacearum* and effect on tomato seedling growth. *Fitopatologia Brasileira* 20, 605–612.

De Cleene, M. (1979) Crown gall: economic importance and control. *Zentralblatt fur Bacteriologie and Parasiten Kunde Infektionskrankheiten und Hygiene, II Abteilung* 134, 551–554.

De Cleene, M. and De Ley, J. (1976) The host range of crown gall. *Botanical Review* 42, 389–466.

Denny, T. and Beak, S. (1991) Genetic evidence that extracellular polysaccharide is avirulence factor of *Pseudomonas solanacearum*. *Molecular Plant–Microbe Interactions* 4, 198–206.

Ellis, J.G. and Murphy, P.J. (1981) Four new opines from crown gall tumours – their detection and properties. *Molecular and General Genetics* 181, 36–43.

El Shanshoury, A.R., Abu El Sououd, S.M., Awadalla, O.A. and El-Bandy, N.B. (1996) Effects of *Streptomyces corchorusii*, *Streptomyces mutabilis*, pendimethalin, and metribuzin on the control of bacterial and *Fusarium* wilt of tomato. *Canadian Journal of Botany* 74, 1016–1022.

Etchebar, C., Trigalet-Demery, D., van Gijsegem, F., Vasse, J. and Trigalet, A. (1998) Xylem colonization by an *HrcV*$^-$ mutant of *Ralstonia solanacearum* is a key factor for the efficient biological control of tomato bacterial wilt. *Molecular Plant–Microbe Interactions* 11, 869–877.

Fellay, R., Frey, J. and Kirsch H. (1987) Intersposon mutagenesis of soil and water bacteria: a family of DNA fragments designed for *in vitro* insertional mutagenesis of Gram-negative bacteria. *Gene* 52, 147–154.

Fravel, D. (1988) Role of antibiosis in the biocontrol of plant diseases. *Annual Review of Phytopathology* 26, 75–91.

French, E.R. (1994) Strategies for integrated control of bacterial wilt of potatoes. In: Hayward, A.C. and Hartman, G.L. (eds) *Bacterial Wilt: the Disease and its Causative Agent,* Pseudomonas solanacearum. CAB International, Wallingford, UK, pp. 199–207.

Frey, P., Prior, P., Marie, C., Kotoujansky, A., Trigalet-Demery, D. and Trigalet, A. (1994) *Hrp* mutants of *Pseudomonas solanacearum* as potential biocontrol agents of tomato bacterial wilt. *Applied and Environmental Microbiology* 60, 3175–3181.

Frey, P., Smith, J.J., Albar, L., Prior, P., Saddler, G.S., Trigalet-Demery, D. and Trigalet, A. (1996) Bacteriocin typing of *Burkholderia* (*Pseudomonas*) *solanacearum* race 1 of the French West Indies and correlation of genomic variation of the pathogen. *Applied and Environmental Microbiology* 62, 473–479.

Furaya, N., Kushima, Y., Tsuchiya, K, Matsuyama, N. and Wakimoto, S. (1991a) Protection of tomato seedlings by pretreatment with *Pseudomonas glumae* from infection with *Pseudomonas solanacearum* and its mechanisms. *Annals of the Phytopathological Society of Japan* 57, 363–370.

Furaya, N., Okamoto, T., Kori, Y., Matsuyama, N. and Wakimoto, S. (1991b) Control of bacterial seedling rot of rice by avirulent strains of *Pseudomonas glumae*. *Annals of the Phytopathological Society of Japan* 57, 371–376.

Furaya, N., Yamasaki, S., Nishioka, M., Shiraishi, I., Iiyama, K. and Matsuyama, N. (1997) Antimicrobial activities of pseudomonads against plant pathogenic organisms and efficacy of *Pseudomonas aeruginosa* ATCC7700 against bacterial wilt of tomato. *Annals of the Phytopathological Society of Japan* 63, 417–424.

Gallardo, P., Ciampi-Panno, L. and Guichaquelem, V. (1989) Inhibition *in vitro* of *Pseudomonas solanacearum* E.F. Smith by using the antagonist BC8 strain of *Pseudomonas fluorescens*. *Revista de Microbiologica* 20, 27–33.

Gough, C.L., Genin, S., Lopes, V. and Boucher, C.A. (1993) Homology between the *HrpO* protein of *Pseudomonas solanacearum* and bacterial proteins implicated in a signal peptide-independent secretion mechanism. *Molecular and General Genetics* 239, 378–392.

Grimault, V. and Prior, P. (1994) Invasiveness of avirulent strains of *Pseudomonas solanacearum* in tomato cultivars, resistant or susceptible to bacterial wilt. *Journal of Phytopathology* 141, 195–201.

Hara, H. and Ono, K. (1991) Effect of weakly-virulent bacteriocin-producing strain of *Pseudomonas solanacearum* on the protection of tobacco plant from bacteria wilt. *Annals of the Phytopathological Society of Japan* 57, 24–31.

Hartati, S.Y., Boa, E.R., Supriadi, Adhi, E.M. and Karyani, N. (1991) Biological control of Sumatra disease bacterium (*Pseudomonas syzygii*) with its avirulent strains and *Pseudomonas solanacearum*. *Industrial Crops Research Journal* 4, 1–4.

Hayward, A.C. (1964) Characteristics of *Pseudomonas solanacearum*. *Journal of Applied Bacteriology* 27, 265–277.

Hayward, A.C. (1991) Biology and epidemiology of bacterial wilt caused by *Pseudomonas solanacearum*. *Annual Review of Phytopathology* 29, 65–87.

Jones, D., Ryder, M., Clare, B., Farrand, S. and Kerr, A. (1988) Construction of a Tra$^-$ deletion mutant of pAgK84 to safeguard the biological control of crown gall. *Molecular and General Genetics* 212, 207–214.

Kao, C.C. and Sequeira, L. (1994) The function and regulation of genes required for extracellular polysaccharide synthesis and virulence in *Pseudomonas solanacearum*. In: Kado, C.I. and Crosa, J.H. (eds) *Molecular Mechanisms of Bacterial Virulence*. Kluwer Academic Publishers, Dordrecht, pp. 93–108.

Kelman, A. (1954) The relationship of pathogenicity in *Pseudomonas solanacearum* to colony appearance on a tetrazolium medium. *Phytopathology* 44, 693–695.

Kelman, A. and Hruschka, J. (1973) The role of motility and aerotaxis in the selective increase of avirulent bacteria in still broth cultures of *Pseudomonas solanacearum*. *Journal of General Microbiology* 76, 177–188.

Kempe, J. and Sequeira, L. (1983) Biological control of bacterial wilt of potatoes: attempts to induce resistance by treating tubers with bacteria. *Plant Disease* 67, 499–503.

Kerr, A. (1980) Biological control of crown gall through production of Agrocin 84. *Plant Disease* 64, 28–30.

Kerr, A. (1991) The genus *Agrobacterium*. In: Balows, A., Trüper, H.G., Dworkin, M., Harder, W. and Schleifer, K.-H. (eds) *The Prokaryotes*, 2nd edn. *A Handbook on the Biology of Bacteria: Ecophysiology, Isolation, Identification, Applications*. Springer-Verlag, Berlin, pp. 2214–2235.

Kerr, A. and Htay, K. (1974) Biological control of crown gall through bacteriocin production. *Physiological Plant Pathology* 4, 37–44.

Kerr, A. and Tate, M.E. (1984) Agrocins and the biological control of crown gall. *Microbiological Sciences* 1, 1–4.

Laferriere, L.T., Helgeson, J.P. and Allen, C. (1998) *Solanum tuberosum – S. commersonii* somatic hybrids are resistant to brown rot caused by *Ralstonia solanacearum*. In: Prior, P., Allen, C. and Elphinstone, J. (eds) *Bacterial Wilt Disease: Molecular and Ecological Aspects*. Springer-Verlag, Berlin, pp. 316–320.

Leong, J. (1986) Siderophores: their biochemistry and possible role in the biocontrol of plant pathogens. *Annual Review of Phytopathology* 24, 187–209.

Liu, F.Q., Wang, J.S. and Wang, J.S. (1998) A preliminary study on controlling rice bacterial leaf blight with a virulent gene deleted strain DU728. *Chinese Journal of Biological Control* 14, 115–118.

Liu, Q.G., Li, Z., Tang, Z. and Zeng, X.M. (1999) Biological control of tobacco bacterial wilt by soil amendments and antagonistic bacteria. *Chinese Journal of Biological Control* 15, 94–95.

Lozano, J.C. and Sequeira, L. (1970) Differentiation of races of *Pseudomonas solanacearum* by a leaf infiltration technique. *Phytopathology* 60, 833–838.

Luo, K. and Wang, Z. (1983) Study of bacterial wilt (*Pseudomonas solanacearum*) controlled by antagonistic and avirulent *P. solanacearum*. *Acta Phytopathologica Sinica* 13, 51–56.

Macrae, S., Thomson, J.A. and van Staden, J. (1988) Colonization of tomato plants by two agrocin-producing strains of *Agrobacterium tumefaciens*. *Applied and Environmental Microbiology* 54, 3133–3137.

Marenda, M., Van Gijsegem, F., Arlat, M., Zichek, C., Barberis, P., Camus, J.C., Castello, P. and Boucher, C.A. (1996) Genetic and molecular dissection of the *hrp* region of *Ralstonia* (*Pseudomonas*) *solanacearum*. In: Stacey, G., Mullin, B. and Gresshoff, P.M. (eds) *Biology of Plant-Microbe Interactions*. International Society for Molecular–Plant–Microbe Interactions, St Paul, Minnesota, pp. 165–172.

McLaughlin, R. and Sequeira, L. (1988) Evaluation of an avirulent strain of *Pseudomonas solanacearum* for biological control of bacterial wilt of potato. *American Potato Journal* 65, 255–268.

McLaughlin, R., Sequeira, L. and Weingartner, D.P. (1990) Biocontrol of bacterial wilt of potato with an avirulent strain of *Pseudomonas solanacearum*: interactions with root-knot nematodes. *American Potato Journal* 67, 93–107.

Moore, L.W. and Warren, G. (1979) *Agrobacterium radiobacter* strain 84 and biological control of crown gall. *Annual Review of Phytopathology* 17, 163–179.

Mulya, K., Watanabe, M., Goto, M., Takikawa, Y. and Tsuyumu, S. (1996) Suppression of bacterial wilt disease of tomato by root-dipping *with Pseudomonas fluorescens* PfG32 – the role of antibiotic substances and siderophore production. *Annals of the Phytopathological Society of Japan* 62, 134–140.

Murphy, P.J. and Roberts, W.P. (1979) A basis for agrocin 84 and sensitivity in *Agrobacterium radiobacter*. *Journal of General Microbiology* 114, 207–213.

Nesmith, W. and Jenkins, W. (1985) Influence on antagonists and controlled matrix potential on the survival of *Pseudomonas solanacearum* in four Carolina soils. *Phytopathology* 75, 1182–1187.

New, P.B. and Kerr, A. (1972) Biological control of crown gall: field measurements and glasshouse experiments. *Journal of Applied Bacteriology* 35, 279–287.

Nyangeri, J.B., Gathuru, E.M. and Mukunya, D.M. (1984) Effect of latent infection on the spread of bacterial wilt of potatoes in Kenya. *Tropical Pest Management* 30, 163–165.

Panagopoulos, C.G., Psallidas, P.G. and Alivizatos, A.S. (1979) Evidence of a breakdown in the effectiveness of biological control, of crown gall. In: Schippers, B. and Gams, W. (eds) *Soil Borne Pathogens*. Academic Press, London, pp. 569–578.

Phae, C.G., Shoda, M., Kita, N., Nakano, M. and Ushiyama, K. (1992) Biological control of crown and root rot and bacterial wilt of tomato by *Bacillus subtilis* NB22. *Annals of the Phytopathological Society of Japan* 58, 329–339.

Prior, P. and Steva, H. (1990) Characteristic strains of *Pseudomonas solanacearum* from the French West Indies. *Plant Disease* 74, 13–17.

Prior, P., Grimault, V. and Schmidt, J. (1994) Resistance to bacterial wilt (*Pseudomonas solanacearum*) in tomato: present status and prospects. In: Hayward, A.C. and Hartman, G.L. (eds) *Bacterial Wilt: the Disease and its Causative Agent,* Pseudomonas solanacearum. CAB International, Wallingford, UK, pp. 209–223.

Quezado-Soares, A.M. and Lopes, C.A. (1994) Lack of biological control of potato bacterial wilt by avirulent mutants of *Pseudomonas solanacearum* and by fluorescent pseudomonads. *Bacterial Wilt Newsletter* No. 11, 3–5.

Ren, X.Z., Shen, D.L. and Fang, Z.D. (1988) Preliminary studies on control of bacterial wilt of tomato (*Pseudomonas solanacearum*) by avirulent bacteriocin-producing strain MA-7. *Chinese Journal of Biological Control* 4, 62–64.

Ryder, M.H. and Jones, D.A. (1990) Biological control of crown gall. In: Hornby, D. (ed.) *Biological Control of Soil-borne Plant Pathogens*. CAB International, Wallingford, UK, pp. 45–63.

Schell, J., Van Montagu, M., de Beuckeleer, M., de Block, M., Depicker, A., de Wilde, M., Engler, G., Genetello, C., Hernalsteens, J.P., Holsters, M., Seurinck, J., Silva, B., Van Vliet, F. and Villarroel, R. (1979) Interactions and DNA transfer between *Agrobacterium tumefaciens*, the Ti-plasmid and the plant host. *Proceedings of the Royal Society of London B* 204, 251–266.

Schmidt, D. (1988) Prevention of bacterial wilt of grasses by phylloplane bacteria. *Journal of Phytopathology* 122, 253–260.

Sequeira, L. (1982) Determinants of plant response to bacterial infection. In: Wood, R.K.S. (ed.) *Active Defense Mechanisms in Plants*. Plenum Press, New York, pp. 85–102.

Sequeira, L. (1994) Epilogue: life with a 'mutable and treacherous tribe'. In: Hayward, A.C. and Hartman, G.L. (eds) *Bacterial Wilt: the Disease and its Causative Agent,* Pseudomonas solanacearum. CAB International, Wallingford, UK, pp. 235–247.

Sequeira, L. and Hill, L. (1974) Induced resistance in tobacco leaves: the growth of *Pseudomonas solanacearum* in protected tissues. *Physiology and Plant Pathology* 4, 447–455.

Sequeira, L., Gaard, G. and de Zoeten, G. (1977) Attachment of bacteria to host cell walls: its relation to mechanisms of induced resistance. *Physiology and Plant Pathology* 10, 43–50.

Shekhawat, G.S., Singh, R. and Chakrabarti, S.K. (1993) Possibility of biological management of potato bacterial wilt with microbial antagonists and latent potato viruses. *Journal of the Indian Potato Association* 20, 219–222.

Smith, J.J., Offord, L.C., Holderness, M. and Saddler, G.S. (1995a) Pulsed-field gel electrophoresis analysis of *Pseudomonas solanacearum*. *EPPO Bulletin* 25, 163–167.

Smith, J.J., Offord, L.C., Holderness, M. and Saddler, G.S. (1995b) Genetic diversity of *Burkholderia solanacearum* (syn. *Pseudomonas solanacearum*) race 3 in Kenya. *Applied and Environmental Microbiology* 61, 4263 – 4268.

Smith, J.J., Offord, L.C., Kibata, G.N., Murimi, Z.K., Trigalet, A. and Saddler, G.S. (1998a) The development of a biological control against *Ralstonia solanacearum* race 3 in Kenya. In: Prior, P., Allen, C. and Elphinstone, J. (eds) *Bacterial Wilt Disease: Molecular and Ecological Aspects*. Springer-Verlag, Berlin, pp. 337–342.

Smith, J.J., Kibata, G.N., Murimi, Z.K., Lum, K.Y., Fernandez-Northcote, E., Offord, L.C. and Saddler, G.S. (1998b) Biogeographic studies of *Ralstonia solanacearum* race 1 and 3 by genomic fingerprinting. In: Prior, P., Allen, C. and Elphinstone, J. (eds) *Bacterial Wilt Disease: Molecular and Ecological Aspects*. Springer-Verlag, Berlin, pp. 50–55.

Surico, G., Comai, L. and Kosuge, T. (1984) Pathogenicity of strains of *Pseudomonas syringae* pv. *savastanoi* and their indoleacetic acid-deficient mutants on olive and oleander. *Phytopathology* 74, 490–493.

Tanaka, H., Negishi, H. and Maeda, H. (1990) Control of tobacco bacterial wilt by an avirulent strain of *Pseudomonas solanacearum* M4S and its bacteriophage. *Annals of the Phytopathological Society of Japan* 56, 243–246.

Terblanche , J. and de Villiers, D.A. (1998) The suppression of *Ralstonia solanacearum* by marigolds. In: Prior, P., Allen, C. and Elphinstone, J. (eds) *Bacterial Wilt Disease: Molecular and Ecological Aspects*. Springer-Verlag, Berlin, pp. 325–331.

Tharaud, M., Laurent, J., Faize, M. and Paulin, J.P. (1997) Fire blight protection with avirulent mutants of *Erwinia amylovora*. *Microbiology* 143, 625–632.

Trigalet, A. and Demery, D. (1986) Invasiveness in tomato plants of Tn5-induced avirulent mutants of *Pseudomonas solanacearum*. *Physiology and Molecular Plant Pathology* 28, 423–430.

Trigalet, A. and Trigalet-Demery, D. (1990) Use of avirulent mutants of *Pseudomonas solanacearum* for the biological control of bacterial wilt of tomato plants. *Physiology and Molecular Plant Pathology* 36, 27–38.

Trigalet, A., Frey, P. and Trigalet-Demery, D. (1994) Biological control of bacterial wilt caused by *Pseudomonas solanacearum*: state of the art and understanding. In: Hayward, A.C. and Hartman, G.L. (eds) *Bacterial Wilt: the Disease and its Causative Agent,* Pseudomonas solanacearum. CAB International, Wallingford, UK, pp. 225–233.

Trigalet, A., Trigalet-Demery, D. and Prior, P. (1998) Elements of biocontrol of tomato bacterial wilt. In: Prior, P., Allen, C. and Elphinstone, J. (eds) *Bacterial Wilt Disease: Molecular and Ecological Aspects*. Springer-Verlag, Berlin, pp. 332–336.

Trigalet-Demery, D., Montrozier, H., Orgambide, G., Patry, V., Adam, O., Navarro, V., Cotelle, V. and Trigalet, A. (1993) Exopolysaccharides of *Pseudomonas solanacearum*: relation to virulence. In: Hartman, G.L. and Hayward, A.C. (eds) *Bacterial Wilt*. Proceedings of an International Conference held in Kaohsiung, Taiwan, 28–30 October 1992 (ACIAR Proceedings No. 45). Australian Centre for International Agricultural Research, Canberra, pp. 312–315.

Tsai, J.-W., Hu, S.-T. and Chen, L.-C. (1985) Bacteriocin-producing strains of *Pseudomonas solanacearum* and their effect on development of bacterial wilt of tomato. *Plant Protection Bulletin (Taiwan)* 27, 267–278.

Varvaro, L. and Martella, L. (1993) Virulent and avirulent isolates of *Pseudomonas syringae* subsp. *savastanoi* as colonisers of olive leaves: evaluation of possible biological control of the olive knot pathogen. *EPPO Bulletin* 23, 423–427.

Vasse, J., Frey, P. and Trigalet, A. (1995) Microscopic studies of the intercellular infection and protoxylem invasion of tomato roots by *Pseudomonas solanacearum*. *Molecular Plant–Microbe Interactions* 8, 241–251.

Wakimoto, S. (1987) Biological control of bacterial wilt of tomato by non-pathogenic strains of *Pseudomonas glumae*. *Korean Journal of Plant Pathology* 3, 300–303.

Weller, D. (1988) Biological control of soilborne plant pathogens in the rhizosphere with bacteria. *Annual Review of Phytopathology* 26, 379–407.

Yao, G., Zhang, F. and Li, Z. (1994) Control of bacterial wilt with soil amendment. *Chinese Journal of Biological Control* 10, 106–109.

Xie, D., Fan, Y. and He, L. (1989) Purification and some properties of bacteriocin produced by *Pseudomonas solanacearum*. *Acta Microbiologica Sinica* 29, 284–292.

Cross-protection: Interactions Between Strains Exploited to Control Plant Virus Diseases

10

Hervé Lecoq[1] and Benjamin Raccah[2]

[1]*INRA, Station de Pathologie Végétale, Domaine Saint Maurice, BP 94, 84143 Montfavet Cédex, France;*
[2]*ARO, Department of Virology, The Volcani Center, PO Box 6, 50250 Bet Dagan, Israel*

Introduction

Virus diseases have a major impact on the productivity of cultivated plants, especially vegetable and fruit crops, in very diverse agricultural ecosystems. Suppression of their impact is almost impossible because of the systemic and non-curable nature of virus infections in the field. In many important crops virus-resistant or -tolerant cultivars of good agronomic quality are lacking. When available, resistances may break down due to the variability and the diversity of plant viruses. In this context, cross-protection appears to be a potentially useful component of an integrated virus management strategy.

The principle of cross-protection was discovered by McKinney in 1929, when he observed that a tobacco plant systemically infected by a green strain of tobacco mosaic virus (TMV, *Tobamovirus*) was protected from infection by another strain of this virus inducing a yellow mosaic. This phenomenon was subsequently observed for many other plant viruses and considered to be of interest for protecting plants in the field. The principle was that an 'infection of a plant with a strain of virus causing only mild disease symptoms may protect it from infection with severe strains' (Matthews, 1991). Gonsalves and Garnsey (1989) introduced a more applied definition of cross-protection as 'the use of a mild virus isolate to protect plants against economic damage caused by infection with a severe challenge strain of the same virus'. In this case cross-protection is mainly evaluated by the farmer's economical benefits rather than defined by a specific mode of action or by a type of interaction between virus strains.

The terminology commonly used refers to the virus strain which induces cross-protection, as the 'protective strain' and to the strain which is used to evaluate cross-protection efficiency as the 'challenging strain'. Protective strains may also be referred to as 'mild', 'attenuated' or 'hypovirulent' strains, while challenging strains may be referred to as 'severe' strains. In laboratory experiments challenging strains are often chosen because they induce either severe or easily recognizable (i.e. white or yellow) symptoms; they can be inoculated mechanically, by grafting or by their natural vectors. In nature, challenging strains are the indigenous severe strains prevailing in a region.

When field application is considered, the protective strain is generally an isolate which induces mild symptoms and does not affect the marketable yield of the crop. In this chapter, protective strains will be referred to as 'mild' strains or isolates. Cross-protection using mild strains has been investigated or commercially used against viruses belonging to various genera (*Badnavirus, Closterovirus, Cucumovirus, Nepovirus, Potyvirus* and *Tobamovirus*), mostly infecting vegetable or fruit crops, probably because it is on these hosts that viruses have the most deleterious effects on the cash value of the crop.

Mechanisms of Cross-protection

Although cross-protection has been used in practice in a number of crops to control different viruses, still little is known about its mode(s) of action. Early theories, developed with only limited experimental support, involved depletion of specific host components required for virus multiplication, occupation of replication sites or synthesis of inhibitors by the cross-protected plants (Palutaikis and Zaitlin, 1984; Fulton, 1986; Sherwood, 1987; Urban *et al.*, 1990). Two other hypotheses were subsequently developed. One suggested that the nucleic acid of the challenging strain could be sequestrated by the coat protein of the protecting strain, preventing subsequent translation and replication of the challenging virus RNA (de Zoeten and Fulton, 1975; Sherwood, 1987; Urban *et al.*, 1990). This was supported by observations using tobamoviruses for which a diversity of strains was available including coat protein-deficient strains. First, cross-protection was overcome by using viral RNA as challenge inoculum instead of virus particles. This suggested that the uncoating of the challenging strain was an important step in cross-protection (Sherwood and Fulton, 1982). Secondly, a strain of TMV not producing coat protein did not confer protection against a regular TMV isolate (Sherwood, 1987). However, these observations were not confirmed using other hosts or strains, leaving some uncertainty on the exact role played by the coat protein in the cross-protection mechanism (Gerber and Sarkar, 1989). The second theory was proposed to include cross-protection phenomenon between naked RNAs (such as viroids or some satellite RNAs) that do not code for their own coat protein. It involved hybridization of (+) and (–) RNA strands of protective and challenging strains preventing replication and/or translation of the challenging strain (Palutaikis and Zaitlin, 1984).

Recently, the development of a potato virus X (PVX, *Potexvirus*) vector that enables the expression of native or mutated TMV coat proteins in *Nicotiana benthamiana* has allowed a complete re-evaluation of the role of this protein in cross-protection (Culvert, 1996; Lu *et al.*, 1998). In a first series of experiments it was demonstrated that TMV coat protein expressed by the PVX vector provided a similar level of protection to TMV (Culvert, 1996). Challenging with increased levels of inoculum, with TMV RNA instead of virions or with a TMV construct in which the coat protein gene had been replaced by that of a distantly related *Tobamovirus*, overcame the protection conferred by the PVX vector. Similar treatments would have also overcome the protection conferred by TMV itself (Culvert, 1996). To further investigate the role of TMV coat protein on cross-protection, the effect of different mutations applied to the coat protein has been evaluated (Lu *et al.*, 1998). The protection was lost when mutations prevented formation of virions and of coat protein helical aggregates or when the coat protein was unable to bind viral RNA. But the cross-protection was maintained when mutations prevented the formation of virions but not of helical aggregates. Finally, mutations which enhanced subunit interactions and favoured coat protein helical aggregation increased the protective effect (Lu *et al.*, 1998). So the capacity of the TMV coat protein to bind to viral RNA and to self-associate in helical structures appears to be essential to confer protection. This confirms the possibility that RNA sequestration of the challenging strain by the protective strain coat protein is a mode of action of cross-protection for *Tobamoviruses* (Lu *et al.*, 1998).

Meanwhile, gene silencing, a phenomenon involving sequence-specific RNA degradation in infected cells, has been shown to be operating in cross-protection for viruses belonging to different genera (Ratcliff *et al.*, 1997, 1999). The hypothesized mechanism involves antisense RNAs as determinants of cross-protection that would bind or degrade challenging strain RNAs. These antisense RNAs could be synthesized by the viral replicase as negative strand replication intermediates or be produced by a plant-encoded RNA-dependent RNA polymerase using viral RNA as a template (Ratcliff *et al.*, 1999).

However, it is quite possible that different mechanisms may be involved in different virus–host combinations or at different stages of the virus interactions. The observation that some level of protection against TMV still occurred when the PVX vector contained the TMV coat protein gene sequence in a non-expressed form (Culvert, 1996) supports the idea that several distinct mechanisms might contribute simultaneously or sequentially to cross-protection.

Interestingly, it appears now that similar mechanisms might be involved in classical cross-protection and in transgenic plants expressing pathogen-derived resistance. Indeed, in the case of *Tobamoviruses*, coat protein expression is necessary to observe resistance, and it seems also to be necessary for cross-protection (Lu *et al.*, 1998). For other viruses, one of the most frequent mechanisms involved in transgenic plant resistance is post-transcriptional gene silencing; gene silencing is now also considered to be a major component of cross-protection (Ratcliff *et al.*, 1999).

Mild Strain Selection

An 'ideal' mild isolate to be used in the field for cross-protection should possess the following characteristics (Lecoq, 1998) :

1. It should induce mild symptoms (not altering the marketable yield and the quality) on the target crop. It should also be mild in cultivated hosts which are not targets for the cross-protection.
2. It should be fully systemic since cross-protection effectiveness depends on the presence of the mild strain in all tissues to be protected from severe strain inoculation.
3. It should be genetically stable and not revert to more severe forms.
4. It should be protective against the widest possible range of severe isolates.
5. The protective inoculum should be easy to produce, easy to check for purity, stable in storage and readily inoculated.
6. It should not be easily disseminated by vectors, in order to limit any non-intentional spread to other crops or fields.

Mild strains have been obtained in various empirical ways (Table 10.1). Some were selected as naturally occurring variants from plants observed in the field with mild symptoms amongst plants with severe symptoms (Muller, 1980). Other mild strains have been obtained in the laboratory, either after single

Table 10.1. Methods to obtain mild strains used for cross-protection.

Field or greenhouse spontaneous variants		
Citrus tristeza virus	*Closterovirus*	Muller and Costa, 1987
Cocoa swollen shoot virus	*Badnavirus*	Hughes and Ollenu, 1994
Tobacco mosaic virus	*Tobamovirus*	Fletcher, 1978
Zucchini yellow mosaic virus	*Potyvirus*	Lecoq *et al.*, 1991
Heat treatment		
Tobacco mosaic virus	*Tobamovirus*	Oshima, 1975
Low temperature treatment		
Soybean mosaic virus	*Potyvirus*	Kosaka and Fukunishi, 1993
Random mutagenesis (nitrous acid)		
Papaya ringspot virus	*Potyvirus*	Yeh and Gonsalves, 1984
Tobacco mosaic virus	*Tobamovirus*	Rast, 1972
Site-directed mutagenesis (on a cDNA)		
Zucchini yellow mosaic virus	*Potyvirus*	Gal-On and Raccah, 2000

local lesion isolation from samples with severe symptoms, or from plants inoculated by severe isolates, but which developed spontaneously axillary branches with mild symptoms (Lecoq *et al.*, 1991). Heat or cold treatments may also yield mild isolates (Oshima, 1975; Kosaka and Fukunishi, 1993), and mild variants were obtained in the laboratory after mutagenesis treatment (generally with nitrous acid) followed by single local lesion selection (Rast, 1972; Yeh and Gonsalves, 1984). More recently, the knowledge gained on the molecular determinants of virus pathogenicity allowed the development of a more rational way to produce mild strains. Mild variants may be obtained through site-directed mutagenesis of infectious cDNA clones at locations known to be involved in symptom mildness (Gal-On and Raccah, 2000).

At this stage, special attention should be paid to ensure that the protective isolate does not contain a minor contaminant of a severe isolate or of another virus. Various control procedures may be used, such as single local lesion transfers, differential hosts analysis, serological assays and electron microscopic examination. When the molecular determinants of severity/mildness are known, it is conceivable that specific probes or primers to recognize severe from mild strains may be developed.

A Case Story: Cross-protection Against Zucchini Yellow Mosaic Virus (ZYMV)

The agronomic context

Zucchini yellow mosaic virus (ZYMV, *Potyvirus*) is one of the major pathogens of cucurbit crops worldwide. Discovered in southern Europe in the mid-1970s, it spread within a decade to the major cucurbit production areas in the world (Lisa *et al.*, 1981; Desbiez and Lecoq, 1997). ZYMV is very efficiently transmitted by several aphid species in a non-persistent manner. In zucchini squash symptoms are particularly severe and include mosaic and distortions of leaves and fruits. Infected plants generally do not produce marketable fruits, resulting in important economical losses particularly when epidemics start early after planting. Symptoms are also very severe in melons, cucumbers and watermelons (Desbiez and Lecoq, 1997). It is not uncommon for complete crop failures to be observed in cucurbit crops in the case of early and severe ZYMV epidemics.

Control measures by cultural practices, including the use of plastic mulches or oil sprays, may provide a temporary protection to the crops, but this is generally not sufficient to prevent significant economic losses. Breeding programmes are in progress, either through classical breeding or using transgenic plants expressing ZYMV coat protein. However, high and durable levels of resistance are not yet available in most commercial cultivars (Desbiez and Lecoq, 1997).

Isolation of a mild ZYMV strain

A mild variant of ZYMV (ZYMV-WK) was obtained from a melon axillary branch showing very attenuated symptoms that appeared on a plant infected by a severe and aphid non-transmissible isolate in a greenhouse trial (Fig. 10.1) (Lecoq *et al.*, 1991). ZYMV-WK produced very mild symptoms in various cucurbits (melon, cucumber, watermelon or squash) following mechanical inoculation. ZYMV-WK only slightly affected the production of some zucchini squash hybrids but not all, by delaying fruit setting for a few days rather than decreasing the production potential of individual plants (Lecoq *et al.*, 1991; Ginoux and Lecoq, 1994, unpublished; Spence *et al.*, 1996). Recently, Gal-On and Raccah (2000) have demonstrated that the mild symptoms induced by ZYMV-WK were due to a single mutation R→I in a FRNK conserved motif present in most severe potyvirus isolates, in the helper component protease (HC-Pro) gene. In addition, ZYMV-WK is not transmissible by aphids as the severe strain (ZYMV-PAT) from which it is derived (Fig. 10.1) (Lecoq *et al.*,

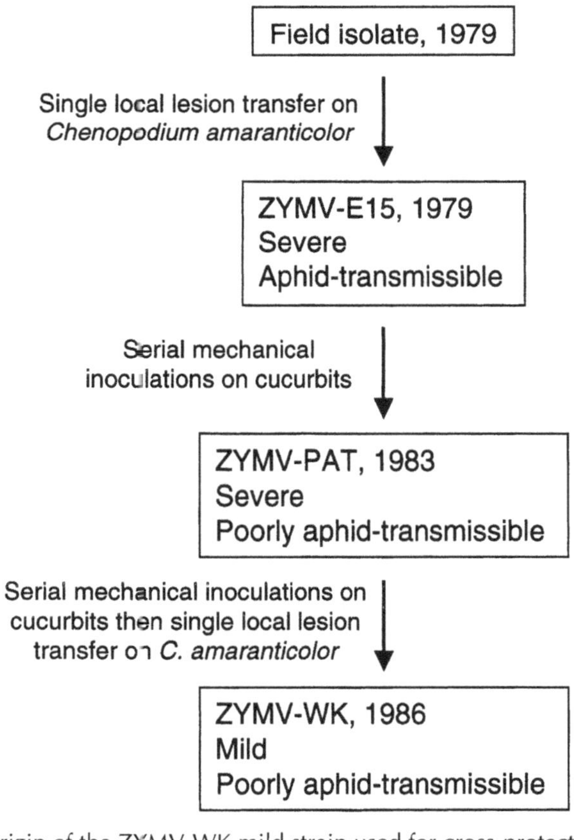

Fig. 10.1. Origin of the ZYMV-WK mild strain used for cross-protection.

1991). This deficiency in aphid transmission is due to a T→A mutation in a PTK conserved motif in the same HC-Pro gene (Huet *et al.*, 1994).

Greenhouse evaluation of the protection specificity

In laboratory experiments, ZYMV-WK was observed to be efficient in protecting plants against severe ZYMV isolates from the Mediterranean Basin, northern Europe, Africa, Asia, northern and Central America (Lecoq *et al.*, 1991; Wang *et al.*, 1991; Desbiez and Lecoq, 1997; Lecoq, unpublished). However ZYMV-WK was not efficient in controlling severe isolates originating from the islands of Réunion, Mauritius and Madagascar. These isolates are serologically distantly related to the type strain of ZYMV, and represent an important molecular divergence (Desbiez and Lecoq, 1997). For this reason, it seems advisable to achieve preliminary cross-protection experiments in controlled environmental conditions with local ZYMV isolates, before considering the use of ZYMV-WK in the field in new locations.

Pilot tests

In order to validate the practical interest of cross-protection to control severe ZYMV, pilot tests have been conducted in different regions of the world, in covered crops as well as in open fields with different artificial or natural inoculation procedures (Table 10.2). In the fields, under severe ZYMV natural epidemics or artificial inoculum pressures, ZYMV-WK provided a very effective protection to squash, melon, watermelon or cucumber, in temperate, Mediterranean and sub-tropical environments (Lecoq *et al.*, 1991; Wang *et al.*,

Table 10.2. Documented pilot tests conducted in the fields in order to evaluate the protective effect of ZYMV-WK against severe local ZYMV isolates.

Location	Challenge inoculation	Crops	Reference
California	Mechanical inoculation	Melon	Perring *et al.*, 1995
England	Aphid transmission	Squash	Walkey *et al.*, 1992
	Natural infections	Squash (four cultivars)	Spence *et al.*, 1996
France	Natural infections	Squash (two cultivars)	Lecoq *et al.*, 1991
Hawaii	Natural infections	Squash	Cho *et al.*, 1992
Israel	Mechanical inoculation	Melon, squash, watermelon	Yarden *et al.*, 2000
Jersey	Aphid transmission	Squash	Walkey *et al.*, 1992
Lebanon	Natural infections	Squash	Abou-Jawdah *et al.*, 2000
Taiwan	Natural infections	Squash	Wang *et al.*, 1991
Turkey	Aphid transmission	Squash, cucumber	Yilmaz *et al.*, 1994

1991; Walkey *et al.*, 1992; Yilmaz *et al.*, 1994; Spence *et al.*, 1996; Abou-Jawdah *et al.*, 2000; Yarden *et al.*, 2000). Yield increases reached up to 40 times that of unprotected crops. In Israel, on a basis of nine experiments conducted in different periods of the year and at different locations, the mean income increase in watermelon crops was estimated to be near to US\$3000 ha^{-1} (Yarden *et al.*, 2000).

Increases in cucumber mosaic virus (CMV, *Cucumovirus*), watermelon mosaic virus (WMV, *Potyvirus*) and papaya ringspot virus (PRSV, *Potyvirus*) were very similar in control and cross-protected plots suggesting that an infection by ZYMV-WK did not specifically interfere with the dissemination of these non-persistently aphid-transmitted viruses (Lecoq *et al.*, 1991; Wang *et al.*, 1991). Although ZYMV-WK is not transmitted by aphids, it was transmitted to a significant extent from plants grown in the field and co-infected by WMV. In fact, WMV provided a functional helper component to assist transmission of the mild isolate which is deficient for this function (Lecoq *et al.*, 1991). This may explain why ZYMV-WK might occasionally be observed in non-inoculated plants in the fields (Spence *et al.*, 1996; Lecoq, unpublished). Alternatively, plant-to-plant contact may assist spread to a limited number of plants.

Pilot tests also revealed the feasibility of using cross-protection within an integrated virus control strategy. For instance, it can be successfully applied with plastic mulches (that repel aphids and delay virus spread) or to CMV-resistant zucchini–squash hybrids (such as Supremo) (Lecoq *et al.*, 1991; Cho *et al.*, 1992).

Factors affecting cross-protection effectiveness

The effectiveness of cross-protection will depend upon the speed at which the mild strain will become systemic in the plants. Indeed, it is generally accepted that cross-protection is directly correlated with the presence of the mild strain in the tissues to be protected. ZYMV-WK was shown to move from inoculated cotyledons to the rest of the plant within 48 h (Yarden *et al.*, 2000). The incubation time after inoculation, necessary for sufficient mild strain multiplication to provide protection, was estimated to be between 4 and 14 days (Walkey *et al.*, 1992; Yarden *et al.*, 2000). Therefore, it is advised to inoculate the mild strain in nurseries, at the cotyledon stage, before transplanting plantlets in the field several days later. Some correlation was observed between the length of the incubation time and the level of nucleotide identity between ZYMV-WK and challenging strains. The more related the strains, the shorter is the incubation time (Desbiez *et al.*, 1997; Desbiez and Lecoq, unpublished).

Inefficient cross-protection has been occasionally observed with plants suffering from various physiological stresses. Therefore, an appropriate irrigation and fertilization should be applied to the cross-protected plants in order to obtain optimal protection.

In the later stage of crop development, a very small number of plants may occasionally develop more severe symptoms on leaves and moderate symptoms on fruits (Lecoq *et al.*, 1991; Walkey *et al.*, 1992). It is not yet known whether the apparent partial breakdown of the cross-protection is due to high inoculum pressure, to physiological changes in the plants (such as senescence) or to the mutation of the mild isolate. However, most of the time, severe symptoms observed in mild strain-inoculated plants in the field are due to infections by other viruses (Lecoq *et al.*, 1991; Wang *et al.*, 1991).

No major synergetic effects have been observed with any of the other viruses commonly infecting cucurbits. However, a slight increase in the multiplication rate of CMV, or of Cucurbit aphid-borne yellows virus (CABYV, *Polerovirus*), has been observed without affecting significantly the symptomatology (Bourdin and Lecoq, 1994; A. Gal-On, Israel, 2000, personal communication). This suggests that ZYMV-WK mild isolate may be used in protected crops as well as in crops grown in the open, even when other virus epidemics occur.

Mild strain production and application

Before contemplating any commercial application of ZYMV-WK, specific agreements have to be obtained from the local authorities responsible for registration of biological control methods. Specific methods should also be developed for mild strain production and inoculation. Mild strain increase must be conducted in a highly protected environment under very strict phytosanitary supervision, in order to eliminate risks of contamination by undesirable viruses, bacteria or fungi. This entails the use of certified seeds and sterilized potting soils for plant production, disinfection of the insect-proof greenhouses or growing cabinet structures, regular pesticide treatments, and random serological tests in order to detect any possible virus contamination.

It is also essential to provide farmers (or nurseries) with an easy and an efficient inoculation methodology. Various techniques have been investigated. Mechanical inoculation (by hands or with sponge pads screwed in inoculum-containing bottles) is labour intensive and time consuming; it may also favour the non-intentional spread of severe viruses which are mechanically transmissible. Another alternative is the use of spray guns (with adapted air pressures and nozzle sizes) (Ginoux *et al.*, 1994; Perring *et al.*, 1995); this technique has the advantage of allowing, if needed, inoculation of mild strains to adult plants. Recently, a machine has been developed especially for the inoculation of a complete tray of seedlings. This machine allows rapid and efficient inoculation of large quantities of seedlings (Yarden *et al.*, 2000).

Mild strain detection

It is necessary to have specific diagnostic tools to differentiate the mild strain from severe strains, for field surveys or inoculum quality controls. Immunocapture polymerase chain reaction (IC-PCR) followed by restriction fragment length polymorphism (RFLP) analysis or triple antibody sandwich enzyme-linked immunosorbent assay (TAS-ELISA) with monoclonal antibodies have been shown to differentiate ZYMV-WK from some (but not all) severe ZYMV isolates (Barbara *et al.*, 1995; Desbiez and Lecoq, 1997). Recently, Gal-On and Raccah (2000) have introduced the mutation responsible for symptom mildness into an infectious cDNA clone of a severe ZYMV isolate. This has led to a new mild isolate, ZYMV-AG. In addition, a specific restriction site has been introduced which will allow easy detection of ZYMV-AG.

Commercialization of cross-protection

ZYMV-WK has been used on a pre-commercial and commercial basis in several countries with different technology transfer approaches. In Hawaii, Cho *et al.* (1992) have successfully implemented the commercialization of cross-protection in the island of Maui which regularly suffers from severe ZYMV epidemics. The source of ZYMV-WK is maintained at the University of Hawaii; it is given to cooperative nurseries which have the responsibility of increasing the mild strain (for a limited number of times) and inoculating the seedlings for the farmers. The success of the programme is demonstrated by the fact that up to 90% of the farmers have utilized cross-protection in squash production (Cho *et al.*, 1992; Fuchs *et al.*, 1997).

In Israel, a cooperation between the Agriculture Research Organization and a kibbutz led to the creation of a start-up company (Bio-Oz) which is now producing and inoculating the mild strain. Upon request from the farmers, the mild strain is inoculated in nurseries with a specially designed machine (Yarden *et al.*, 2000). The frequency and severity of the ZYMV epidemics have been estimated in Israel for the different cucurbit crops and the different periods of the year. These indications are of major importance for farmers in order to decide whether they should cross-protect their crops or not. More than 2500 ha of cucurbits have been successfully cross-protected in Israel since 1996 (Yarden *et al.*, 2000).

Other Examples of Field Application of Cross-protection

Tomato mosaic virus

Tomato mosaic virus (ToMV, *Tobamovirus*) induces a very severe disease in tomato crops. It provokes severe mosaic on leaves, flower abortion and necrotic symptoms on the fruits (internal browning) which render the crop unmarketable. ToMV is seed-transmitted in tomatoes and readily spread by cultural operations (pruning, disbudding, etc.) (Broadbent, 1976).

Mild strains were obtained either from naturally occurring strains or by culturing a severe strain in tissue held above 34°C (Table 10.1). Mild strain MII-16, which was extensively used in Western Europe, was obtained by nitrous acid mutagenesis followed by single local lesion selection (Rast, 1972). Cross-protection against ToMV in tomatoes has been widely used in Europe, Canada, New Zealand and in Japan (Oshima, 1975; Broadbent, 1976; Fletcher, 1978).

In France, a subculture of MII-16 was used commercially from 1972 to 1983 (Lecoq, 1998). Pilot tests demonstrated that it did not reduce yield and provided a very efficient protection against severe strains. However, the risk of synergism with CMV, a virus frequent in the fields in southern France, led to the use of ToMV cross-protection restricted to protected crops, where CMV is rarely observed. The inoculum (purified virus preparation) to be used by farmers was prepared under very strict sanitary control measures. It was subsequently submitted to a series of quality control tests for specific infectivity, absence of severe strain contaminants and cross-protective potential (Lecoq, 1998).

The ToMV cross-protection scheme has been a real success in many countries in the world (Broadbent, 1976). In France alone, more than 13 million tomato plants were still cross-protected in 1981, but progressively this number decreased with the release of ToMV-resistant tomato cultivars. This illustrates the interest in cross-protection as a temporary control method which can be rapidly implemented in the fields and which can be used until easier or better control methods are developed such as resistant cultivars of good agronomic quality.

Papaya ringspot virus

PRSV is an aphid-borne potyvirus which causes important economical losses to papaya in most tropical and sub-tropical regions. Symptoms on papaya include plant stunting, severe mosaic on leaves and spotting on fruits which reduce their quality or cash value. PRSV also causes a severe disease in cucurbits (Wang *et al.*, 1987; Yeh *et al.*, 1988; Gonsalves and Garnsey, 1989).

Two mild variants were obtained by mutagenesis with nitrous acid treatment of crude virus preparations and after selection from 663 single local lesion isolations (Yeh and Gonsalves, 1984). These mild strains proved to be very effective both in laboratory tests and in pilot field tests in Hawaii. Only a low percentage of plants were observed with severe symptoms 2.5 years after plantings (Fuchs *et al.*, 1997). Some level of protection was also observed in Taiwan, but severe symptoms were noticed only 6 months after planting on some cross-protected plants, particularly under severe inoculum pressure.

In Thailand, no protection was observed (Yeh *et al.*, 1988). The PRSV mild isolates were derived from an Hawaii severe strain and may not completely protect against the Taiwan and Thailand isolates. A search for mild variants derived from a Taiwan severe strain has yielded some promising new mild isolates (Yeh *et al.*, 1988). A study was done to compare the specificity of the protection conferred by classical cross-protection using a mild variant of PRSV and by transgenic papaya expressing the coat protein gene of the same isolate (Tennant *et al.*, 1994). In both cases, a high level of protection was observed against severe PRSV isolates from Hawaii. In contrast, a range of reactions varying from delayed and attenuated symptoms to severe symptoms was observed with a series of other isolates from different geographical origins (Tennant *et al.*, 1994).

Cucumber mosaic virus

CMV has a wide host range including many wild plants and several vegetable crops. CMV is efficiently transmitted non-persistently by several aphid species. An original form of cross-protection has been investigated for controlling CMV. It associates the use of a mild CMV strain and a non-necrogenic satellite RNA that attenuates symptoms (Montasser *et al.*, 1998). This type of inoculum has a double protective effect: protection against severe CMV strains (through classical cross-protection between virus strains) and protection against necrogenic satellites (through cross-protection between satellite RNAs). This strategy has been widely used in China to control CMV in tobacco, pepper and tomato crops (Tien and Wu, 1991) and preliminary tests conducted in Europe and in the USA were quite promising (Jacquemond and Tepfer, 1998; Montasser *et al.*, 1998). However, the use of satellite RNAs in cross-protection raises specific biosafety issues which should be considered before large-scale field application (Jacquemond and Tepfer, 1998).

Citrus tristeza virus

Citrus tristeza virus (CTV, *Closterovirus*) is prevalent in many regions where citrus is grown; it is transmitted by aphids in the semi-persistent mode. It causes very important losses particularly in certain scion/rootstock combinations, where it may provoke reduction in fruit size and yield, stem-pitting, decline and the premature death of the plant. Mild isolates were obtained in Brazil, Australia, South Africa and Réunion island, from trees noticed as showing no or very discrete symptoms in the field, whereas all other trees were severely infected (Muller, 1980; Muller and Costa, 1987).

A specificity of this system is that it is necessary to select a mild variant for each scion/rootstock combination or for each type of symptom. For instance in Brazil, the best isolates for Pera orange were not convenient for lime or grapefruit and consequently, specific mild isolates have to be found for each citrus type. Similarly in Florida, three mild isolates were used to cross-protect sweet orange or Ruby Red grapefruit grafted on to sour orange. While cross-protection breakdown was observed in sweet orange in all treatments 8 years after planting, a good level of protection was still observed in grapefruit 16 years after planting (Powell *et al.*, 1999).

Cross-protection has been (and still is) widely used with great success in Brazil: more than 8 million Pera orange trees were cross-protected in 1980 (Muller, 1980), increasing to more than 50 million in 1987 (Urban *et al.*, 1990).

Biosafety Issues and Discussion

More than for any other control method, cross-protection has been associated with potential hazards (Fulton, 1986; Matthews, 1991). Several limitations or possible risks associated with this control method have been emphasized. An incomplete protection may occasionally be observed, and indeed some apparent breakdowns of cross-protection have been reported, especially in senescing plants (Fulton, 1986; Lecoq *et al.*, 1991; Fuchs *et al.*, 1997). Amplified disease symptoms caused by a synergism with other viruses has been rarely observed, except in the case of ToMV and CMV. ZYMV-WK aphid transmission efficiency was improved through heteroassistance in mixed infection with WMV. The possibilities of genetic recombination between the protecting strain and another virus in mixed infection and of mutation of the protecting virus into a more severe form that would cause a destructive disease have both been raised as risks associated with cross-protection, but the occurrence of such events has not yet been demonstrated.

Despite these anticipated difficulties, the method has been, and continues to be, effective with a number of virus diseases and crops. Nevertheless, due to the constraints linked to its implementation in the field, cross-protection should be contemplated only for viruses which present a real threat to a crop, and for which no other alternative control method is presently available. It should be considered only if farmers will gain a significant economical benefit from its use and there are no risks for the environment. Under these conditions, the use of cross-protection should not introduce additional specific risks in comparison to those which may arise by allowing a severe strain to spread naturally in the fields, without appropriate control measures.

The versatility of this method makes it easy to use as new problems arise. It can be applied easily across a range of different cultivars, contributing in part to the maintenance of crop biodiversity. Cross-protection should be integrated into the crop management system; indeed the use of specific cultural practices along with cross-protection may enhance its field efficiency and minimize any slight reduction in yield sometimes observed after mild strain inoculation.

References

Abou-Jawdah, Y., Sobh, H., El-Zammar, S., Fayyad, A. and Lecoq, H. (2000) Incidence and management of virus diseases of cucurbits in Lebanon. *Crop Protection* 19, 217–224.

Barbara, D.J., Morton, A., Spence, N. and Miller, A. (1995) Rapid differentiation of closely related isolates of two plant viruses by polymerase chain reaction and restriction fragment length polymorphism analysis. *Journal of Virological Methods* 55, 121–131.

Bourdin, D. and Lecoq, H. (1994) Increase in cucurbit aphid-borne yellows virus concentration by co-infection with sap-transmissible viruses does not increase its aphid transmissibility. *Journal of Phytopathology* 141, 143–152.

Broadbent, L. (1976) Epidemiology and control of tomato mosaic virus. *Annual Review of Phytopathology* 14, 75–96.

Cho, J.J., Ullman, D.E., Wheatley, E., Holly, J. and Gonsalves, D. (1992) Commercialization of ZYMV cross protection for zucchini production in Hawaii. *Phytopathology* 82, 1073.

Culvert, J.N. (1996) Tobamovirus cross protection using a potexvirus vector. *Virology* 226, 228–235.

Desbiez, C. and Lecoq, H. (1997) Zucchini yellow mosaic virus. *Plant Pathology* 46, 809–829.

Desbiez, C., Gal-On, A., Raccah, B. and Lecoq H. (1997) Characterization of epitopes on zucchini yellow mosaic potyvirus permits studies on the interactions between strains. *Journal of General Virology* 78, 2073–2076.

de Zoeten, G.A. and Fulton, R.W. (1975) Understanding generates possibilities. *Phytopathology* 65, 221–222.

Fletcher, J.T. (1978) The use of avirulent strains to protect plants against the effects of virulent strains. *Annals of Applied Biology* 89, 110–114.

Fuchs, M., Ferreira, S. and Gonsalves D. (1997) Management of virus diseases by classical and engineered cross protection. *Molecular Plant Pathology On-Line* [http://www.bspp.org.uk/mppol/] 1997/0116fuchs.

Fulton, R.W. (1986) Practices and precautions in the use of cross protection for plant virus disease control. *Annual Review of Phytopathology* 24, 67–81.

Gal-On, A. and Raccah, B. (2000) A point mutation in the FRNK motif of the potyvirus helper component-protease gene alters symptom expression in cucurbits and elicits protection against the severe homologous virus. *Phytopathology* 90, 467–473.

Gerber, M. and Sarkar, S. (1989) The coat protein of tobacco mosaic virus does not play a significant role for cross protection. *Journal of Phytopathology* 124, 323–331.

Ginoux, G., Wipf-Scheibel, C. and Lecoq, H. (1994) Mosaïque jaune de la courgette: lutte biologique par prémunition. *PHM Revue Horticole* 346, 15–18.

Gonsalves, D. and Garnsey, S.M. (1989) Cross-protection techniques for control of plant virus diseases in the Tropics. *Plant Disease* 73, 592–597.

Huet, H., Gal-On, A., Meir, E., Lecoq, H. and Raccah, B. (1994) Mutations in the helper component protease gene of zucchini yellow mosaic virus affect its ability to mediate aphid transmissibility. *Journal of General Virology* 75, 1407–1414.

Hughes, J. d'A. and Ollenu, A.A. (1994) Mild strain protection of cocoa in Ghana against cocoa swollen shoot virus – a review. *Plant Pathology* 43, 442–457.

Jacquemond, M. and Tepfer, M. (1998) Satellite RNA-mediated resistance to plant viruses: are the ecological risks well assessed? In: Hadidi, A., Kheterpal, R.K. and

Koganezawa, H. (eds) *Plant Virus Disease Control.* APS Press, St Paul, Minnesota, pp. 94–120.

Kosaka, Y. and Fukunishi, T. (1993) Attenuated isolates of soybean mosaic virus derived at low temperature. *Plant Disease* 77, 882–886.

Lecoq, H. (1998) Control of plant virus diseases by cross protection. In: Hadidi, A., Kheterpal, R.K. and Koganezawa, H. (eds) *Plant Virus Disease Control.* APS Press, St Paul, Minnesota, pp. 33–40.

Lecoq, H., Lemaire, J.M. and Wipf-Scheibel, C. (1991) Control of zucchini yellow mosaic virus in squash by cross protection. *Plant Disease* 75, 208–211.

Lisa, V., Boccardo, G., D'Agostino, G., Dellavalle, G. and D'Aquilio, M. (1981) Characterization of a potyvirus that causes zucchini yellow mosaic. *Phytopathology* 71, 667–672.

Lu, B., Stobbs G. and Culver, N. (1998) Coat protein interactions involved in tobacco mosaic tobamovirus cross-protection. *Virology* 248, 188–198.

Matthews, R.E.F. (1991) *Plant Virology*, 3rd edn. Academic Press, San Diego, California.

McKinney, H.H. (1929) Mosaic diseases in the Canary Islands, West Africa and Gibraltar. *Journal of Agricultural Research* 39, 557–578.

Montasser, M.S., Tousignant, M.E. and Kaper, J.M. (1998) Viral satellite RNAs for the prevention of cucumber mosaic virus (CMV) disease in field-grown pepper and melon plants. *Plant Disease* 82, 1298–1303.

Muller, G.W. (1980) Use of mild strains of citrus tristeza virus (CTV) to reestablish commercial production of 'Pera' sweet orange in Sao Paulo, Brazil. *Proceedings of the Florida State Horticultural Society* 93, 62–64.

Muller, G.W. and Costa, A.S. (1987) Search for outstanding plants in tristeza infected citrus orchards: the best approach to control the disease by preimmunization. *Phytophylactica* 19, 197–198.

Oshima, N. (1975) The control of tomato mosaic virus disease with attenuated virus of tomato strain of TMV. *Review of Plant Protection Research* 8, 126–135.

Palutaikis, P. and Zaitlin, M. (1984) A model to explain the 'cross protection' phenomenon shown by plant viruses and viroids. In: Kosuge, T. and Nester, E.W. (eds) *Plant–Microbe Interactions: Molecular and Genetics Perspectives.* Macmillan, New York, pp. 420–429.

Perring, T.M., Farrar, C.A., Blua, M.J., Wang, H.L. and Gonsalves D. (1995) Cross protection of cantaloupe with a mild strain of zucchini yellow mosaic virus: effectiveness and application. *Crop Protection* 14, 601–606.

Powell, C.A., Pelosi, R.R., Rundell, P.A., Stover, E. and Cohen, M. (1999) Cross-protection of grapefruit from decline-inducing isolates of citrus tristeza virus. *Plant Disease* 83, 989–991.

Rast, A.T.B. (1972) MII-16, an artificial symptomless mutant of tobacco mosaic virus for seedling inoculation of tomato crops. *Netherlands Journal of Plant Pathology* 78, 110–112.

Ratcliff, F.G., Harrison, B.D. and Baulcombe, D.C. (1997) A similarity between viral defense and gene silencing in plants. *Science* 276, 1558–1560.

Ratcliff, F.G., MacFarlane, S.A. and Baulcombe, D. (1999) Gene silencing without DNA: RNA-mediated cross-protection between viruses. *The Plant Cell* 11, 1207–1215.

Sherwood, J.L. (1987) Mechanisms of cross-protection between plant virus strains. In: Evered, D. and Harnett, S. (eds) *Plant Resistance to Viruses.* John Wiley & Sons, Chichester, pp. 136–150.

Sherwood, J.L. and Fulton, R.W. (1982) The specific involvement of the coat protein in tobacco mosaic virus cross protection. *Virology* 119, 150–158.

Spence, N.J., Mead, A., Miller, A., Shaw, E.D. and Walkey, D.G.A. (1996) The effect on yield in courgette and marrow of the mild strain of zucchini yellow mosaic virus used for cross-protection. *Annals of Applied Biology* 129, 247–259.

Tennant, P.F., Gonsalves, C., Ling, K.S., Fitch, M., Manshardt, R., Slightom, J.L. and Gonsalves, D. (1994) Differential protection against papaya ringspot virus isolates in coat protein gene transgenic papaya and classically cross-protected papaya. *Phytopathology* 84, 1359–1366.

Tien, P. and Wu, G. (1991) Satellite RNA for the biocontrol of plant diseases. *Advances in Virus Research* 39, 321–339.

Urban, L.A., Sherwood, J.L., Rezende, J.A.M. and Melcher, U. (1990) Examination of mechanisms of cross protection with non-transgenic plants. In: Fraser, R.S.S. (ed.) *Recognition and Response in Plant–Virus Interactions*. Springer-Verlag, Berlin, pp. 415–426.

Walkey, D.G.A., Lecoq, H., Collier, R. and Dobson, S. (1992) Studies on the control of zucchini yellow mosaic virus in courgettes by mild strain protection. *Plant Pathology* 4, 762–771.

Wang, H.L., Yeh, S.D., Chiu, R.J. and Gonsalves, D. (1987) Effectiveness of cross protection by mild mutants of papaya ringspot virus for control of ringspot disease of papaya in Taiwan. *Plant Disease* 71, 491–497.

Wang, H.L., Gonsalves, D., Provvidenti, R. and Lecoq, H. (1991) Effectiveness of cross-protection by a mild strain of zucchini yellow mosaic virus in cucumber, melon and squash. *Plant Disease* 75, 203–207.

Yarden, G., Hemo, R., Livne, H., Maoz, E., Lev, E., Lecoq, H. and Raccah, B. (2000) Cross-protection of Cucurbitaceae from zucchini yellow mosaic virus. In: Katzir, N. and Paris, H.S. (eds) *Proceedings of 7th EUCARPIA Meeting on Cucurbits Genetics and Breeding, Acta Horticulturae 510*. ISHS, Leuven, pp. 349–356.

Yeh, S.D. and Gonsalves, D. (1984) Evaluation of induced mutants of papaya ringspot virus for control by cross protection. *Phytopathology* 74, 1086–1091.

Yeh, S.D., Gonsalves, D., Wang, H.L., Namba, R. and Chiu, R.J. (1988) Control of papaya ringspot virus by cross protection. *Plant Disease* 72, 375–380.

Yilmaz, M.A., Abak, K., Lecoq, H., Baloglu, S., Sari, N., Kesici, S., Ozaslan, M. and Guldur, M.E. (1994) Control of zucchini yellow mosaic virus (ZYMV) in cucurbits by ZYMV-WK strain. *Proceedings of the 9th Congress of the Mediterranean Phytopathological Union*, Turkish Phytopathological Society, Izmir, pp. 353–356.

Plant Pathogen–Herbivore Interactions and Their Effects on Weeds

11

Paul E. Hatcher[1] and Nigel D. Paul[2]

[1]Department of Agricultural Botany, School of Plant Sciences, The University of Reading, 2 Earley Gate, Whiteknights, Reading RG6 6AU, UK; [2]Division of Biology, Institute of Environmental and Natural Sciences, Lancaster University, Lancaster LA1 4YQ, UK

Introduction

Fungi and insects form two of the most numerous groups of living organisms in the world. Many individual species of each group utilize higher plants as a food source and thus it is not surprising that they should interact. Insects and fungi also form the major groups used in weed biological control. Although weed biocontrol using combinations of insects and fungi was suggested at the onset of utilizing plant pathogens in this field (Wilson, 1969), it has rarely been put into practice. However, these herbivore–pathogen interactions have been implicated in some of the first successful weed biocontrol projects. For example, in the biocontrol programme against *Opuntia* in Australia, plant pathogens including *Gloesporium lunatum* and bacterial soft rots followed damage by the introduced insect agent, *Cactoblastis cactorum*, and completed the eradication of the weed (Dodd, 1940).

The study of insect herbivore–pathogen–plant (or tripartite) interactions is complicated in that not only must the traditional two-way interactions of the plant pathologist (pathogen–plant) and entomologist (insect–plant) be studied but a new, three-way indirect interaction must be studied. This is the interaction between insect and pathogen mediated through the host plant and needs the combined attention of plant pathologists, entomologists and plant physiologists. Direct herbivore–pathogen interactions can also occur: for example, insects consuming spores or acting as vectors for pathogens. This is not discussed in detail here as it has been well reviewed elsewhere (Carter, 1973; Agrios, 1980; Wheeler and Blackwell, 1984; Wilding *et al.*, 1989).

Indirect interactions between insects and fungi cannot be studied in isolation, without considering the effect of the insect and pathogen together on the plant. This is a time-consuming process that has been carried out for few systems, leaving little in the way of generalization concerning the outcomes of different interactions in terms of biological weed control. However, in reviewing the existing literature, we will discuss whether certain herbivore–pathogen combinations may hold greater promise for weed control than others.

Even if traditional biocontrol using either insects or fungi is being attempted, the study of herbivore–pathogen–plant relationships can still be important. This will particularly be so where native, or long-established, weeds are being controlled. Here it is likely that the biocontrol agent will not be colonizing or infecting a healthy plant but one that either is or has been attacked or infected to some extent. Thus, we review the ecological and biochemical evidence for interactions between plant pathogens and insect herbivores.

In this chapter we confine ourselves to plant-pathogenic fungi. Other fungi, for example vesicular-arbuscular mycorrhiza (Gange and West, 1994; Gange and Bower, 1997) and endophytes (Clay, 1997; Prestidge and Ball, 1997; Saikkonen *et al.*, 1998), also interact with insects, and the biocontrol implications of insect interactions with the former group have been discussed recently by Harris and Clapperton (1997).

Indirect Interactions Between Insects and Fungi

Currently, the study of interactions between insects and fungi is proceeding along two research lines. In one, the effects of insects on fungi, and *vice versa*, and their effects on their host plant are investigated in the field and laboratory. The second is fundamentally laboratory-based and is focused on investigating the mechanisms by which signalling pathways for systemic acquired resistance (SAR) against pathogens and induced resistance (IR) against insects interact. Just as the first of these often has little regard for the *mechanisms* of the interactions, so the second route often has little regard for the *effects* of these interactions. A combination of these two research approaches is long overdue (Hatcher and Paul, 2000b; Paul *et al.*, 2000) but until this is achieved, evidence from each approach has still to be considered largely in isolation.

Most of the studies reported below are laboratory investigations or manipulative field experiments. Little research has been carried out on field interactions. However, Ericson and Wennström (1997) reported more scale insects, *Arctorthezia cataphracta* (Hemiptera), on *Trientalis europaea* infected with the smut *Urocystis trientalis*, and Lappalainen and Helander (1997) found that *Phratora polaris* (Coleoptera) performance was worse on *Betula pubescens* infected with a low level of the rust *Melampsoridium betulinum* rather than no or a heavy infection. These field interactions are important but remain hard to analyse in the absence of manipulative experiments investigating the nature of the interactions.

How insects may affect fungi

Positive interactions

Insects may facilitate fungal infection by two main actions: (i) acting as vectors for fungal propagules, and (ii) providing wound sites for fungal entry (Leach, 1940; Carter, 1973; Agrios, 1980). These direct interactions were among the first herbivore–pathogen interactions discovered (Leach, 1940) and may be very important in weed biological control. For example, the beetle *Chrysolina hyperici*, used as a biological control agent against *Hypericum perforatum*, can vector the fungus *Colletotrichum gloesporioides*, and Morrison *et al.* (1998) have shown that the combination of these agents can be very effective in causing weed mortality. Likewise, *Colletotrichum orbiculare*, a potential mycoherbicide for control of Bathurst burr (*Xanthium spinosum*), had enhanced infection after plant wounding (Klein and Auld, 1996). Here, not only may the wound site have offered easier access to the fungus, but also wounding of plants ruptures cells and releases moisture and possibly nutrients which may enhance infection.

The role of insect feeding wounds in providing a suitable site of entry for phytophages should not be overstated: it is far from an inevitable effect. Indeed, it may be a safer generalization to state that feeding wounds are unsuitable infection sites for many fungal pathogens, notably biotrophs such as rusts and powdery mildews. Even for necrotrophic pathogens, the role of feeding wounds as infection sites may be rather complex. For example, although Dillard and Cobb (1995) found that injured tissue of cabbage, caused by both artificial damage and insect feeding (by a variety of Lepidoptera larvae), could be colonized by *Sclerotinia sclerotiorum*, the frequency of colonization was low except when bruising injuries resulted in extensive cell damage. Their results suggest that only a low proportion of insect feeding wounds would be colonized by *S. sclerotiorum* in the field.

It is often difficult to disentangle the positive effects of these direct interactions on fungal infection from the effects of more indirect interactions. For example, larval fungus gnats (*Sciara* and *Bradysia* spp.) actively feed on both fungi and plant roots (Harris *et al.*, 1996) which may weaken the roots, predisposing them to infection by pathogens (Leath and Newton, 1969). Along with these indirect interactions, fungus gnats also vector phytopathogens.

In one study that went further than most to account for direct effects (Virtanen *et al.*, 1997), it was noted that necrosis of *Pinus sylvestris* seedlings caused by *Gremmeniella abietina* was more pronounced in seedlings that were infested with the aphid *Cinara pinea* 1 month before infection. Firstly, they established that the aphid was not a vector for the fungus, secondly they also suggested that the insect does not facilitate fungal entry because *G. abietina* infects through buds or current season's shoots (especially through stomata situated in bracts subtending the needle base), whereas the aphid feeds from the current or previous year's shoots by penetration of the needle axils

(Virtanen *et al.*, 1997). In this case the most parsimonious explanation is that aphids predisposed *P. sylvestris* to *G. abietina* by causing a general growth disruption or weakening of the seedlings.

Just as abiotic stresses such as drought and freezing can predispose plants to fungal attack, so can biotic stresses such as defoliation (Schoeneweiss, 1981). This is particularly evident in woody plants: for example, maples defoliated by the gypsy moth often succumb more rapidly to attack by the root rot *Armillaria mellea* (Wargo and Houston, 1973); and defoliation of *Eucalyptus* spp. enhanced infection by *Endothia gyrosa* and *Botryosphaeria ribis* (Old *et al.*, 1990). The latter pathogen also infects the invasive tree *Melaleuca quinquenervia* which is the subject of biocontrol efforts in Florida (Rayachhetry *et al.*, 1999). Again, defoliation enhances the ability of the fungus to infect (Rayachhetry *et al.*, 1996). The infection of some non-woody plants with fungi, for example *Fusarium* spp., is also enhanced by plant stress, including that caused by aphids (Leath and Byers, 1977) and homopteran leaf hoppers (Moellenbeck *et al.*, 1992).

Negative interactions

Several of the positive responses of pathogens to herbivores described above are widely known, for example vector associations form part of the mainstream of plant-pathological research. However, most insect herbivores do not serve as vectors for plant pathogens (Vega *et al.*, 1995). For example, of 53 species of insect reported as pests of bean in the UK, 94% were non-vectors and of 54 pest insect species on tomato, 83% were also non-vectors (Hill, 1987).

By contrast, systems in which insect feeding has an inhibitory effect on fungal infection of their host plants are less well known. The most direct negative effect of herbivores on pathogenic fungi is caused by mycophagy, the consumption of pathogen tissue by herbivores. Circumstantial evidence suggests that this might be important in some cases. For example, Powell (1971) found 137 species of insect and 23 species of mite and spider associated with the pine stem rust *Cronartium comandrae*. These arthropods damaged up to 62% of rust cankers during 1 year. Larvae of *Bradysia* spp. (fungus gnats) may play an important role in the rate of destruction and lysis of sclerotia of *Sclerotinia sclerotiorum* (Anas and Reeleder, 1987). Recently, some experimental studies have demonstrated this negative interaction. For example, the millipede *Aniulus bollmani* had a significant negative impact on the reproductive output of *Epichlöe typhina* by feeding on reproductive tissues (Bultman and Mathews, 1996). Likewise, Guevara *et al.* (2000) demonstrated that two specialist ciid beetle species reduced the reproductive potential of the fungus *Coriolus versicolor* by up to 50% alone and 64% in combination. This was also supported by field data. There is a continuum here, some vectors are also mycophagous, but vectoring outweighs the negative effect while some mycophages might be casual vectors, even if others are pure predators of fungal tissue.

Indirect effects may be local (around the site of feeding damage or on the same leaf) or systemic (other, undamaged leaves on the plant). For example, prior herbivory of *Rumex crispus* and *R. obtusifolius* by the chrysomelid beetle *Gastrophysa viridula* induced an 80% reduction in the pustule density of the rust fungus *Uromyces rumicis* within and around the feeding site within 1 day of feeding damage (Hatcher *et al.*, 1994a). Feeding damage also induced resistance throughout the undamaged portion of the leaf and a systemic induced resistance in undamaged leaves (Fig. 11.1). Herbivory by *G. viridula* mainly affected the early stages of uredinial development: once intercellular hyphae had started to develop, there was no difference between rate of development of uredinia on damaged and undamaged plants (Hatcher *et al.*, 1995a).

In the field, we found that artificial damage to leaves (caused by a hole punch) produced a reduction in *U. rumicis* pustule density of 74% and 86% on *R. crispus* and *R. obtusifolius*, respectively, when they were exposed to natural rust infection for 1 month (Hatcher *et al.*, 1994a). Recent experiments have shown that beetle grazing causes an overall reduction in the amount of natural infection on *R. obtusifolius* by three pathogenic fungi in the field: *U. rumicis*; the hemibiotrophic Ascomycete *Venturia rumicis;* and the necrotrophic Ascomycete *Ramularia rubella* (Hatcher and Paul, 2000a). The results also suggest that beetle grazing induces a systemic resistance to these fungi in the field (Table 11.1).

Other examples include a reduction of infection of barley by *Erysiphe graminis* when plants were infested by the aphid *Macrosiphum avenae* (Sipos

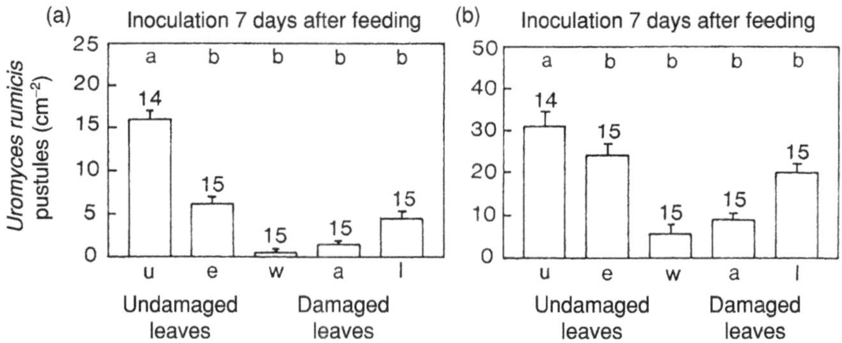

Fig. 11.1. The effect of prior feeding by *Gastrophysa viridula* on *Uromyces rumicis* establishment 7 days after feeding finished on (a) *Rumex crispus,* (b) *Rumex obtusifolius.* Categories: u, leaf on an undamaged plant; e, undamaged leaf on a plant with feeding damage; w, within the area of feeding damage; a, within a 1-cm wide area circling the circumference of the feeding damage; l, undamaged portion of the damaged leaf. Mean + SE given, numbers above indicate sample size. The same letters above each column indicate no significant difference ($P > 0.05$), Tukey-HSD multiple-range test from ANOVA. (From Hatcher *et al.*, 1994a.)

Table 11.1. The effect of herbivory by *Gastrophysa viridula* on the per-plant mean pustule density of *Ramularia rubella, Venturia rumicis* and *Uromyces rumicis* on undamaged leaves. (From Hatcher and Paul, 2000a.)

Pathogen	Experiment	Without beetles		With beetles		Mann–Whitney
Ramularia rubella	Spring	0.0	(54)	0.0	(23)	*
	Autumn	5.8	(74)	4.5	(76)	ns
Venturia rumicis	Spring	6.5	(54)	0.0	(23)	**
	Autumn	119.9	(74)	72.7	(76)	*
Uromyces rumicis	Autumn	38.0	(74)	21.0	(75)	***

Experiments carried out in the field in spring and autumn with natural infection. All values are medians (pustules per 100 cm^2) with sample sizes in parentheses. For Mann–Whitney test: ns, $P > 0.05$; $*P < 0.05$; $**P < 0.01$; $***P < 0.001$.

and Sagi, 1987) and an 84% reduction in total canker area of *Diaporthe phaseolorum* infection of soybean after defoliation by *Pseudoplusia includens* (Lepidoptera) (Padgett *et al.*, 1994). Similarly, infestation of watermelon (*Citrullus lanatus*) with thrips or aphids reduced the subsequent necrotic area of anthracnose, caused by *Colletotrichum orbiculare* (Russo *et al.*, 1997).

How fungi may affect insects

Positive interactions

While some arthropods may specialize in consuming fungi, many phytophagous insects practice facultative mycophagy (e.g. Barbe, 1964; Lewis, 1979). The selective grazing of leaves infected by fungi, a direct interaction, involves invertebrates removing spores or pustules with only the minimum of leaf material, and with often a tendency for uninfected leaves to be avoided (Hatcher, 1995). This is distinct from indirect interactions between fungi and herbivores that are mediated through host chemistry. Such indirect effects have been noted in several systems.

Plant infection by necrotrophic pathogens can facilitate the development of phytophagous insects For example, the European corn borer, *Ostrinia nubilalis* (Lepidoptera), had a faster larval development when reared on maize infected with *Fusarium graminearum* (Chiang and Wilcoxson, 1961), compared to those reared on uninfected maize. Plants infected with *Colletotrichum graminicola* alone (Carruthers *et al.*, 1986) and in combination with other fungi (Jarvis *et al.*, 1990) had a similar effect on these larvae. Recently, Moran (1998) demonstrated that spotted cucumber beetles, *Diabrotica undecimpunctata*, preferred feeding on leaf discs from cucumber with necrotic lesions from *Cladosporium cucumerinum* infection, rather than healthy young leaves.

These effects may be due to predigestion of complex carbohydrates by enzymes released from the fungus (Carruthers *et al.*, 1986). A similar mechanism may have important effects on insects consuming plant tissues infected by *Botrytis*, since both aphids (Zebitz and Kehlenbeck, 1991) and Lepidoptera (Savopoulou-Soultani and Tzanakakis, 1988; Mondy *et al.*, 1998a) had increased development rates when reared on *Botrytis*-infected tissue. In the latter case the larvae of the moth *Lobesia botrana* were also significantly attracted to grape berries infected with *B. cinerea* (Mondy *et al.*, 1998b).

Positive indirect interactions could also occur as infection of plants by pathogenic fungi can lead to an increase in nutritional quality via increases in carbohydrate and nitrogen concentrations. The increases in nitrogen can be considerable: for example up to 30% in *Tussilago farfara* infected with *Puccinia poarum* (Ramsell and Paul, 1990) and between 28 and 50% in *Allium porrum* infected with *P. allii* (Roberts and Walters, 1989).

In addition, pathogen infection can cause a reduction in the concentration of secondary metabolites that are inhibitory to some Lepidoptera. For example, infection of *Senecio vulgaris* by the rust *Puccinia lagenophorae* results in a substantial reduction in the concentration of pyrolizidine alkaloids, especially in the uninfected tissues of the root and capitula (Tinney *et al.*, 1998b). In this case, rust infection appeared to affect the distribution of alkaloids between different tissues, perhaps as the result of overall changes in allocation patterns, rather than alkaloid synthesis.

Negative interactions

The potential for fungal infection to affect insects negatively is great. Apart from any effects of resistance mechanisms (see below), plant-pathogenic fungi often alter the nutritional quality of the plant for insects. Hatcher (1995) reviewed these fungal-induced indirect effects on the plant which include alterations in carbohydrate composition and concentration, increases in leaf toughness, lignification and fibre content, and reduction in nitrogen content. Nitrogen quality can also be affected, with increases in nitrate (Piening, 1972), potentially poisonous to insect herbivores, and changes in total free and protein amino acid concentration and composition (Reddy and Rao, 1976) being reported following fungal infection.

Gastrophysa viridula was inhibited during all life stages when it was reared upon *Rumex* leaves infected with *U. rumicis* (Hatcher *et al.*, 1994b). The larvae reared upon infected food took 15% longer to complete development and had a relative growth rate up to 25% lower than beetles feeding on uninfected leaves. The fecundity of adults reared on infected food was also reduced, with these adults laying up to 55% fewer eggs (also with a reduced viability when reared on infected *R. crispus*) than those reared on healthy food (Fig. 11.2).

These effects were consistent with the chemical changes occurring in *Rumex* spp. after infection with the rust *U. rumicis*. This included a reduction in

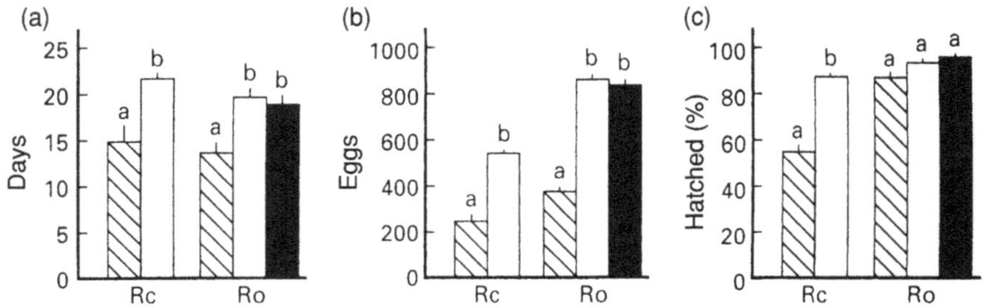

Fig. 11.2. The effect of infection of *Rumex crispus* (Rc) and *R. obtusifolius* (Ro) with *Uromyces rumicis* on *Gastrophysa viridula*: (a) duration of egg laying, (b) number of eggs laid by individuals, (c) percentage of eggs that hatched. Beetles reared on infected leaves (▧), uninfected leaves (□), or on uninfected leaves as an adult but infected leaves as a larva (■). $n = 15$ except infected *R. crispus* ($n = 9$) and uninfected leaves/infected leaves as a larva ($n = 13$). Mean + SE given. The same letter above columns within species indicates no significant difference ($P > 0.05$), Tukey-HSD multiple-range test from ANOVA. (Data from Hatcher *et al.*, 1994c.)

leaf nitrogen by up to 50% and a small rise in carbohydrate concentrations (Hatcher *et al.*, 1995b, 1997c; Hatcher and Ayres, 1998), leading to an increase in the carbohydrate:nitrogen ratio from 2.5 and 4.0 in the healthy leaves of both species to maxima of 6.8 and 7.7 in infected *R. crispus* and *R. obtusifolius* leaves, respectively (Hatcher *et al.*, 1995b).

Few other studies have considered the effect of plant pathogen infection on chewing insect herbivores. Kingsley *et al.* (1983) reported that the efficiency of conversion of digested food to biomass was significantly reduced in larvae of *Spodoptera eridania* (Lepidoptera) when they were fed on lucerne infected with *Verticillium albo-atrum*. Similarly, *Tyria jacobaeae* (Lepidoptera) larvae were smaller, fed for longer and had lower growth rates when fed on *Tussilago farfara* leaves infected with *Coleosporium tussilaginis* rather than healthy leaves (Tinney *et al.*, 1998a).

In both the above studies, the measured nutritional changes in the host plant with fungal infection did not explain the observed effects on the larvae (Kingsley *et al.*, 1983; Tinney *et al.*, 1998a). Apart from gross nutritional changes and induced defence reactions (discussed below), mycotoxins from plant-pathogenic fungi may affect insect herbivores. For example, *Heliothis virescens* (Lepidoptera) was negatively affected (reduced larval weight, inhibited pupation, increased larval development time) by feeding on plants infected with *Alternaria alternata*, *Fusarium moniliforme* and *F. oxysporum* (Abbas and Mulrooney, 1994). Toxins were isolated from these fungi and proved effective against *H. virescens* larvae, reducing larval weight by up to 87% and pupal weight by 33% (Abbas and Mulrooney, 1994).

Several studies have reported that the growth and reproduction of aphids was inhibited both by biotrophic fungi such as *Uromyces viciae-fabae* and

Erysiphe graminis, and necrotrophs such as *Phytophthora erythroseptica* and *Fusarium* sp. (Leath and Byers, 1977; Pratt *et al.*, 1982; Sipos and Sagi, 1987; Pesel and Poehling, 1988; Prüter and Zebitz, 1991).

In the above examples, insects were usually given leaves or plants that had recently been infected with fungi, and were usually still infected. However, the effect of fungal infection on insect herbivores may be more long-lasting. For example, the pupal weight of *Epirrita autumnata* (Lepidoptera), larvae of which feed on young leaves in the spring, was reduced when feeding on mountain birch, *Betula pubescens*, infected with the foliar rust *Melampsoridium betulinum* the previous autumn (Lappalainen *et al.*, 1995). Likewise, infection of *Larix decidua* by the needle cast fungus, *Mycosphaerella laricinia*, reduced larch sawfly, *Pristiphora erichsonii*, consumption rates 1 year later, and larvae rapidly abandoned seedlings previously defoliated by *M. laricinia* (Krause and Raffa, 1992).

Induced resistance mechanisms

The preceding section has demonstrated that insects and fungi can negatively affect each other. Apart from the correlation between reduction in insect performance and the reduction in host plant quality by fungal infection, there is little experimental evidence of the mechanisms involved in these interactions.

There is considerable evidence, however, that induced defence mechanisms that were thought to be induced solely by fungi (e.g. phytoalexins, the hypersensitive response) can be induced by or affect insects; and that induced defences thought to be induced solely by insects (e.g. proteinase inhibitors) could be induced by or affect fungi. Earlier literature has been reviewed by Hatcher (1995) and Hatcher and Ayres (1997).

Systems acquired resistance (SAR) against pathogens and the induced immunization of plants against pathogens by inoculation with non-pathogenic races of the organism has been demonstrated repeatedly (e.g. Kuc, 1982; Weete, 1992). This immunization is effective against a range of pathogens. For example, infection with *Colletotrichum lagenarium* or tobacco mosaic virus, followed 2–3 weeks later by a booster inoculation of the two pathogens, immunized cucumber against ten viral, bacterial and fungal pathogens, including *C. lagenarium* and *Fusarium oxysporum* (Kuc, 1982).

Several investigators have examined whether this SAR is also effective against insects. McIntyre *et al.* (1981) not only achieved whole-plant resistance of tobacco to *Phytophthora parasitica* after the plant was immunized with TMV, but reproduction of the aphid *Myzus persicae* was significantly reduced when placed on these plants 7 days after immunization. However, other studies have not found this effect. For example, although cucumber inoculated with TMV became systemically protected from *Colletotrichum lagenarium*, simultaneous tests found no effect on the survival of the mite *Tetranychus urticae*, larval growth and survival of *Spodoptera frugiperda*

(Lepidoptera), oviposition of *Trialeurodes vaporariorum* (Hemiptera) or food consumption by *Acalymma vittatum* (Coleoptera) (Apriyanto and Potter, 1990). Similarly Ajlan and Potter (1991, 1992) found no effect of the *C. lagenarium* or TMV induction of cucumber and tobacco, respectively, on insect herbivores; in reciprocal experiments the herbivores had no systemic effect on the pathogens.

Recently, however, the greatest advance in this area has come from the elucidation of the signalling pathways involved in the pathogen-induced salicylic acid (SA) SAR and in the insect-induced octadecanoid or jasmonate-induced (JA) resistance pathway. Furthermore, recent studies have demonstrated cross-talk between these pathways which could explain many of the insect–pathogen interactions described (Bostock, 1999; Pieterse and van Loon, 1999; Stout and Bostock, 1999).

The biochemistry and molecular biology of such interactions remain controversial. Some studies have suggested that stimulation of one pathway facilitates the other pathway. For example, wounding of one leaf of young rice plants caused both a strong accumulation of jasmonic acid and SAR to infection by the rice blast fungus *Magnaporthe grisea* (Schweizer *et al.*, 1998).

However, there are now several examples of the SA and JA pathways inhibiting each other. Early *in vitro* studies suggested that the two pathways might be mutually inhibitory (e.g. Doares *et al.*, 1995), and this has received recent support *in planta* by Felton *et al.* (1999) who showed that tobacco, in which pathogen-induced SAR was suppressed, expressed greater grazing-induced resistance to larvae of *Heliothis virescens*. Likewise, Thaler *et al.* (1999) demonstrated in field-grown tomato that a salicylate mimic, benzothiadiazole, reduces the jasmonate-induced expression of the enzyme polyphenol oxidase, linked to anti-herbivore defence, and also reduces host plant resistance to larvae of *Spodoptera exigua*; while treating plants with JA reduces resistance to a bacterial speck disease caused by *Pseudomonas syringae*. In addition, treatment with SA prevented wounded plants from accumulating proteinase inhibitors and polyphenol oxidase, possible defence mechanisms triggered by jasmonic acid (Stout *et al.*, 1998, 1999).

The reported incidences of the 'wrong pathway' being induced, i.e. insects inducing the SA pathway and fungi inducing the JA pathway (Inbar *et al.*, 1998; Stout *et al.*, 1998, 1999; Thomma *et al.*, 1998; Stout and Bostock, 1999) could explain some of the negative interactions between insects and fungi reported above, although sequential studies, in which either the insect or pathogen is placed first on the plant, rather than together, are needed. However, other molecular information is more consistent with positive interactions between insects and fungi. For example, the inhibition of the JA pathway by the SA pathway, and *vice versa*, could enhance fungal entry after insect attack and could enhance insect success after fungal attack.

These interactions are counterintuitive in terms of host adaptation, and recent suggestions propose that they are exploited by herbivores. Karban and Kuc (1999) suggested that perhaps insects that stimulate SA and pathogens that stimulate JA are indirectly depressing the plant's responses that would be the most effective against each individually, thereby increasing their own fitness. This suggestion is supported by the work of Korth and Dixon (1997) and Felton and Eichenseer (1999) who found that glucose oxidase from the saliva of *Heliothis zea* stimulated the SA pathway (thus inhibiting the JA pathway) in soybean and induced resistance against *Pseudomonas* and *Cercospora sojinae*.

The SA- and JA-mediated resistance pathways cannot be considered in isolation, however, and are likely not to be the only resistance pathways present in plants: these may be many and varied, and may mediate interactions between insects and fungi. For example, apart from resource depletion as a possible resistance mechanism, *Rumex obtusifolius* has the following resistance mechanisms:

1. Age-specific constitutive resistance: undamaged and uninfected young developing leaves are resistant to infection by *U. rumicis* (Hatcher *et al.*, 1995a) – as these leaves age, they become susceptible to the pathogen.
2. Localized induced resistance against rust has been demonstrated following artificial damage, insect herbivory and rust infection (Hatcher *et al.*, 1994a, 1995a).
3. Systemic induced resistance against rust has been demonstrated following insect herbivory and rust infection (Hatcher *et al.*, 1994a, 1995a).

Whereas interactions between the SA and JA pathways are mediated at the biochemical level, other interactions may be mediated entirely at the ecological level, and between resistance mechanisms that may share no biochemical similarity. For example, in *Rumex obtusifolius*, resource depletion by the rust *U. rumicis* induces *G. viridula* to move from the infected leaves to healthy younger leaves that are normally protected against the pathogen by age-specific resistance, leading to a greater amount of damage to the plant (Fig. 11.4). This negative effect, i.e. the feeding of *G. viridula* on younger leaves, must affect fitness to a smaller extent than the main effect of age-specific resistance towards the pathogen, which enables the plant always to outgrow a fungal infection.

Effects on the Plant

The preceding section has demonstrated that both negative and positive interactions are possible between insects and fungi. It is also possible for insects and fungi to coexist on plants without discernible effect on one another or in an idiosyncratic fashion so that no patterns are detectable, possibly due to low densities (Tipping, 1993; Moran, 1998; Moran and Schultz, 1998).

The interactions between insects and fungi can also be classified as reciprocal (i.e. both agents benefit or are harmed by the relationship) or non-reciprocal (i.e. one agent benefits, the other does not benefit). Based on the nature of the herbivore–pathogen relationships discussed above, it is possible to construct a classification of the effects of the interaction on the host plant

(Hatcher, 1995; Table 11.2). The effects range from synergistic, where the effect of the interaction is greater than the sum of the individual effects, through to inhibitory, where the effect of the interaction on the host is less than the effect of the least damaging agent alone.

In the context of weed biocontrol, an important question is whether the nature of the herbivore–pathogen interaction can be used to predict the effect of the interaction on the host plant. This question also highlights the lack of ecological understanding of such interactions since few studies that have examined the nature of the interactions between fungi and insects have looked at their effects on the host plant. Conversely, few studies that have investigated the effect of insect and pathogen combinations on the host plant have also investigated the nature of the insect–pathogen interaction. To investigate this we have examined some systems in which both aspects – the nature of the herbivore–pathogen interaction and its effect on the host – have been studied.

Rumex

Over the last decade we have carried out a number of experiments designed to investigate the effects of combined insect (the chrysomelid beetle *Gastrophysa viridula*) and fungal (the rust fungus *Uromyces rumicis*) attack on their host plants *Rumex crispus* and *R. obtusifolius*. These experiments have ranged from simple microplot designs over one summer (Hatcher *et al.*, 1994c), through overwintering experiments (Hatcher, 1996), field experiments with varying nitrogen fertilizer concentrations (Hatcher *et al.*, 1997a) to long-term field experiments (Hatcher, 1999) over the life-time of the plant.

Since our previous studies had quite consistently indicated a reciprocal negative relationship between the beetle and the fungus in this system, the expectation was an equivalent effect on the host at best, and an inhibitory effect at worse. There was some evidence of inhibition in one experiment, where the predicted effect (calculated from the effect of the insect and fungus alone, and the length of time each agent was on the plant) of a treatment on

Table 11.2. Classification of the effects of insect–fungus interactions on their host plants. (From Hatcher, 1995.)

Synergistic: Interaction causes a reduction in a plant variable significantly greater than that obtained from adding the damage from insect and fungus alone
Additive: Interaction causes a reduction in a plant variable equivalent to that obtained from adding the damage from insect and fungus alone
Equivalent: Interaction causes a reduction in a plant variable equivalent to the damage obtained from either insect or fungus alone (usually the agent causing the greater damage)
Inhibitory: Interaction causes a reduction in a plant variable significantly less than that caused by the weaker of the two agents alone

R. crispus of the beetle followed by the rust was 40% greater than the actual effect of the treatment (Hatcher *et al.*, 1994c). This result was explained by the inhibition of rust infection by beetle grazing (Hatcher *et al.*, 1994a, 1995a). However, this is the only example of a less than equivalent effect of the combined insect and rust treatment on the host, and in some cases (Hatcher *et al.*, 1994c, 1997a; Hatcher, 1996) the effect has been additive (Fig. 11.3).

One possible explanation for the greater than predicted effects of the herbivore–pathogen combination came from an experiment designed to investigate whether rust infection altered host choice in *G. viridula*. Although *G. viridula* will feed and egg-lay upon both *R. crispus* and *R. obtusifolius*, when given the choice between healthy plants of both species it prefers the latter (Bentley and Whittaker, 1979; Fig. 11.4). When given a choice between infected *R. obtusifolius* and healthy *R. crispus* in laboratory choice tests, *G. viridula* gravid females chose *R. crispus*, a host shift induced by the infection. However, when

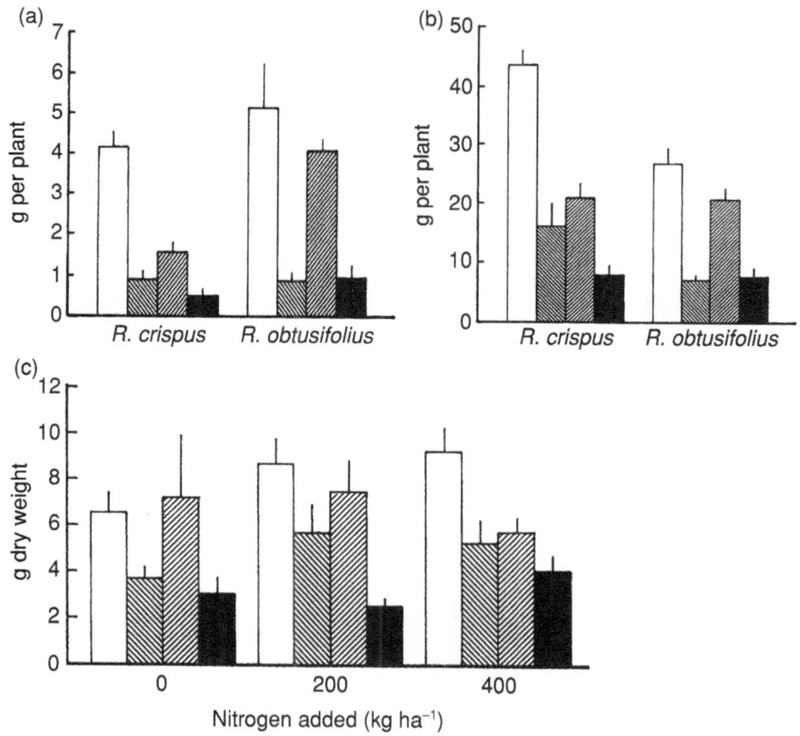

Fig. 11.3. The effects of *Gastrophysa viridula* and *Uromyces rumicis*, alone and combined, on *Rumex*: (a) shoot dry weight; (b) root dry weight of seedlings sown in the field in August and harvested in February; (c) *Rumex obtusifolius* shoot dry weight grown in the field with 0, 200, or 400 kg ha^{-1} nitrogen fertilization added. Mean + SE given. Treatments: (□) control, healthy plants; (▧) herbivory by *G. viridula*; (▨) infection by *U. rumicis*; (■) herbivory by *G. viridula* and infection by *U. rumicis*. n = 11–15. (From Hatcher, 1996; Hatcher *et al.*, 1997c.)

this effect was investigated in the field, gravid females feeding on infected plants switched not between species but rather from feeding predominantly on the older to feeding on the younger leaves (Fig. 11.4). Even though the majority of leaves on the plants were infected in this experiment, young leaves of *R. crispus* and *R. obtusifolius* are resistant to rust infection (Hatcher *et al.*, 1995a). The beetles responded to rust infection of their host by selecting the younger, rust-free leaves, and we believe that this spatial separation of insect and pathogen leads to the additive amount of damage observed.

We believe that this spatial 'displacement' of one organism because of host damage by the other is important for at least two reasons. Firstly, it will tend to minimize many of the potential interactions between the two organisms, the probable exception being indirect interactions mediated through systemic induced changes in the host. Secondly, in terms of effects on the host it also reduces the potential effects of facultative mycophagy: the herbivore will not simply remove tissue that is already 'lost' in terms of its contribution to host

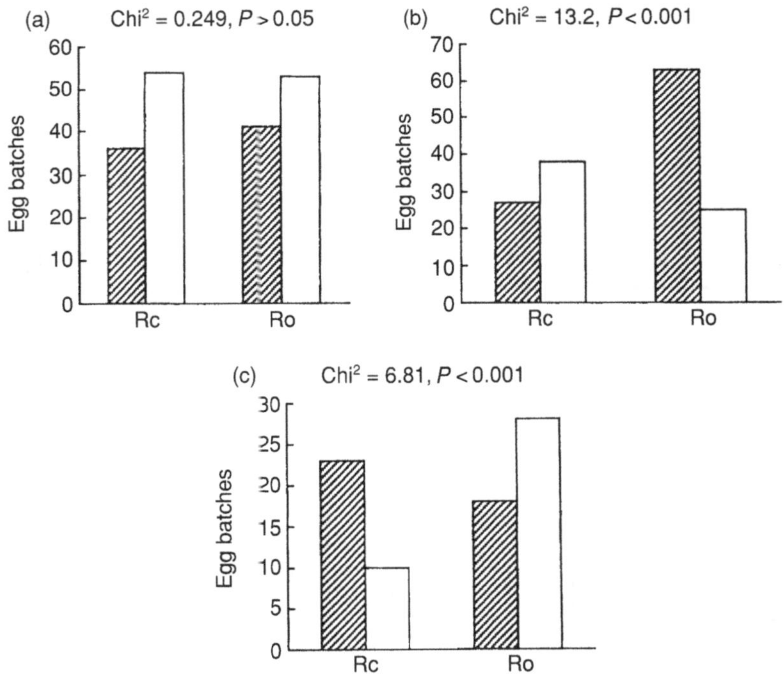

Fig. 11.4. Within-plant host-shifting by *Gastrophysa viridula*. Egg batches per plant laid over 1 week when a population of *G. viridula* was given a choice, in the field, between: (a) healthy *R. crispus* (Rc) and healthy *R. obtusifolius* (Ro); (b) healthy *R. crispus* and infected *R. obtusifolius*; and (c) infected *R. crispus* and healthy *R. obtusifolius*. Hatched bars indicate young leaves (resistant to *U. rumicis* infection); open bars indicate older leaves (not resistant to *U. rumicis* infection). Summary statistics of Chi^2 contingency tests given. $n = 36$. (From Hatcher, unpublished.)

assimilation. Spatial separation of insects and fungi on the host plant could be very important for weed biocontrol: not least because combinations of agents that look initially unpromising from studies of their interactions with one another could have unexpected effects on the host, perhaps leading to good biocontrol. While we are unaware of other studies reporting spatial separation between insects and fungi as described above, such responses would be consistent with the avoidance of fungal-infected tissues by herbivores that has been shown in several studies (Kok *et al.*, 1996; Tinney *et al.*, 1998a; Wilson *et al.*, 2000).

Another effect that could uncouple host responses from negative interactions between insects and fungi is the observation that when confronted with a nutritionally inferior food, phytophagous insects often eat more of it in order to compensate for its poor quality (Scriber and Slansky, 1981). This also happens with insects feeding on some fungal-infected tissue (Tinney *et al.*, 1998a). In the *Rumex* system, *G. viridula* feeding on *U. rumicis*-infected *R. crispus* consumed up to 2.5 times as much plant material as those feeding on healthy *R. crispus* (Hatcher *et al.*, 1994b). As well as short-term effects on the host, this increased food consumption could ameliorate some of the negative effects of the lowered food quality on the populations of the insect discussed above, and thus have long-term effects on the target weed.

Carduus

The musk thistle, *Carduus thoermeri*, is an introduced weed in many parts of the USA. Several species of insect have been introduced as biocontrol agents against it, including the weevils, *Rhinocyllus conicus* and *Trichosirocalus horridus*, and the chrysomelid, *Cassida rubiginosa* (all Coleoptera) (Kok *et al.*, 1996). In addition, the rust fungus *Puccinia carduorum* was introduced as an agent in 1987. The rust infects most parts of the plant, and thus all three insects are likely to be exposed to it: *C. rubiginosa* most, as it is a leaf-feeder.

The effect of the rust and introduced insects was generally additive, except in 1 year of a 3-year field experiment, when the effect of the insects and pathogen on seed production was synergistic (Fig. 11.5) (Baudoin *et al.*, 1993). This is particularly noteworthy, as only seed production would be expected to have an important effect on the population development of the thistle.

This effect was achieved by the insects and fungi having no negative effect on each other. All three insects have been shown to carry rust uredospores (Kok and Abad, 1994) while there was no overall effect of the insects on the rust (Baudoin *et al.*, 1993). Likewise, the rust did not affect any part of the life cycle of the insects (Kok *et al.*, 1996). In laboratory experiments, *C. rubiginosa* and *T. horridus* adults chose healthy in preference to infected leaves for feeding and ovipositing; and tended to avoid feeding and ovipositing on the pustules while feeding on infected leaves (Kok *et al.*, 1996). Again, this dispersal will minimize interference between agents.

Senecio vulgaris

In a controlled-environment study, Tinney (1997) studied the effect of herbivory by *Tyria jacobaeae* (Lepidoptera) followed by infection by the rust *Puccinia lagenophorae* on growth of the annual weed *Senecio vulgaris*. Herbivory alone significantly reduced all measured components of plant growth, including root and shoot weight, leaf area and capitula number (Fig. 11.6). Rust infection alone had no significant effect, apart from a reduction in capitula number. The damage caused by the combination of insect and pathogen was

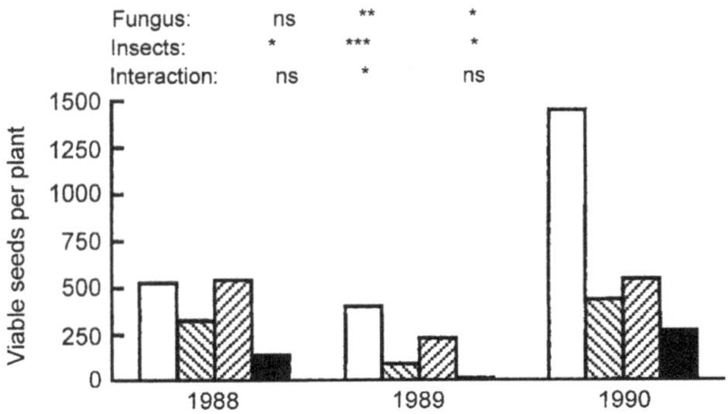

Fig. 11.5. The effects of herbivores and the rust *Puccinia carduorum* alone and together on number of viable seeds per plant of *Carduus thoermeri* in a 3-year field exclusion experiment. The insects included *Cassida rubiginosa, Rhinocyllus conicus* and *Trichosirocalus horridus*. Treatments: (□) control, healthy plants; (▧) insect herbivory; (▨) infection by *P. carduorum*; (■) combined insect herbivory and *P. carduorum* infection. Summary statistics of two-factor ANOVA given for each year: ns, $P > 0.05$; $*P < 0.05$; $**P < 0.01$; $***P < 0.001$. (Data from Baudoin *et al.*, 1993.)

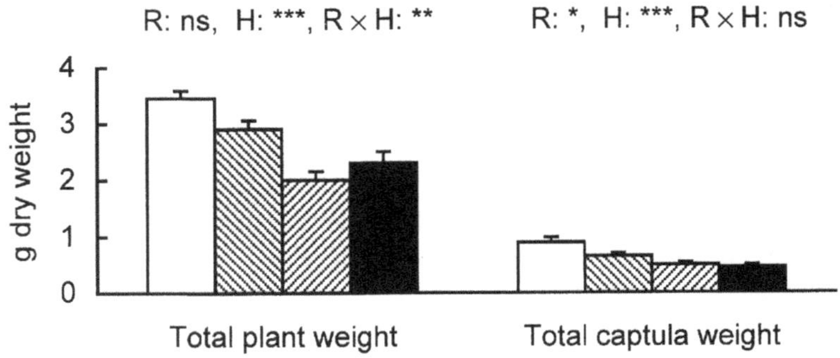

Fig. 11.6. The effect of herbivory by *Tyria jacobaeae* and infection by *Puccinia lagenophorae* alone and together on *Senecio vulgaris*. Treatments: (□) control, undamaged plants; (▧) infection by *P. lagenophorae*; (▨) herbivory by *T. jacobaeae*; (■) combined *P. lagenophorae* infection and *T. jacobaeae* herbivory. Mean + SE given. Summary statistics of two-factor ANOVA between rust (R) and herbivore (H) given: ns, $P > 0.05$; $*P < 0.05$; $**P < 0.01$; $***P < 0.001$. $n = 12–15$. (Data from Tinney, 1997.)

equivalent to that caused by the insect alone, including the effect on capitula number (Fig. 11.6). This is largely what was expected from the study of the interactions between the insect and pathogen. The insect was not affected by feeding on rust-infected *S. vulgaris* (Tinney *et al.*, 1998a) and herbivore damage had little effect on subsequent colonization by the pathogen, although it is possible that either herbivore-damaged leaves became more susceptible or undamaged leaves became less susceptible (Tinney, 1997).

Emex australis

Emex australis is an annual weed of crops and pastures in Australia (Gilbey and Weiss, 1980). Several biocontrol agents have been proposed for this weed; one, the weevil *Perapion antiquum*, has been introduced and released at sites throughout southern Australia since 1974 (Shivas and Scott, 1993). This weevil reduced the stem length and number of fruits in mature plants by 68% (Fig. 11.7). Another potential biocontrol agent, the fungus *Phomopsis emicis*, alone also reduced stem length and fruit number. However, when the fungus and weevil were combined the effect on dry weight of stem, leaves and roots, and fruit number was equivalent and the effect on stem length and fruit weight was inhibitory (Fig. 11.7).

These effects were due to the negative insect–pathogen interaction: it was suggested that attack by the weevils induced a host response that slowed the development of the fungus (stems that had been attacked by *P. antiquum*

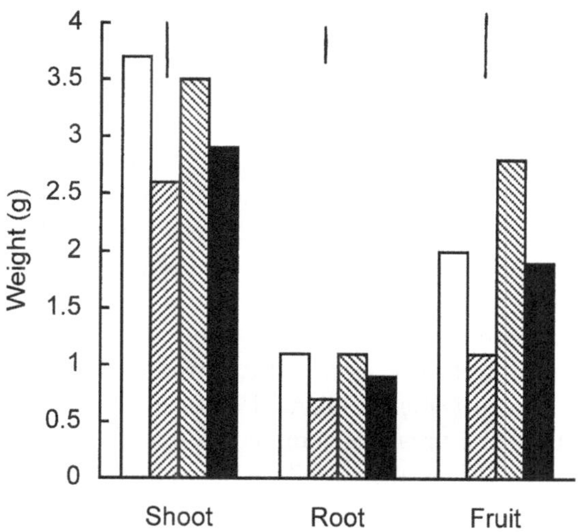

Fig. 11.7. The effect of herbivory by *Perapion antiquum* and infection by *Phomopsis emicis* alone and together on dry weight of *Emex australis*. Treatments: (□), control, undamaged plants; (▨) infection by *P. emicis*; (▧) herbivory by *P. apion*; (■) combined infection by *P. emicis* and herbivory by *P. apion*. Bars indicate SE (32 df). (Data from Shivas and Scott, 1993.)

appeared thicker and more woody than those of non-infected plants). Similarly, the ability of the fungus to cause stem collapse meant that there were fewer sites where the weevil could lay eggs and feed (Shivas and Scott, 1993).

In this example, a reciprocally negative herbivore–pathogen interaction led to at best an equivalent effect on the host, unlike in the *Rumex* example above. We suggest that this is because there was no room for escape within the *Emex* system; both the pathogen and the weevil predominately utilize the stem of the plant and thus cannot fail to interact. Thus, such systems should be avoided for weed biocontrol unless the herbivore–pathogen interaction is reciprocally positive. An example of the latter in a crop system is the maize–*Ostrinia nubilalis* (Lepidoptera)–*Fusarium* system where insect damage facilitates entry of the pathogen and pathogen infection enhances larval development.

Proposed uses in weed biocontrol

There are a few examples in the literature of the use of non-vector herbivore–pathogen interactions in weed biocontrol. The most widely cited example is on water-hyacinth, *Eichhornia crassipes*, control in southern USA. Several insects, including the weevils *Neochetina eichhorniae* and *N. bruchi*, and the pyralid moth *Sameodes albiguttalis* have been introduced to control the weed (Center *et al.*, 1990). It was noted that plants damaged by these insects had also come under attack by pathogens (Charudattan *et al.*, 1978), and that parts of water-hyacinth with larval tunnels usually became necrotic and rotted as a result of secondary microbial infection (Charudattan, 1986). One of the most virulent pathogens isolated from these plants, *Acremonium zonatum*, caused severe damage to the plant, but neither it nor the weevils alone killed the plants. However, combined insect and pathogen application resulted in the death of all 12 trial plants (Charudattan *et al.*, 1978). Likewise, a combination of the pathogen *Cercospora rodmanii* and the weevils caused an additive amount of damage to the water-hyacinth, and 99% of the plants with the combined herbivore–pathogen treatment died after 7 months (Charudattan, 1986).

A herbivore–pathogen interaction may be occurring on water-hyacinth elsewhere: with the weevil *N. eichhorniae*, introduced into South Africa, and the fungus *Cercospora piaropi* (Cilliers, 1990). In Egypt, *Alternaria alternata* infected over twice the area on wounded as opposed to undamaged plants (Elwakil *et al.*, 1990). The same weevil was introduced into southern Thailand and Malaysia. Here, water-hyacinth inoculated with the indigenous pathogen *Myrothecium roridum* at field levels and which were subsequently exposed to the weevil were significantly reduced in growth compared to those with the weevil alone (Caunter and Mohamed, 1990; Caunter and Lee, 1996). It appears that herbivore–pathogen interactions are ripe for development for water-hyacinth control.

However, any wider consideration of weed biocontrol strategies begs the question of whether the apparent potential for exploiting herbivore–pathogen interactions in water-hyacinth is anything more than a function of the ecology of this particular weed? If we consider weed–herbivore–pathogen interactions as a special case of the 'disease triangle', it might be argued that the water-hyacinth system is almost unique in that all its elements are optimum for maximum host damage, i.e.:

1. It is a floating water weed, with large fleshy leaf bases which are vulnerable to attack and is largely clonal, resulting in genetically uniform host populations.
2. The herbivores that have been used in biocontrol create feeding tunnels that represent wounds and possibly infection sites for the facultative pathogens that are involved in this system.
3. Interactions take place in an aquatic environment, providing conditions ideal for pathogen infection processes and dispersal.

The ecology of a few other weeds, notably those of aquatic environments, may similarly predispose them to insect–herbivore interactions comparable to those in water-hyacinth. However, we would argue that water-hyacinth is a poor model for the majority of weeds, certainly those of terrestrial habitats.

None the less, other workers have proposed using herbivore–pathogen combinations for weed biocontrol. These include: Srikanth and Pushpalatha (1991) who proposed using a pathogen–insect combination to control *Parthenium hysterophorus* in India; Alber *et al.* (1984) who proposed combining the rust *Puccinia expansa* and *Tyria jacobaeae* and *Longitarsus jacobaeae* to control *Senecio jacobaea*; Leen *et al.* (1996) who proposed releasing *Septoria passiflorae* along with a variety of Lepidoptera against *Passiflora mollissima*; and Wilson *et al.* (1996) who proposed introducing the coelomycete pathogen *Phloeospora mimosae-pigrae* along with weevils and Lepidoptera against *Mimosa pigra*.

Many of these accounts of herbivore–pathogen combinations give no indication that any investigation into the nature of the interaction is being or has been, carried out. This is particularly worrying when introducing an exotic pathogen, as the diversity of interactions evident in studies with *Rumex*, *Carduus*, *Emex* and *Senecio* suggest that it is quite possible to introduce a pathogen that, while having no effect on controlling the weed, could have a seriously detrimental effect on established biocontrol agents.

Effects of the Environment – the Systems Management Approach

The systems management approach to weed biological control, as developed by Müller-Schärer and Frantzen (1996), is based on the management of a weed pathosystem in order to maximize the natural spread and the severity of a pathogen. The aim of this system is to shift the balance between weed host and pathogen in favour of the pathogen, mainly by stimulating the build-up of a disease epidemic. It excludes disruptive events, such as the introduction of exotic control organisms or the mass release of inoculum (Müller-Schärer and Frantzen, 1996).

This approach, therefore, seeks to manipulate the environment favourably for the pathogen. As the pathogen is assumed to be indigenous, this approach is suited to the control of native or long-established weeds, which are also likely to be attacked by insect herbivores. Thus, this approach should also consider manipulating the environment so as to increase/decrease herbivores, depending on their effect on the pathogen. This has been suggested for the control of *Senecio vulgaris* (Frantzen and Hatcher, 1997).

The effects of the environment should also be considered on other potential herbivore–pathogen combinations for weed biological control. For example, the effect of soil nitrogen fertilization on insects (Scriber, 1984) and fungi (Huber and Watson, 1974; Anderson and Dean, 1986) is an environmental factor that has received considerable attention in two-way interactions, but very little attention in herbivore–pathogen interactions. In the interaction between infection of soybean by *Diaporthe phaseolorum* and defoliation by *Pseudoplusia includens* (Lepidoptera), the reduction in stem canker length due to fungal infection in response to defoliation was diminished in plants that were fertilized by ammonium or a commercial inoculant of an N_2-fixing bacterium (Russin *et al.*, 1989). Similarly, the inhibitory effect of *G. viridula* grazing on *Venturia rumicis* infection in the spring was inhibited by increasing nitrogen fertilization, although the effect of grazing on *V. rumicis* in the autumn and *R. rubella* in the spring was enhanced by increasing nitrogen fertilization (Hatcher and Paul, 2000a).

The density and percentage of *U. rumicis* pustules sporulating 8 days after infection and the leaf area consumed and number of eggs laid by *G. viridula* decreased as the concentration of nitrate given to *R. obtusifolius* increased in laboratory experiments (Hatcher *et al.*, 1997b) (Fig. 11.8). Likewise, first instar mortality of *G. viridula* was increased and adult fecundity was decreased additively by the combination of *U. rumicis* infection and decreasing the nitrate concentration fed to plants from 10 to 1 mmol l^{-1} (Hatcher *et al.*, 1997b). However, in field experiments the effects of the beetle and the rust on the plant were consistently additive over a range of nitrogen fertilization concentrations (Hatcher *et al.*, 1997a).

Other aspects of the weed's environment, such as mowing, important to weeds of grassland, are only just being applied to the study of herbivore–pathogen interactions. Regrowth foliage following complete defoliation can differ considerably in suitability for insect herbivores than primary foliage (Meijden *et al.*, 1988; Vrieling *et al.*, 1996). However, in *Rumex obtusifolius* the effects of cutting on the interaction between herbivore and pathogen are likely to be small or masked by other effects. In a long-term experiment on the effect of herbivores and fungi on *R. obtusifolius*, plants with added insects and fungi were still being severely, and at least equivalently damaged by the agents, 3 years into a management regime consisting of mowing every

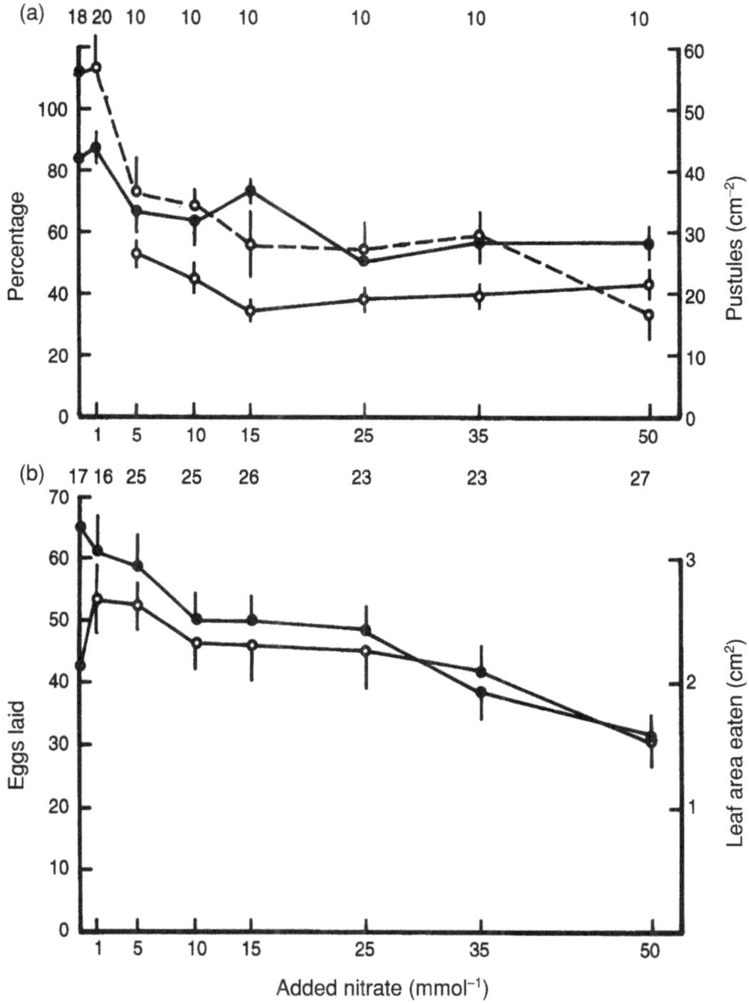

Fig. 11.8. The effects of nitrate fertilization of *Rumex obtusifolius* on: (a) *Uromyces rumicis* infection and (b) *Gastrophysa viridula* egg laying and feeding in a laboratory experiment. Plants were artificially inoculated with *U. rumicis* spores, and pustules were examined 8 days later. Gravid female *G. viridula* were fed on excised leaves for 24 h and number of eggs laid and leaf area eaten was recorded. (a) (●—●) percentage of *U. rumicis* pustules sporulating; (○—○) percentage dry weight of leaves infected; (○---○) pustules cm^{-2}. (b) (●—●) leaf area eaten per beetle; (○—○) mean number of eggs laid per beetle. Means ± SE given, although SE omitted sometimes for clarity. Numbers indicate sample size. (From Hatcher *et al.*, 1997b.)

6 weeks during the growing season (Hatcher, 1999). Similarly, preliminary experiments demonstrate that *G. viridula* larval development and survival is unaffected by feeding on regrowth leaves from a variety of types of defoliation. However, regrowth foliage is altered in some way – both *Uromyces rumicis* pustule density and colonization by *Aphis fabae* were reduced on the regrowth foliage of plants that had been defoliated twice by a variety of means (Fig. 11.9).

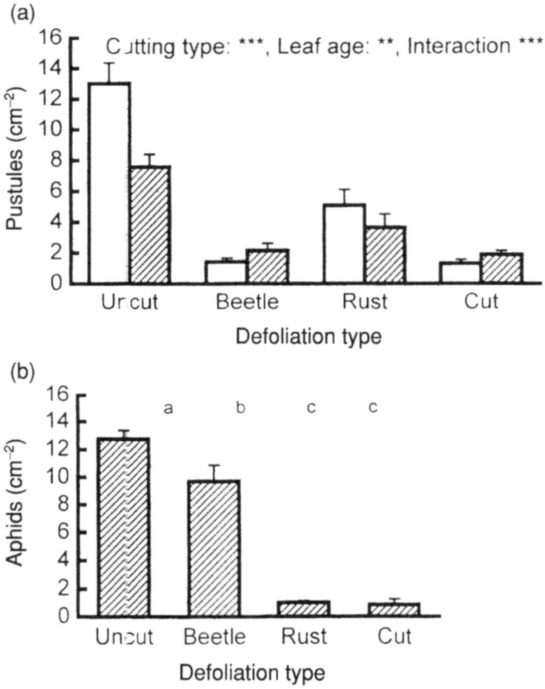

Fig. 11.9. The effect of different types of defoliation of *Rumex obtusifolius* on: (a) subsequent infection by *Uromyces rumicis* (b) colonization by *Aphis fabae*. Plants were grown for 6 weeks before defoliation, they were defoliated again by the same agent 6 weeks later, and 3 weeks later infected with *U. rumicis* on a mature leaf (□) and younger leaf (▨) using our established methods. Pustule density was recorded after 9 days. (b) Another set of previously defoliated plants became colonized naturally by *Aphis fabae*: aphid density was recorded 3 weeks after the final defoliation. Defoliation types: 'uncut', plants remained uncut; 'beetle', defoliated (over 75% leaf area removed) by third-instar *Gastrophysa viridula* larvae; 'rust', plants defoliated (approx. 50% leaf area removed) by *Uromyces rumicis* infection; 'cut', plants cut at ground level (100% leaf area removed). $n = 15$. Summary statistics: (a) two-factor ANOVA; $**P < 0.01$; $***P < 0.001$. (b) Tukey-HSD multiple-range test from ANOVA, different letters indicate significant difference ($P < 0.05$) between means. (From Hatcher and Bloxsom, unpublished.)

Conclusions

In one of the first reviews of the use of plant pathogens in weed control, Wilson (1969) suggested that: (i) more investigation of the combined effects of insects and diseases on weeds was needed; (ii) the combined effect of insects and pathogen damage on weeds had not been measured; and (iii) that the predisposition of plants to insect attack by disease and vice versa needed consideration. Although advances in molecular biology have shed light on the third point and have shown that predisposition is one possibility from a wide range of interactions, much the same comments could be made on the basis of the current literature. Indeed, reviewers over the past 30 years have often re-phrased Wilson's (1969) key points, and highlighted the lack of integration between biocontrol based on herbivores and that based on pathogens. For example, Hill (1996) noted that '. . . despite the common interest of plant pathologists and entomologists in the biological control of weeds, the exploitation of such synergies [herbivore–pathogen interactions] continue to be a neglected field' and Cullen (1996) observed that 'we might expect the business of this symposium to be concerned with the integration of insects and pathogens, though there has been virtually no mention of this as an area of interest in its own right'. Even more recently, in the book edited by Hawkins and Cornell (1999) on the theory of biological control, there is no mention of the use of herbivore–pathogen interactions.

Yet, understanding of herbivore–pathogen interactions has moved forward in the past 30 years, even if only in a few well-defined areas. It is, perhaps, clearer that herbivore–pathogen interactions make a substantial contribution to effective biological control of weeds in some systems. Water-hyacinth can now be added to *Opuntia* as a 'classic' example of such interactions, with the notable difference that the former involved deliberate consideration of possible synergy between two agents. These two examples share certain features in common, notably that both involve necrotrophic pathogens exaggerating initial herbivore damage by an insect that causes gross feeding wounds. We anticipate that new examples of this type of interaction will continue to appear, especially perhaps for aquatic weeds. New examples may be as likely to result from chance combinations of agents, for example following the introduction of an effective herbivore, as from the planned use of herbivores and pathogens. Indeed, the current state of knowledge provides a very limited foundation for predicting how herbivores and pathogens might interact in any particular biocontrol system.

It has been noted before that one of the major shortcomings of classical biological control of weeds is its lack of predictability: Crawley (1990), in noting that about 85% of the introduced agents fail to control the target weed, stated that 'ecological theory has made virtually no contribution to the way in which biological control schemes are planned or executed'. Since our understanding of most herbivore–pathogen interactions falls very far short of 'ecological theory', it is not surprising that planned combinations of agents remain a rather distant hope except, perhaps, at the most general level.

In broader terms, the possible benefits of herbivore–pathogen interactions for weed control relates to the question of whether combinations of agents are necessarily superior to the use of single agents. Myers (1985) has pointed out the frequent success from a single agent and suggests that each introduction of a new agent species is a 'lottery' as regards its chances of reducing the density of the host plant. However, given the high failure rate of introduced agents (Crawley, 1990) the odds in this particular lottery seem rather poor. The reasons for the failure of an agent to produce significant biocontrol of the target weed are manifold, but include:

1. Failure of initial establishment or dispersal.
2. Failure to establish or maintain an adequate population density due, for example, to sub-optimal environmental conditions or host genetic diversity.
3. Failure to induce sufficient damage to the target to reduce its population significantly, which may be due to disease escape, tolerance or resistance, as well as inadequate population density of the biocontrol agent.

Clearly, only the last of these constraints could be overcome by the use of combinations of herbivores and pathogens. Harris (1984) has hypothesized that successful biocontrol is achieved through cumulative stress on the host plant and that the more species that are introduced, the greater the stress on the target weed. One clear message from studies of pathogen–herbivore interactions is that 'more species' does not necessarily equal 'greater stress', due to potential negative interactions between organisms. This should encourage a cautious approach to the use of pathogen–herbivore combinations. Without an understanding of the nature of the interactions between specific organisms being considered for combined use, there is a significant risk that one organism may reduce the effect of the other, compromising the overall success of weed control. The results obtained for *Emex australis* show that this caveat applies even to interactions such as a necrotroph invading feeding wounds, that might intuitively be expected to be synergistic. Nevertheless, in the short to medium term it may be interactions of this type that are most readily developed for biocontrol: the essential first question is whether there is evidence that agents may act together within the native range? Here, the key is appropriate ecological experimentation in the earliest stages of developing a biocontrol programme; and above all the need for observations that are not too confined by the traditional split between entomology and pathology.

Field studies are probably less likely to detect more subtle interactions, including those based on 'cumulative stress'. We think there is scope for the exploitation of such interactions, as is evident from the data for interactions between herbivores and rust fungi in *Carduus* and to some extent, for *Rumex*. In the short to medium term, exploitation of these interactions in biocontrol will rely on careful experimentation to establish the nature of interactions. In the longer term, we hope that the challenge of understanding how plants cope with multiple enemies will stimulate ecologists to address herbivore–pathogen interactions, leading to a rational foundation for planning combinations of agents for future biocontrol programmes.

Acknowledgements

We thank the Natural Environment Research Council for funding our work on *Rumex*, and the Nuffield Foundation which supported the work on interactions between consumers and regrowth foliage reported in Fig. 11.9.

References

Abbas, H.K. and Mulrooney, J.E. (1994) Effect of some phytopathogenic fungi and their metabolites on growth of *Heliothis virescens* (F.) and its host plants. *Biocontrol Science and Technology* 4, 77–87.

Agrios, G.N. (1980) Insect involvement in the transmission of fungal pathogens. In: Harris, K.F. and Maramorosch, K. (eds) *Vectors of Plant Pathogens*. Academic Press, New York, pp. 293–324.

Ajlan, A.M. and Potter, D.A. (1991) Does immunization of cucumber against anthracnose by *Colletotrichum lagenarium* affect host suitability for arthropods? *Entomologia Experimentalis et Applicata* 58, 83–91.

Ajlan, A.M. and Potter, D.A. (1992) Lack of effect of tobacco mosaic virus-induced systemic acquired resistance on arthropod herbivores in tobacco. *Phytopathology* 82, 647–651.

Alber, G., Défago, G., Sedlar, L. and Kern, H. (1984) Damage to *Senecio jacobaea* by the rust fungus *Puccinia expansa*. In: Delfosse, E.S. (ed.) *Proceedings of the VIth International Symposium on Biological Control of Weeds*. Agriculture Canada, Ottawa, pp. 587–592.

Anas, O. and Reeleder, R.D. (1987) Recovery of fungi and arthropods from sclerotia of *Sclerotinia sclerotiorum* in Quebec muck soils. *Phytopathology* 77, 327–331.

Anderson, D.L. and Dean, J.L. (1986) Relationship of rust severity and plant nutrients in sugarcane. *Phytopathology* 76, 581–585.

Apriyanto, D. and Potter, D.A. (1990) Pathogen-activated induced resistance of cucumber: response of arthropod herbivores to systemically protected leaves. *Oecologia* 85, 25–31.

Barbe, G.D. (1964) Relation of the European earwig to snapdragon rust. *Phytopathology* 54, 369–371.

Baudoin, A.B.A.M., Abad, R.G., Kok, L.T. and Bruckart, W.L. (1993) Field evaluation of *Puccinia carduorum* for biological control of musk thistle. *Biological Control* 3, 53–60.

Bentley, S. and Whittaker, J.B. (1979) Effects of grazing by a chrysomelid beetle *Gastrophysa viridula* on competition between *Rumex obtusifolius* and *Rumex crispus*. *Journal of Ecology* 67, 79–90.

Bostock, R.M. (1999) Signal conflicts and synergies in induced resistance to multiple attackers. *Physiological and Molecular Plant Pathology* 55, 99–109.

Bultman, T.L. and Mathews, P.L. (1996) Mycophagy by a millipede and its possible impact on an insect–fungus mutualism. *Oikos* 75, 67–74.

Carruthers, R.I., Bergstrom, G.C. and Haynes, P.A. (1986) Accelerated development of the European corn borer, *Ostrinia nubilalis* (Lepidoptera: Pyralidae), induced by interactions with *Colletotrichum graminicola* (Melanconiales: Melanconiaceae), the causal fungus of maize anthracnose. *Annals of the Entomological Society of America* 79, 385–389.

Carter, W. (1973) *Insects in Relation to Plant Disease*, 2nd edn. John Wiley & Sons, New York.

Caunter, I.G. and Lee, K.C. (1996) Initiating the use of fungi for biocontrol of weeds in Malaysia. In: Moran, V.C. and Hoffmann, J.H. (eds) *Proceedings of the IX International Symposium on Biological Control of Weeds.* University of Cape Town, South Africa, pp. 249–252.

Caunter, I.G. and Mohamed, S. (1990) The effect of *Neochetina eichhorniae* and *Myrothecium roridum* alone and in combination for the initial biological control of waterhyacinth. *Journal of Plant Protection in the Tropics* 7, 133–140.

Center, T.D., Cofrancesco, A.F. and Balciunas, J.K. (1990) Biological control of aquatic and wetland weeds in the southeastern United States. In: Delfosse, E.S. (ed.) *Proceedings of the VIIth International Symposium on Biological Control of Weeds.* Istituto Sperimentale per la Patalogia Vegetale, Ministero dell'Agricoltura e delle Foreste, Rome, pp. 239–262.

Charudattan, R. (1986) Integrated control of waterhyacinth (*Eichhornia crassipes*) with a pathogen, insects, and herbicides. *Weed Science* 34 (Suppl. 1), 26–30.

Charudattan, R., Perkins, B.D. and Littell, R.C. (1978) Effects of fungi and bacteria on the decline of arthropod-damaged waterhyacinth (*Eichhornia crassipes*) in Florida. *Weed Science* 26, 101–107.

Chiang, H.C. and Wilcoxson, R.D. (1961) Interactions of the European corn borer and stalk rot in corn. *Journal of Economic Entomology* 54, 850–852.

Cilliers, C.J. (1990) Biological control of aquatic weeds in South Africa – an interim report. In: Delfosse, E.S. (ed.) *Proceedings of the VIIth International Symposium on Biological Control of Weeds.* Istituto Sperimentale per la Patalogia Vegetale, Ministero dell'Agricoltura e delle Foreste, Rome, pp. 263–267.

Clay, K. (1997) Fungal endophytes, herbivores and the structure of grassland communities. In: Gange, A.C. and Brown, V.K. (eds) *Multitrophic Interactions in Terrestrial Systems.* Blackwell Science, Oxford, pp. 151–169.

Crawley, M.J. (1990) Plant life-history and the success of weed biological control projects. In: Delfosse, E.S. (ed.) *Proceedings of the VIIth International Symposium on Biological Control of Weeds.* Istituto Sperimentale per la Patalogia Vegetale, Ministero dell'Agricoltura e delle Foreste, Rome, pp. 17–26.

Cullen, J.M. (1996) Integrated control and management: synthesis of session 6. In: Moran, V.C. and Hoffmann, J.H. (eds) *Proceedings of the IX International Symposium on Biological Control of Weeds.* University of Cape Town, South Africa, pp. 483–486.

Dillard, H.R. and Cobb, A.C. (1995) Relationship between leaf injury and colonization of cabbage by *Sclerotinia sclerotiorum. Crop Protection* 14, 677–682.

Doares, S.H., Narvaez-Vasquez, J., Conconi, A. and Ryan, C.A. (1995) Salicylic acid inhibits synthesis of proteinase inhibitors in tomato leaves induced by systemin and jasmonic acid. *Plant Physiology* 108, 1741–1746.

Dodd, A.P. (1940) *The Biological Campaign Against Prickly Pear*. Prickly Pear Board, Brisbane.

Elwakil, M.A., Sadik, E.A., Fayzalla, E.A. and Shabana, Y.M. (1990) Biological control of waterhyacinth with fungal plant pathogens in Egypt. In: Delfosse, E.S. (ed.) *Proceedings of the VIIth International Symposium on Biological Control of Weeds*. Istituto Sperimentale per la Patalogia Vegetale, Ministero dell'Agricoltura e delle Foreste, Rome, pp. 483–497.

Ericson, L. and Wennström, A. (1997) The effect of herbivory on the interaction between the clonal plant *Trientalis europaea* and its smut fungus *Urocystis trientalis*. *Oikos* 80, 107–111.

Felton, G.W. and Eichenseer, H. (1999) Herbivore saliva and its effects on plant defense against herbivores and pathogens. In: Agrawal, A.A., Tuzun, S. and Bent, E. (eds) *Induced Plant Defenses Against Pathogens and Herbivores*. APS Press, St Paul, Minnesota, pp. 19–36.

Felton, G.W., Korth, K.L., Bi, J.L., Wesley, S.V., Huhman, D.V., Mathews, M.C., Murphy, J.B., Lamb, C. and Dixon, R.A. (1999) Inverse relationships between systemic resistance of plants to microorganisms and to insect herbivory. *Current Biology* 9, 317–320.

Frantzen, J. and Hatcher, P.E. (1997) A fresh view on the control of the annual plant *Senecio vulgaris*. *Integrated Pest Management Reviews* 2, 77–85.

Gange, A.C. and Bower, E. (1997) Interactions between insects and mycorrhizal fungi. In: Gange, A.C. and Brown, V.K. (eds) *Multitrophic Interactions in Terrestrial Systems*. Blackwell Science, Oxford, pp. 115–132.

Gange, A.C. and West, H.M. (1994) Interactions between arbuscular mycorrhizal fungi and foliar-feeding insects in *Plantago lanceolata* L. *New Phytologist* 128, 79–87.

Gilbey, D.J. and Weiss, P.W. (1980) The biology of Australian weeds. 4. *Emex australis* Steinh. *Journal of the Australian Institute of Agricultural Science* 46, 221–228.

Guevara, R., Rayner, A.D.M. and Reynolds, S.E. (2000) Effects of fungivory by two specialist ciid beetles (*Octotemnus glabriculus* and *Cis boleti*) on the reproductive fitness of their host fungus, *Coriolus versicolor*. *New Phytologist* 145, 137–144.

Harris, M.A., Gardner, W.A. and Oetting, R.D. (1996) A review of the scientific literature on fungus gnats (Diptera: Sciaridae) in the genus *Bradysia*. *Journal of Entomological Science* 31, 252–276.

Harris, P. (1984) Current approaches to biological control of weeds. In: Kelleher, J.S. and Hulme, M.A. (eds) *Biological Control Programmes Against Insects and Weeds in Canada 1969–1980*. Commonwealth Agricultural Bureaux, Farnham, pp. 95–103.

Harris, P. and Clapperton, M.J. (1997) An exploratory study on the influence of vesicular-arbuscular mycorrhizal fungi on the success of weed biological control with insects. *Biocontrol Science and Technology* 7, 193–201.

Hatcher, P.E. (1995) Three-way interactions between plant pathogenic fungi, herbivorous insects and their host plants. *Biological Reviews* 70, 639–694.

Hatcher, P.E. (1996) The effect of insect–fungus interactions on the autumn growth and over-wintering of *Rumex crispus* and *R. obtusifolius* seedlings. *Journal of Ecology* 84, 101–109.

Hatcher, P.E. (1999) A long-term study of the effects of insects and fungi on *Rumex obtusifolius* – the first two years. *Proceedings of the 11th EWRS Symposium 1999*, Basel, p. 95.

Hatcher, P.E. and Ayres, P.G. (1997) Indirect interactions between insect herbivores and pathogenic fungi on leaves. In: Gange, A.C. and Brown, V.K. (eds) *Multitrophic Interactions in Terrestrial Systems*. Blackwell Science, Oxford, pp. 133–149.

Hatcher, P.E. and Ayres, P.G. (1998) The effect of fungal infection and nitrogen fertilization on the carbohydrate composition of *Rumex obtusifolius* leaves. *European Journal of Plant Pathology* 104, 553–559.

Hatcher, P.E. and Paul, N.D. (2000a) Beetle grazing reduces natural infection of *Rumex obtusifolius* by fungal pathogens. *New Phytologist* 146, 325–333.

Hatcher, P.E. and Paul, N.D. (2000b) On integrating molecular and ecological studies of plant resistance: variety of mechanisms and breadth of antagonists. *Journal of Ecology* 88, 702–706.

Hatcher, P.E., Paul, N.D., Ayres, P.G. and Whittaker, J.B. (1994a) Interactions between *Rumex* spp., herbivores and a rust fungus: *Gastrophysa viridula* grazing reduces subsequent infection by *Uromyces rumicis*. *Functional Ecology* 8, 265–272.

Hatcher, P.E., Paul, N.D., Ayres, P.G. and Whittaker, J.B. (1994b) The effect of a foliar disease (rust) on the development of *Gastrophysa viridula* (Coleoptera: Chrysomelidae). *Ecological Entomology* 19, 349–360.

Hatcher, P.E., Paul, N.D., Ayres, P.G. and Whittaker, J.B. (1994c) The effect of an insect herbivore and a rust fungus individually, and combined in sequence, on the growth of two *Rumex* species. *New Phytologist* 128, 71–78.

Hatcher, P.E., Ayres, P.G. and Paul, N.D. (1995a) The effect of natural and simulated insect herbivory, and leaf age, on the process of infection of *Rumex crispus* L. and *R. obtusifolius* L. by *Uromyces rumicis* (Schum.) Wint. *New Phytologist* 130, 239–249.

Hatcher, P.E., Paul, N.D., Ayres, P.G. and Whittaker, J.B. (1995b) Interactions between *Rumex* spp., herbivores and a rust fungus: the effect of *Uromyces rumicis* infection on leaf nutritional quality. *Functional Ecology* 9, 97–105.

Hatcher, P.E., Paul, N.D., Ayres, P.G. and Whittaker, J.B. (1997a) Added soil nitrogen does not allow *Rumex obtusifolius* to escape the effects of insect–fungus interactions. *Journal of Applied Ecology* 34, 88–100.

Hatcher, P.E., Paul, N.D., Ayres, P.G. and Whittaker, J.B. (1997b) Nitrogen fertilization affects interactions between the components of an insect–fungus–plant tripartite system. *Functional Ecology* 11, 537–544.

Hatcher, P.E., Paul, N.D., Ayres, P.G. and Whittaker, J.B. (1997c) The effect of nitrogen fertilization and rust fungus infection, singly and combined, on the leaf chemical composition of *Rumex obtusifolius*. *Functional Ecology* 11, 545–553.

Hawkins, B.A. and Cornell, H.V. (eds) (1999) *Theoretical Approaches to Biological Control*. Cambridge University Press, Cambridge.

Hill, D.S. (1987) *Agricultural Pests of Temperate Regions and their Control*. Cambridge University Press, Cambridge.

Hill, R.L. (1996) Agent performance: synthesis of session 5. In: Moran, V.C. and Hoffmann, J.H. (eds) *Proceedings of the IX International Symposium on Biological Control of Weeds*. University of Cape Town, South Africa, pp. 19–26.

Huber, D.M. and Watson, R.D. (1974) Nitrogen form and plant disease. *Annual Review of Phytopathology* 12, 139–165.

Inbar, M., Doostdar, H., Sonoda, R.M., Leibee, G.L. and Mayer, R.T. (1998) Elicitors of plant defensive systems reduce insect densities and disease incidence. *Journal of Chemical Ecology* 24, 135–149.

Jarvis, J.L., Ziegler, K.E., Webber, D.F. and Guthrie, W.D. (1990) Effect of second-generation European corn borers and stalk rot fungi on popcorn hybrids. *Maydica* 35, 259–265.

Karban, R. and Kuc, J. (1999) Induced resistance against pathogens and herbivores: an overview. In: Agrawal, A.A., Tuzun, S. and Bent, E. (eds) *Induced Plant Defenses Against Pathogens and Herbivores*. APS Press, St Paul, Minnesota, pp. 1–16.

Kingsley, P., Scriber, J.M., Grau, C.R. and Delwiche, P.A. (1983) Feeding and growth performance of *Spodoptera eridania* (Noctuidae: Lepidoptera) on 'vernal' alfalfa, as influenced by *Verticillium* wilt. *Protection Ecology* 5, 127–134.

Klein, T.A. and Auld, B.A. (1996) Wounding can improve efficacy of *Colletotrichum orbiculare* as a mycoherbicide for Bathurst burr. *Australian Journal of Experimental Agriculture* 36, 185–187.

Kok, L.T. and Abad, R.G. (1994) Transmission of *Puccinia carduorum* by the musk thistle herbivores, *Cassida rubiginosa* (Coleoptera: Chrysomelidae), *Trichosirocalus horridus* and *Rhinocyllus conicus* (Coleoptera: Curculionidae). *Journal of Entomological Science* 29, 186–191.

Kok, L.T., Abad, R.G. and Baudoin, A.B.A.M. (1996) Effects of *Puccinia carduorum* on musk thistle herbivores. *Biological Control* 6, 123–129.

Korth, K.L. and Dixon, R.A. (1997) Evidence for chewing insect-specific molecular events distinct from a general wound response in leaves. *Plant Physiology* 115, 1299–1305.

Krause, S.C. and Raffa, K.F. (1992) Comparison of insect, fungal, and mechanically induced defoliation of larch: effects on plant productivity and subsequent host susceptibility. *Oecologia* 90, 411–416.

Kuc, J. (1982) Induced immunity to plant disease. *BioScience* 32, 854–860.

Lappalainen, J.H. and Helander, M.J. (1997) The role of foliar microfungi in mountain birch–insect herbivore relationships. *Ecography* 20, 116–122.

Lappalainen, J., Helander, M.J. and Palokangas, P. (1995) The performance of the autumnal moth is lower on trees infected by birch rust. *Mycological Research* 99, 994–996.

Leach, J.G. (1940) *Insect Transmission of Plant Diseases*. McGraw-Hill, London.

Leath, K.T. and Byers, R.A. (1977) Interaction of *Fusarium* root rot with pea aphid and potato leafhopper feeding on forage legumes. *Phytopathology* 67, 226–229.

Leath, K.T. and Newton, R.C. (1969) Interaction of a fungus gnat, *Bradysia* sp. (Sciaridae) with *Fusarium* spp. on alfalfa and red clover. *Phytopathology* 59, 257–258.

Leen, R., Friesen, R. and Causton, C. (1996) Biological control of *Passiflora mollissima*: what's up? In: Moran, V.C. and Hoffmann, J.H. (eds) *Proceedings of the IX International Symposium on Biological Control of Weeds*. University of Cape Town, South Africa, p. 230.

Lewis, A.C. (1979) Feeding preference for diseased and wilted sunflower in the grasshopper, *Melanoplus differentialis*. *Entomologia Experimentalis et Applicata* 26, 202–207.

McIntyre, J.L., Dodds, J.A. and Hare, J.D. (1981) Effects of localized infections of *Nicotiana tabacum* by tobacco mosaic virus on systemic resistance against diverse pathogens and an insect. *Phytopathology* 71, 297–301.

Meijden, E. van der, Wijn, M. and Verkaar, H.J. (1988) Defence and regrowth: alternative plant strategies in the struggle against herbivores. *Oikos* 51, 355–363.

Moellenbeck, D.J., Quisenberry, S.S. and Colyer, P.D. (1992) *Fusarium* crown-rot development in alfalfa stressed by threecornered alfalfa hopper (Homoptera: Membracidae) feeding. *Journal of Economic Entomology* 85, 1442–1449.

Mondy, N., Charrier, B., Fermaud, M., Pracros, P. and Corio-Costet, M.-F. (1998a) Mutualism between a phytopathogenic fungus (*Botrytis cinerea*) and a vineyard pest (*Lobesia botrana*). Positive effects on insect development and oviposition behaviour. *Comptes Rendus de l'Académie des sciences Serie III, Sciences de la vie/Life Sciences* 321, 665–671.

Mondy, N., Pracros, P., Fermaud, M. and Corio-Costet, M.-F. (1998b) Olfactory and gustatory behaviour by larvae of *Lobesia botrana* in response to *Botrytis cinerea*. *Entomologia Experimentalis et Applicata* 88, 1–7.

Moran, P.J. (1998) Plant-mediated interactions between insects and a fungal plant pathogen and the role of plant chemical responses to infection. *Oecologia* 115, 523–530.

Moran, P.J. and Schultz, J.C. (1998) Ecological and chemical associations among late-season squash pests. *Environmental Entomology* 27, 39–44.

Morrison, K.D., Reekie, E.G. and Jensen, K.I.N. (1998) Biocontrol of common St. Johnswort (*Hypericum perforatum*) with *Chrysolina hyperici* and a host-specific *Colletotrichum gloeosporioides*. *Weed Technology* 12, 426–435.

Müller-Schärer, H. and Frantzen, J. (1996) An emerging system management approach for biological weed control in crops: *Senecio vulgaris* as a research model. *Weed Research* 36, 483–491.

Myers, J.H. (1985) How many insect species are necessary for successful biocontrol of weeds? In: Delfosse, E.S. (ed.) *Proceedings of the VI International Symposium on Biological Control of Weeds*. Agriculture Canada, Vancouver, pp. 77–82.

Old, K.M., Gibbs, R., Craig, I., Myers, B.J. and Yuan, Z.Q. (1990) Effect of drought and defoliation on the susceptibility of eucalypts to cankers caused by *Endothia gyrosa* and *Botryosphaeria ribis*. *Australian Journal of Botany* 38, 571–581.

Padgett, G.B., Russin, J.S., Snow, J.P., Boethel, D.J. and Berggren, G.T. (1994) Interactions among the soybean looper (Lepidoptera: Noctuidae), threecornered alfalfa hopper (Homoptera: Membracidae), stem canker, and red crown rot in soybean. *Journal of Entomological Science* 29, 110–119.

Paul, N.D., Hatcher, P.E. and Taylor, J.E. (2000) Coping with multiple enemies: an integration of molecular and ecological perspectives. *Trends in Plant Science* 5, 220–225.

Pesel, E. and Poehling, H.-M. (1988) Zum Einfluss von abiotischen (Wassermangel) und biotischen (echter Mehltau, *Erysiphe graminis* f. sp. *tritici*) Stessfaktoren auf die Populationsentwicklung der getreideblattläuse *Metopolophium dirhodum* Walk. und *Sitobion avenae* F. *Mitteilungen der Deutschen Gesellschaft für Allgemeine und Angewandte Entomologie* 6, 531–536 (in German with English summary).

Piening, L.J. (1972) Effects of leaf rust on nitrate in rye. *Canadian Journal of Plant Science* 52, 842–843.

Pieterse, C.M.J. and van Loon, L.C. (1999) Salicylic acid-independent plant defence pathways. *Trends in Plant Science* 4, 52–58.

Powell, J.M. (1971) The arthropod fauna collected from the comandra blister rust, *Cronartium comandrae*, on lodgepole pine in Alberta. *Canadian Entomologist* 103, 908–918.

Pratt, R.G., Ellsbury, M.M., Barnett, O.W. and Knight, W.E. (1982) Interactions of bean yellow mosaic virus and an aphid vector with *Phytophthora* root diseases in arrowleaf clover. *Phytopathology* 72, 1189–1192.

Prestidge, R.A. and Ball, O.J.-P. (1997) A catch 22: the utilization of endophytic fungi for pest management. In: Gange, A.C. and Brown, V.K. (eds) *Multitrophic Interactions in Terrestrial Systems.* Blackwell Science, Oxford, pp. 171–192.

Prüter, C. and Zebitz, C.P.W. (1991) Effects of *Aphis fabae* and *Uromyces viciae-favae* on the growth of a susceptible and an aphid resistant cultivar of *Vicia faba*. *Annals of Applied Biology* 119, 215–226.

Ramsell, J. and Paul, N.D. (1990) Preferential grazing by molluscs of plants infected by rust fungi. *Oikos* 58, 145–150.

Rayachhetry, M.B., Blakeslee, G.M. and Center, T.D. (1996) Predisposition of melaleuca (*Melaleuca quinquenervia*) to invasion by the potential biological control agent *Botryosphaeria ribis*. *Weed Science* 44, 603–608.

Rayachhetry, M.B., Elliott, M.L., Center, T.D. and Laroche, F. (1999) Field evaluation of a native fungus for control of melaleuca (*Melaleuca quinquenervia*) in Southern Florida. *Weed Technology* 13, 59–64.

Reddy, M.N. and Rao, A.S. (1976) Changes in the composition of free and protein amino acids in groundnut leaves induced by infection with *Puccinia arachidis* Speg. *Acta Phytopathologica, Academiae Scientiarum Hungaricae* 11, 167–172.

Roberts, A.M. and Walters, D.R. (1989) Shoot:root interrelationships in leeks infected with the rust, *Puccinia allii* Rud.: growth and nutrient relations. *New Phytologist* 111, 223–228.

Russin, J.S., Layton, M.B., Boethel, D.J., McGawley, E.C., Snow, J.P. and Berggren, G.T. (1989) Severity of soybean stem canker disease affected by insect-induced defoliation. *Plant Disease* 73, 144–147.

Russo, V.M., Russo, B.M., Peters, M., Perkins-Veazie, P. and Cartwright, B. (1997) Interaction of *Colletotrichum orbiculare* with thrips and aphid feeding on watermelon seedlings. *Crop Protection* 16, 581–584.

Saikkonen, K., Faeth, S.H., Helander, M. and Sullivan, T.J. (1998) Fungal endophytes: a continuum of interactions with host plants. *Annual Review of Ecology and Systematics* 29, 319–343.

Savopoulou-Soultani, M. and Tzanakakis, M.E. (1988) Development of *Lobesia botrana* (Lepidoptera: Tortricidae) on grapes and apples infected with the fungus *Botrytis cinerea*. *Environmental Entomology* 17, 1–6.

Schoeneweiss, D.F. (1981) The role of environmental stress in diseases of woody plants. *Plant Disease* 65, 308–314.

Schweizer, P., Buchala, A., Dudler, R. and Métraux, J.-P. (1998) Induced systemic resistance in wounded rice plants. *The Plant Journal* 14, 475–481.

Scriber, J.M. (1984) Host-plant suitability. In: Bell, W.J. and Cardé, R.T. (eds) *Chemical Ecology of Insects*. Chapman & Hall, London, pp. 159–204.

Scriber, J.M. and Slansky, F. (1981) The nutritional ecology of immature insects. *Annual Review of Entomology* 26, 183–211.

Shivas, R.G. and Scott, J.K. (1993) Effect of the stem blight pathogen, *Phomopsis emicis*, and the weevil, *Perapion antiquum*, on the weed *Emex australis*. *Annals of Applied Biology* 122, 617–622.

Sipos, L. and Sagi, G. (1987) Tavaszi árpán károsító gabonalevéltetu (*Macrosiphum avenae*) és lisztharmatgomba (*Eryisphe graminis* f. sp. *hordei*) kölcsönhatásának vizsgálata üvegházban. *Növenytermelés* 36, 31–34 (in Hungarian with English summary).

Srikanth, J. and Pushpalatha, N.A. (1991) Status of biological control of *Parthenium hysterophorus* L. in India: a review. *Insect Science and its Application* 12, 347–359.

Stout, M.J. and Bostock, R.M. (1999) Specificity of induced responses to arthropods and pathogens. In: Agrawal, A.A., Tuzun, S. and Bent, E. (eds) *Induced Plant Defenses Against Pathogens and Herbivores*. APS Press, St Paul, Minnesota, pp. 183–209.

Stout, M.J., Workman, K.V., Bostock, R.M. and Duffey, S.S. (1998) Specificity of induced resistance in the tomato, *Lycopersicon esculentum*. *Oecologia* 113, 74–81.

Stout, M.J., Fidantsef, A.L., Duffey, S.S. and Bostock, R.M. (1999) Signal interactions in pathogen and insect attack: systemic plant-mediated interactions between pathogens and herbivores of the tomato *Lycopersicon esculentum*. *Physiological and Molecular Plant Pathology* 54, 115–130.

Thaler, J.S., Fidantsef, A.L., Duffey, S.S. and Bostock, R.M. (1999) Trade-offs in plant defense against pathogens and herbivores: a field demonstration of chemical elicitors of induced resistance. *Journal of Chemical Ecology* 25, 1597–1609.

Thomma, B.P.H.J., Eggermont, K., Penninckx, I.A.M.A., Maunch-Mani, B., Vogelsang, R., Cammue, B.P.A. and Broekaert, W.F. (1998) Separate jasmonate-dependent and salicylate-dependent defense-response pathways in *Arabidopsis* are essential for resistance to distinct microbial pathogens. *Proceedings of the National Academy of Sciences USA* 95, 15107–15111.

Tinney, G.W. (1997) Tripartite interactions of host plant, herbivore, and rust pathogen. PhD thesis, Lancaster University, Lancaster, UK.

Tinney, G.W., Hatcher, P.E., Ayres, P.G., Paul, N.D. and Whittaker, J.B. (1998a) Inter- and intra-species differences in plants as hosts to *Tyria jacobaeae*. *Entomologia Experimentalis et Applicata* 88, 137–145.

Tinney, G.W., Theuring, C., Paul, N. and Hartmann, T. (1998b) Effects of rust infection with *Puccinia lagenophorae* on pyrrolizidine alkaloids in *Senecio vulgaris*. *Phytochemistry* 49, 1589–1592.

Tipping, P.W. (1993) Field studies with *Cassida rubiginosa* (Coleoptera: Chrysomelidae) in Canada thistle. *Environmental Entomology* 22, 1402–1407.

Vega, F.E., Barbosa, P., Kuo-Sell, H.L., Fisher, D.B. and Nelson, T.C. (1995) Effects of feeding on healthy and diseased corn plants on a vector and on a non-vector insect. *Experientia* 51, 293–299.

Virtanen, T., Ranta, H. and Neuvonen, S. (1997) Shoot-feeding aphids promote development of *Gremmeniella abietina*, the fungal pathogen causing scleroderris canker disease in conifers. *Journal of Phytopathology* 145, 245–251.

Vrieling, K., de Jong, T.J., Klinkhamer, P.G.L., van der Meijden, E. and van der Veen-van Wijk, C.A.M. (1996) Testing trade-offs amongst growth, regrowth and anti-herbivore defences in *Senecio jacobaea*. *Entomologia Experimentalis et Applicata* 80, 189–192.

Wargo, P.M. and Houston, D.R. (1973) Infection of defoliated sugar maple trees by *Armillaria mellea*. *Phytopathology* 63, 209.

Weete, J.D. (1992) Induced systemic resistance to *Alternaria cassiae* in sicklepod. *Physiological and Molecular Plant Pathology* 40, 437–445.

Wheeler, Q. and Blackwell, M. (eds) (1984) *Fungus–Insect Relationships, Perspectives in Ecology and Evolution*. Columbia University Press, New York.

Wilding, N., Collins, N.M., Hammond, P.M. and Webber, J.F. (eds) (1989) *Insect–Fungus Interactions*. Academic Press, London.

Wilson, C.L. (1969) Use of plant pathogens in weed control. *Annual Review of Phytopathology* 7, 411–434.

Wilson, C.G., Farrell, G.S. and Forno, I.W. (1996) Biological control of *Mimosa pigra* begins to work. In: Moran, V.C. and Hoffmann, J.H. (eds) *Proceedings of the IX International Symposium on Biological Control of Weeds.* University of Cape Town, South Africa, p. 510.

Wilson, P.A., Room, P.M., Zalucki, M.P. and Chakraborty, S. (2000) Interaction between *Helicoverpa armigera* and *Colletotrichum gloeosporioides* on the tropical pasture legume *Stylosanthes scabra. Australian Journal of Agricultural Research* 51, 107–112.

Zebitz, C.P.W. and Kehlenbeck, H. (1991) Performance of *Aphis fabae* on chocolate spot disease-infected faba bean plants. *Phytoparasitica* 19, 113–119.

The Role of Hyperparasites in Host Plant–Parasitic Fungi Relationships

12

Levente Kiss

Plant Protection Institute, Hungarian Academy of Sciences, H-1525 Budapest, PO Box 102, Hungary

Introduction

Traditionally, the interactions between plant-parasitic fungi and host plants are regarded as closed, two-species systems (Harper, 1990). However, both parasites and their hosts are, in fact, components of complex multitrophic interactions in which parasitic fungi are often attacked and killed by specialized fungal hyperparasites or other antagonists. Parasites, by definition, have a negative effect on host fitness (Jarosz and Davelos, 1995), so hyperparasitism might be favourable for plants infected with parasites. Unfortunately, studies on the possible role of hyperparasites in the natural host–parasite relationships are absent from the literature (Hirsch and Braun, 1992), although their use in biological control is largely based on their supposed importance in the natural control phenomena. This chapter is a first attempt to review our current knowledge in this field.

Interactions Between Plants and Parasitic Fungi: an Evolutionary Approach

Before discussing the role of specialized hyperparasites as components of the third trophic level built on plant–parasite relationships, it is helpful to outline briefly our current knowledge of the plant–parasite interactions themselves. In this respect, Jarosz and Davelos (1995) wrote: 'Perceptions of disease within wild plant populations have been influenced, over the years, by a near-schizophrenic set of stereotypes'. They referred to the two contrasting views existing in the literature: some authors considered that pathogen infections have been a driving force in plant evolution, while others thought that the role of pathogens in the evolution of plants is minor and can be disregarded. It is clear that pathogens will play an important role in plant evolution only if they cause significant reductions in plant fitness. So, the ultimate question is: how important are fungal pathogens, in an evolutionary time scale, in the *natural* control of their host plants?

This issue was discussed in detail by Dinoor and Eshed (1984), Harper (1990) and most recently by Jarosz and Davelos (1995). It was concluded from case studies of natural plant–parasite interactions in wild plant populations (e.g. Alexander and Burdon, 1984; Paul and Ayres, 1986a,b; Paul, 1989; Jarosz and Burdon, 1992), as well as from the widespread occurrence of resistance genes and defence molecules in plants, that pathogen infections represented powerful selection forces during plant evolution.

Interaction Between Fungal Parasites and Their Hyperparasites: a Structural Approach

A hyperparasitic interaction always consists of three trophic levels (Fig. 12.1). This chapter focuses on fungal hyperparasites, although bacteria (Hevesi and Mashaal, 1975) or mycoviruses (Brasier, 1990) could also act as hyperparasites of fungal plant pathogens. Their role will be discussed later.

There are two main scenarios for the activity of fungal hyperparasites. Necrotrophic plant pathogens, which kill host tissues quickly, are attacked by hyperparasites mainly on dead plant materials or even later, during their saprophytic stage in the soil. This means that the 'host plant–parasite' and the 'parasite–hyperparasite' interactions occur during different time and usually space coordinates (Fig. 12.2). Thus, there is, in general, no direct contact between living plants and hyperparasitic fungi. For example, *Sclerotinia* spp., well-known pathogens of many plants, rapidly kill the host plant tissues, then produce sclerotia which survive in the soil and can be attacked by specialized hyperparasites such as *Coniothyrium minitans* (Fig. 12.3) (Whipps and Gerlach, 1992) and *Sporidesmium sclerotivorum* (Adams and Ayers, 1979)

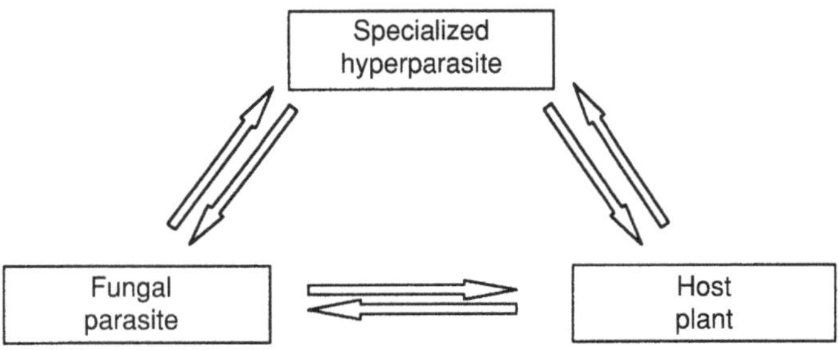

Fig. 12.1. Interactions between host plants, fungal parasites and their specialized hyperparasites.

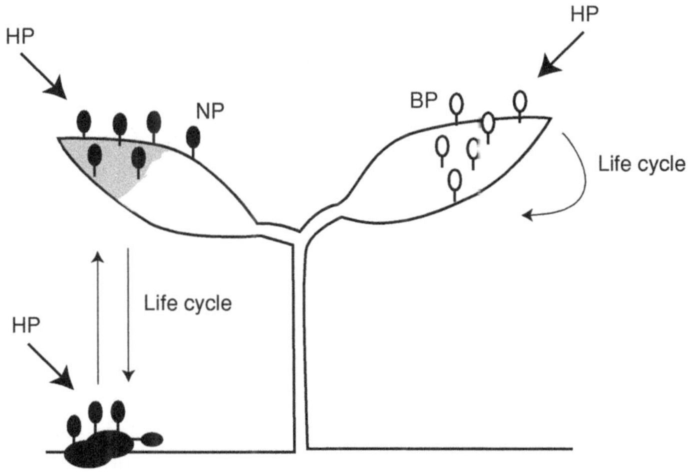

Fig. 12.2. Interactions between plants, necrotrophic fungal parasites (NP), biotrophic parasites (BP) and their specialized hyperparasites (HP).

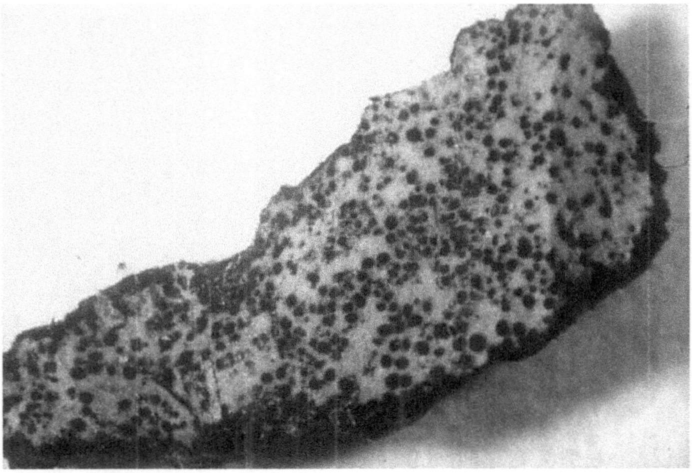

Fig. 12.3. Pycnidia of the hyperparasite *Coniothyrium minitans* in a cross-sectioned sclerotium of *Sclerotinia sclerotiorum*. Courtesy of Dr L. Vajna.

either on dead plant materials or in the soil. Other examples for naturally occurring hyperparasites of plant-pathogenic fungi were reviewed by Jeffries and Young (1994).

In contrast, biotrophic pathogens can survive on infected living plants only. They have no saprophytic stage, so they can only be attacked by hyperparasites in or on living host plant tissues (Fig. 12.2). In these cases, besides the 'host–parasite' interface, there is also a new structural interaction, namely the 'host–hyperparasite' interface. Clearly, the hyperparasites must survive and germinate on living host plant tissues before attacking the pathogens. Examples of these hyperparasites include *Eudarluca caricis* (Kranz, 1981),

Ampelomyces spp. (Fig. 12.4a) (Kiss, 1998) and some ballistosporic yeasts such as *Tilletiopsis* spp. (Fig. 12.4b) (Hoch and Provvidenti, 1979).

Reaction of Infected Plants to the Activity of Fungal Hyperparasites

Obviously, hyperparasitism is favourable for plants infected with parasites because 'the enemy of my enemy is my friend'. The literature on biological control of plant pathogens contains much data showing that diseased plants were greener and taller, and the yield was also higher, when treated artificially with fungal hyperparasites of the relevant pathogen. However, there are few detailed studies of the 'host plant–hyperparasite' interactions in this respect. Recently, Abo-Foul *et al.* (1996) showed that the dramatic decrease in the photosynthetic CO_2-fixation and also in the chlorophyll content of cucumber leaves caused by powdery mildew infection is stopped by the artificial application of *Ampelomyces* hyperparasites. The infected plants regained

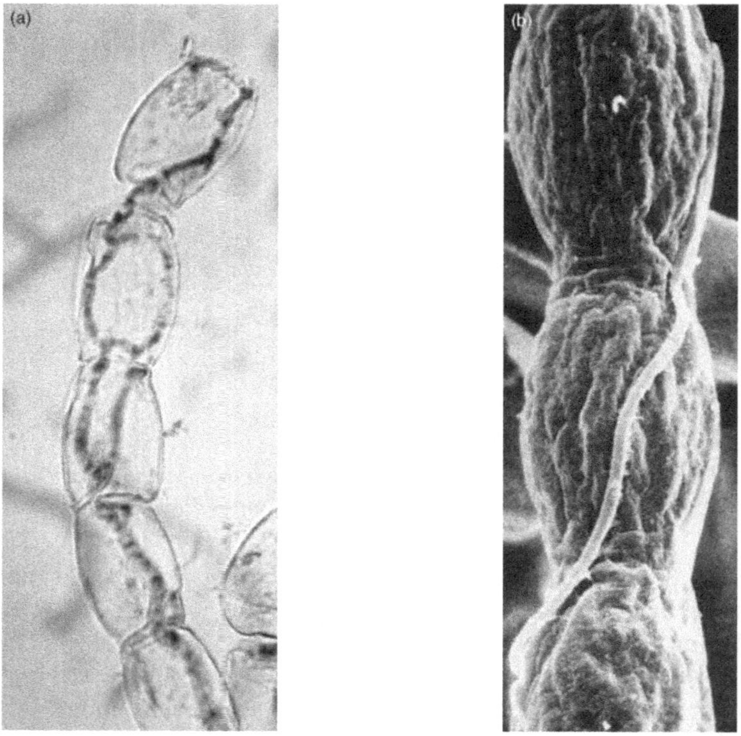

Fig. 12.4. Two hyperparasites of powdery mildew fungi. (a) Intracellular hyphae of an *Ampelomyces* sp. in a conidial chain of *Sphaerotheca fusca*. (Reprinted with permission from Kiss, 1998.) (b) Hyphae of a ballistosporic yeast, *Tilletiopsis* sp., on the surface of a conidial chain of *Sphaerotheca fusca*. (Reprinted with permission from Hoch and Provvidenti, 1979.)

vigour after the hyperparasites killed the pathogen. Eight days after treatment, the chlorophyll content and also the CO_2-fixation of infected leaves were almost the same as with the uninfected controls. Furthermore, Benhamou *et al.* (1997) provided the first convincing evidence that *Pythium oligandrum* induced plant defence reactions in tomato infected with *Fusarium oxysporum* f. sp. *radicis-lycopersici* in addition to acting as a hyperparasite. Due to this dual effect, striking differences in the extent of parasite colonization were observed between *P. oligandrum*-treated and control tomato plants. Both experiments demonstrated the mechanisms through which positive effects of a hyperparasite suppressed a fungal disease.

However, the effects of host plants on the activity of their fungal hyperparasites have never been studied in detail. In this review, I hypothesize that there are feedback mechanisms between the plant tissues infected with fungal parasites and the activity of their specialized hyperparasites, at least in the case of biotrophic pathogens where there is physical contact between them (Fig. 12.2). For example, host plant surfaces might influence the germination or survival of hyperparasites, and, thus to a certain extent, the suppression of the disease. Such effects might exist in the phyllosphere and also in the rhizosphere. Experiments should be carried out to test this hypothesis. It is worth mentioning that interactions between the first and the third trophic level are well known in plant–arthropod relationships. The density of leaf hairs, for instance, influences the predation rate of some predatory mites, particularly when the density of herbivorous mites is low (Krips *et al.*, 1999). The presence of leaf domatia clearly affects the distribution and abundance of many predatory mites (Walter and Odowd, 1992). The interactions between plant allelochemicals and predators of the herbivores are much more studied. It is well known that a number of caterpillar-damaged plants emit chemical signals that guide parasitoid wasps to the caterpillars (e.g. Turlings *et al.*, 1990, 1995; Paré and Tumlinson, 1997; De Moraes *et al.*, 1998). Thus, these predators often prevent plants from being severely damaged by killing the herbivores as they feed on the plants. Similar interactions also exist between plants infected with herbivorous mites and their predatory mites (e.g. Dicke *et al.*, 1993). In all these interactions, which involve host plants, herbivores and predators, there are clear feedback mechanisms between the first and the third trophic level. Similar mechanisms might also exist in the case of 'host plant–fungal parasite–fungal hyperparasite' relationships.

Reaction of Infected Plants to the Activity of Non-fungal Hyperparasites

Bacteria and mycoviruses can also serve as specialized hyperparasites of fungal plant pathogens. However, little is known about their role in suppressing plant diseases. *Erwinia uredovora*, for example, is a natural bacterial hyperparasite of rust fungi. In an experimental system, when broad bean plants susceptible to the rust fungus *Uromyces fabae* were co-inoculated with both rust and *E. uredovora*, the disease did not develop due to a hypersensitive response. This was presumed to be due to suppression of the rust development by the bacterial hyperparasite (Hevesi and Mashaal, 1975).

The role of a double-stranded RNA mycovirus in chestnut blight is probably the best example for the activity of a non-fungal hyperparasite. In some areas, the chestnut blight fungus, *Chryphonectria parasitica*, has naturally become hypovirulent by acquiring this mycovirus that reduces its virulence to the host plant (Brasier, 1990). The viral hyperparasite is transmitted horizontally among colonies of *C. parasitica*, so multiple infections of the same chestnut tree increase its acquisition and, thus, the reduction of pathogen virulence.

The examples mentioned above showed that fungal and non-fungal hyperparasites can act in many different ways against fungal plant pathogens (e.g. eliminating them, inducing plant defence reactions or reducing pathogen virulence). However, the direction of these effects (i.e. top-down or bottom-up) in these tritrophic systems still requires detailed analysis.

The Possible Role of Naturally Occurring Fungal Hyperparasites in the Fitness of Infected Plants

Experimental systems and biological control trials have already provided convincing data on the effects of fungal hyperparasites applied artificially against plant pathogens. However, there are only presumptions about their role in natural host–parasite relationships (Hirsch and Braun, 1992; Cooke and Whipps, 1993; Jeffries and Young, 1994; Jeffries, 1995, 1997). By analogy to the plant–pathogen interactions discussed earlier, two contrasting views could be formulated in this respect. It could be suggested that hyperparasites have no significant impact on plant pathogens in nature, so they are only a source of noise in the system. Alternatively, it might be that hyperparasites have played, and continually play, a role in the evolution of host–pathogen relationships by reducing the negative effects of pathogens and, thus, exerting a positive effect on the fitness of host plants. Thus, the ultimate question that might help to elucidate which of these presumptions is realistic is analogous with the question formulated for the plant–pathogen interactions: how important are the hyperparasites, in an evolutionary time scale, in the *natural* control of plant-parasitic fungi?

Unfortunately, there is a lack of quantitative data on the natural occurrence of hyperparasites that would represent the first step towards evaluating the impact of hyperparasitism on both host fungal and plant populations in nature. First, Kranz (1981) studied in detail the dynamics of the interaction between *E. caricis* and some rust fungi. His data showed that this common hyperparasite spontaneously destroyed many of the rust uredia in the field and its development closely followed the spread of the rust fungi. In general, up to 40% of uredia of many different rust fungi were destroyed by *E. caricis* on different host plants under natural conditions, but in some cases all rust colonies were destroyed (Kranz, 1981). Similarly, Kuhlman and Matthews (1976) found that *E. caricis* parasitized 91% of the uredia produced on oak leaves in Florida and almost totally suppressed the formation of teliospores, the overwintering spores of rusts. Another investigation showed that in a sunflower field naturally infested with both *Sclerotinia sclerotiorum* and its pycnidial hyperparasite, *Coniothyrium minitans*, 59%, 76% and 29% of sclerotia on the root surface, inside the root and inside the stem, respectively, had been killed by *C. minitans* (Huang, 1977). The natural occurrence of *Syncephalis californica*, a hyperparasite of the apricot tree pathogen *Rhizopus oryzae*, was also determined in the soil (Hunter *et al.*, 1977). A recent 4-year study of both the natural incidence of *Ampelomyces* hyperparasites (Fig. 12.4a) and the intensity of hyperparasitism in a total of 27 species of powdery mildew fungi infecting 41 host plant genera showed that in seven of the nine genera of the Erysiphales studied, *Ampelomyces* spp. parasitized and destroyed only less than 20% of powdery mildew mycelia (Kiss, 1998). In contrast, hyperparasitism was extremely high, approximately 65%, in *Arthrocladiella mougeotii* infecting *Lycium* plants. Surprisingly, *Ampelomyces* was rarely found in *Blumeria graminis*, the only powdery mildew fungus that infects monocots, although *B. graminis* is heavily parasitized by *Ampelomyces* in experimental systems (Kiss, 1997, 1998).

These field studies showed that, at least in some cases, the naturally occurring hyperparasites do significantly reduce the inoculum density of their fungal hosts, and thus, the infection pressure on the host plants. However, according to a mathematical model of parasite–hyperparasite interactions (Shaw, 1994; Shaw and Peters, 1994), hyperparasites might cause apparently random fluctuations in the abundance of parasites from year to year, even in an absolutely constant environment. So it is not obvious how environmental factors could be distinguished from intrinsic population instability in the field data presented earlier.

The existence of resistance genes and antifungal compounds in plants were considered as evidence that pathogens had a serious impact on plant populations during evolution. Unfortunately, similar evidence is not available for parasite–hyperparasite interactions. It is known that some fungi produce papillae as a reaction to the attack of mycoparasites (Fig. 12.5) (Vajna, 1985a,b, 1987; Jeffries and Young, 1994). However, little is known about the specificity of this response. Clearly, the mechanisms of resistance of pathogenic fungi to hyperparasites have not been studied enough to be considered as evidence of the impact of hyperparasites on the evolution of parasitic fungi.

Conclusions

In conclusion, our current knowledge on both the natural occurrence of hyperparasitism and functional aspects of parasite–hyperparasite interactions

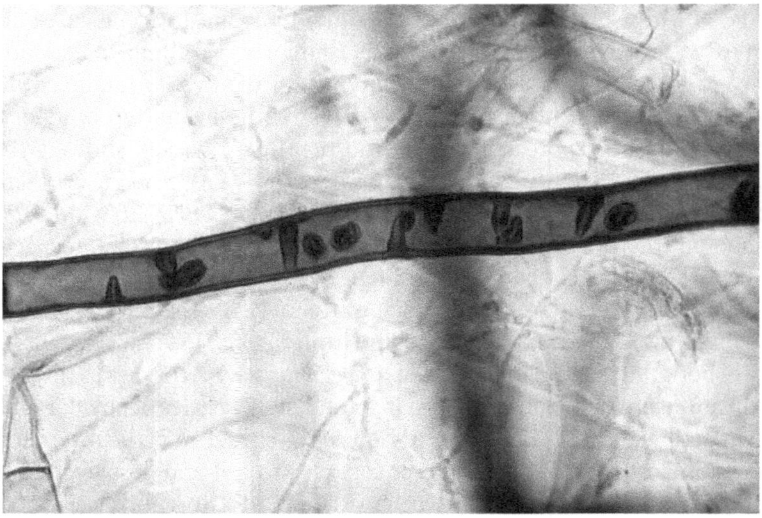

Fig. 12.5. Papillae formation in the hyphae of *Rhizoctonia solani* attacked by *Fusarium oxysporum* f. sp. *dianthi* acting as a mycoparasite. Courtesy of Dr L. Vajna.

is insufficient to answer the intriguing question: how important are the hyperparasites, in an evolutionary time scale, in the *natural* control of plant-parasitic fungi? It seems that the thoughts of John Harper (1990) remain valid, at least for the role of hyperparasites in natural systems. He wrote: 'Ecologists gain great delight from demonstrating how complex the world is. Pests and pathogens contribute extra tangles in the webs of interactions that ecologists love to draw. It is not clear how we ought to set about discovering which among all the forces and interactions that occur in vegetation determine its character and which are minor, superficial and can safely be disregarded'.

Acknowledgements

The author is especially indebted to Peter Jeffries for his comments on a draft of this paper and to John H. Andrews, László Vajna, Tibor Érsek and James H. Tumlinson for helpful suggestions. Financial support was provided by a János Bolyai Research Fellowship, the Royal Society (a UK-CEE joint project grant) and the Hungarian Scientific Research Fund (OTKA F026334).

References

Abo-Foul, S., Raskin, V.I., Sztejnberg, A. and Marder, J.B. (1996) Disruption of chlorophyll organization and function in powdery mildew-diseased cucumber leaves and its control by the hyperparasite *Ampelomyces quisqualis*. *Phytopathology* 86, 195–199.

Adams, P.B. and Ayers, W.A. (1979) Mycoparasitism of sclerotia of *Sclerotinia* and *Sclerotium* species by *Sporidesmium sclerotivorum*. *Canadian Journal of Microbiology* 25, 17–23.

Alexander, H.M. and Burdon, J.J. (1984) The effect of disease induced by *Albugo candida* (white rust) and *Peronospora parasitica* (downy mildew) on the survival and reproduction of *Capsella bursa-pastoris* (shepherd's purse). *Oecologia* 64, 314–318.

Benhamou, N., Rey, P., Chérif, M., Hockenhull, J. and Tirilly, Y. (1997) Treatment with the mycoparasite *Pythium oligandrum* triggers induction of defence-related reactions in tomato roots when challenged with *Fusarium oxysporum* f. sp. *radicis-lycopersici*. *Phytopathology* 87, 108–122.

Brasier, C.M. (1990) The unexpected element: mycovirus involvement in the outcome of two recent pandemics, Dutch elm and chestnut blight. In: Burdon, J.J. and Leather, S.R. (eds) *Pests, Pathogens and Plant Communities*. Blackwell Scientific Publications, Oxford, pp. 289–308.

Cooke, R.C. and Whipps, J.M. (1993) *Ecophysiology of Fungi*. Blackwell Scientific Publications, Oxford.

De Moraes, C.M., Lewis, W.J., Paré, P.W., Alborn, H.T. and Tumlinson, J.H. (1998) Herbivore-infested plants selectively attract parasitoids. *Nature* 393, 570–573.

Dicke, M., Van Baarlen, P., Wessels, R. and Dijkman, H. (1993) Herbivory induces systemic production of plant volatiles that attract predators of the herbivore: extraction of endogenous elicitor. *Journal of Chemical Ecology* 19, 581–599.

Dinoor, A. and Eshed, N. (1984) The role and importance of pathogens in natural plant communities. *Annual Review of Phytopathology* 22, 443–466.

Harper, J.L. (1990) Pests, pathogens and plant communities: an introduction. In: Burdon, J.J. and Leather, S.R. (eds) *Pests, Pathogens and Plant Communities*. Blackwell Scientific Publications, Oxford, pp. 3–14.

Hevesi, M. and Mashaal, S.F. (1975) Contributions to the mechanisms of infection of *Erwinia uredovora*, a parasite of rust fungi. *Acta Phytopathologica Academiae Scientiarum Hungaricae* 10, 275–280.

Hirsch, G. and Braun, U. (1992) Communities of parasitic microfungi. In: Winterhoff, W. (ed.) *Fungi in Vegetation Science*. Kluwer Academic Publishers, Dordrecht, pp. 225–250.

Hoch, H.C. and Provvidenti, R. (1979) Mycoparasitic relationships: cytology of the *Sphaerotheca fuliginea–Tilletiopsis* sp. interaction. *Phytopathology* 69, 359–362.

Huang, H.C. (1977) Importance of *Coniothyrium minitans* in survival of sclerotia of *Sclerotinia sclerotiorum* in wilted sunflower. *Canadian Journal of Botany* 55, 289–295.

Hunter, W.E., Duniway, J.M. and Butler, E.E. (1977) Influence of nutrition, temperature, moisture and gas composition on parasitism of *Rhizopus oryzae* by *Syncephalis californica*. *Phytopathology* 67, 664–669.

Jarosz, A.M. and Burdon, J.J. (1992) Host–pathogen interactions in natural populations of *Linum marginale* and *Melampsora lini*. III. Influence of pathogen epidemics on host survivorship and flower production. *Oecologia* 89, 53–61.

Jarosz, A.M. and Davelos, A.L. (1995) Effects of disease in wild plant populations and the evolution of pathogen aggressiveness. *New Phytologist* 129, 371–387.

Jeffries, P. (1995) Biology and ecology of mycoparasitism. *Canadian Journal of Botany* 73 (Suppl. 1), S1284-S1290.

Jeffries, P. (1997) Mycoparasitism. In: Wicklow, D.T. and Söderström, B.E. (eds) *The Mycota. A Comprehensive Treatise on Fungi as Experimental Systems for Basic and Applied Research*. Springer-Verlag, Berlin, pp. 149–164.

Jeffries, P. and Young, T.W.K. (1994) *Interfungal Parasitic Relationships.* CAB International, Wallingford, UK.

Kiss, L. (1997) Graminicolous powdery mildew fungi as new natural hosts of *Ampelomyces* mycoparasites. *Canadian Journal of Botany* 75, 680–683.

Kiss, L. (1998) Natural occurrence of *Ampelomyces* intracellular mycoparasites in mycelia of powdery mildew fungi. *New Phytologist* 140, 709–714.

Kranz, J. (1981) Hyperparasitism of biotrophic fungi. In: Blakeman, J.P. (ed.) *Microbial Ecology of the Phylloplane.* Academic Press, London, pp. 327–352.

Krips, O.E., Kleijn, P.W., Willems, P.E.L., Gols, G.J.Z. and Dicke, M. (1999) Leaf hairs influence searching efficiency and predation rate of the predatory mite *Phytoseiulus persimilis* (Acari: Phytoseiidae). *Experimental and Applied Acarology* 23, 119–131.

Kuhlman, E.G. and Matthews, F.R. (1976) Occurrence of *Darluca filum* on *Cronartium strobilinum* and *C. fusiforme* infecting oak. *Phytopathology* 66, 1195–1197.

Paré, P.W. and Tumlinson, J.H. (1997) Induced synthesis of plant volatiles. *Nature* 385, 30–31.

Paul, N.D. (1989) The effects of *Puccinia lagenophorae* on *Senecio vulgaris* in competition with *Euphorbia peplus. Journal of Ecology* 77, 552–564.

Paul, N.D. and Ayres, P.G. (1986a) The impact of a pathogen (*Puccinia lagenophorae*) on populations of groundsel (*Senecio vulgaris*) overwintering in the field. I. Mortality, vegetative growth and the development of size hierarchies. *Journal of Ecology* 74, 1069–1084.

Paul, N.D. and Ayres, P.G. (1986b) The impact of a pathogen (*Puccinia lagenophorae*) on populations of groundsel (*Senecio vulgaris*) overwintering in the field. II. Reproduction. *Journal of Ecology* 74, 1085–1094.

Shaw, M.W. (1994) Seasonally induced chaotic dynamics and their implications in models of plant disease. *Plant Pathology* 43, 790–801.

Shaw, M.W. and Peters, J.C. (1994) The biological environment and pathogen population dynamics: uncertainty, coexistence and competition. In: Blakeman, J.P. and Williamson, B. (eds) *Ecology of Plant Pathogens.* CAB International, Wallingford, UK, pp. 17–37.

Turlings, T.C.J., Tumlinson, J.H. and Lewis, W.J. (1990) Exploitation of herbivore-induced plant odors by host-seeking parasitic wasps. *Science* 250, 1251–1253.

Turlings, T.C.J., Loughrin, J.H., McCall, P.J., Röse, U.S.R., Lewis, W.J. and Tumlinson, J.H. (1995) How caterpillar-damaged plants protect themselves by attracting parasitic wasps. *Proceedings of the National Academy of Sciences USA* 92, 4169–4174.

Vajna, L. (1985a) Phytopathogenic *Fusarium oxysporum* Schlecht as a necrotrophic mycoparasite. *Journal of Phytopathology* 114, 338–347.

Vajna, L. (1985b) Mutual parasitism between *Trichoderma hamatum* and *Trichoderma pseudokoningii. Journal of Phytopathology* 113, 300–303.

Vajna, L. (1987) *Fusarium lateritium* Nees ex Link as a parasite and host in interfungal hyphal interactions. *Journal of Phytopathology* 118, 157–164.

Walter, D.E. and Odowd, D.J. (1992) Leaf morphology and predators – effect of leaf domatia on the abundance of predatory mites (Acari, Phytoseiidae). *Environmental Entomology* 21, 478–484.

Whipps, J. and Gerlach, M. (1992) Biology of *Coniothyrium minitans* and its potential for use in disease control. *Mycological Research* 96, 897–907.

Mutualism and Antagonism: Ecological Interactions Among Bark Beetles, Mites and Fungi

13

K.D. Klepzig,[1] J.C. Moser,[1] M.J. Lombardero,[2] M.P. Ayres,[2] R.W. Hofstetter[2] and C.J. Walkinshaw[1]

[1]USDA Forest Service, Pineville, LA 71360, USA;
[2]Dartmouth College, Hanover, NH 03755, USA

Introduction

Insect–fungal complexes provide challenging and fascinating systems for the study of biotic interactions between plants, plant pathogens, insect vectors and other associated organisms. The types of interactions among these organisms (mutualism, antagonism, parasitism, phoresy, etc.) are as variable as the range of organisms involved (plants, fungi, insects, mites, etc.). We focus on bark beetles and their associated organisms, in particular, on the relationship between the southern pine beetle and its associates in coniferous trees of the southern USA. We begin, however, with an attempt to clearly define the terms we use to describe these relationships.

Symbiosis

Zook (1998) stated that 'Defining symbiosis has become something of a life science cliché, an act of verbal, and often verbose, masochism'. Nevertheless, before exploring the manners in which closely associated organisms can interact, we must attempt to arrive at some basic definitions. Perhaps the most widely used, and perhaps widely debated, definition of symbiosis comes from Frank and Debary who defined the term as the 'Living together of unlike organisms'. That definition is useful in that it manages to avoid placing any values on the interaction between organisms (mutualism is not implied here). However, this definition is also vague enough that it might encompass all manner of close relationships between unlike organisms that we might not

 Biotic Interactions in Plant–Pathogen Associations
(eds M.J. Jeger and N.J. Spence)

view as at all symbiotic (e.g. a goldfish and a frog in a bowl, a mouse in a cow barn). A more specific definition, which is value neutral and still broad enough to encompass the variety of symbiotic organisms, is: 'the acquisition and maintenance of one or more organisms by another that results in novel structures and (or) metabolism' (Zook, 1998) (we have added the 'or' to indicate our belief that the existence of modified structures *or* metabolism is sufficient to qualify as symbiosis).

The Continuum from Mutualism to Antagonism – Intersymbiont Interactions

Even with a clearly stated and acceptable definition of symbiosis, problems arise in the classification of interactions between organisms. In particular, attempts to classify a specific relationship as being strictly competitive, or strictly mutualistic, may be frustrated by seemingly contradictory evidence. One group of researchers finds that a particular organism is more successful in the presence of another. Other research may indicate that the two organisms compete for resources and even actively defend against one another. In this case, one might ask, what is the true nature of this relationship? Are the organisms mutualists, or antagonists? Often a satisfactory answer can be arrived at by careful consideration of the developmental and/or resource state being considered in the attempt to classify the relationship. In effect, many studies of symbiotic relationships consider only a limited range of time (or resource conditions). Within a specific window in time it is often possible to characterize a relationship as being primarily mutualistic or antagonistic. However, as noted by Callaway and Walker (1997) most (if not all) studies examining competition and/or facilitation do not measure a long enough period of time. Relationships among closely associated, even symbiotic, organisms may change over the developmental cycles of the organisms (time) as well as over ranges of available nutrient and energy sources (resources). In addition, other organisms may indirectly effect a relationship between two organisms. These third, or even fourth, organisms may become an integral part of the manner in which the two original organisms interact, facilitating and/or interfering as time and resources change (Callaway and Walker, 1997).

Fungal Interactions

Fungi utilizing the same resource may interact in at least three broadly defined ways (Rayner and Webber, 1984). For example, two fungal species may interact mutualistically (in which each facilitates the success of the other), neutralistically (in which each has no discernible effect on the other) or competitively (in which each tries to utilize the resource at the expense of the other). Competitive interactions may be detrimental to either species, and may be further subdivided into primary resource capture and secondary resource capture. In primary resource capture the interacting fungi compete to gain access and influence over an uncolonized resource. At this point, the fungi are not directly challenging one another. However, as the fungi colonize the available resource, they may eventually come into direct contact with one another. The two directly interacting fungi may now engage in defence against one another (e.g. antibiosis), they may intermingle with no discernible effects on one another, or they may attempt to engage in secondary resource capture (in which one fungus attempts to colonize the resource already held by the other). These fungal interactions may be of particular importance when they occur between species which are symbiotically linked to other species. The fungal associates of bark beetles have been extensively studied not only due to their effects on trees, but also as integral parts of complex systems of interacting organisms.

Bark Beetles, Mites and Fungi

The biology and ecology of bark beetle–fungal interactions have been extensively studied, and well reviewed elsewhere (Malloch and Blackwell, 1993; Paine *et al.*, 1997). The interactions among insects that infest the bark, phloem and outer xylem of trees, and fungi that possess varying degrees of virulence within these tissues are complex. Fungi may be carried within specialized cuticular structures termed mycangia (Fig. 13.1), or externally in simple pits or on the exoskeleton. The roles of the associated fungi in the beetle life cycles may be differentiated by the manner in which they are vectored. Fungi carried within mycangia tend to be mutualists of the beetles, those carried externally are more likely to be tree pathogens, or wood-staining fungi. There is substantial taxonomic diversity among the fungi vectored by bark beetles, but many fall within the ascomycete genera *Ophiostoma* or *Ceratocystis* (the term ophiostomatoid is frequently used to refer to this group of fungi (Malloch and Blackwell, 1993)). The details of the interactions among the many species of beetles and fungi vary extensively, making broad generalization problematic (Paine *et al.*, 1997). We will concentrate, below, on the system which we study and which provides examples of the basic types of beetle–fungal interactions.

The Southern Pine Beetle System

Although insect–fungus–mite interactions are important to several bark beetle species, these complex relationships have been extensively studied in the southern pine beetle (SPB). *Dendroctonus frontalis* Zimmermann (Coleoptera: Scolytidae) is among the most damaging of North American forest insects (Thatcher *et al.*, 1980; Drooz, 1985; Price *et al.*, 1992). The SPB is considered a primary bark beetle, in that it is essentially an obligate parasite (Raffa *et al.*,

1993) that attacks and kills healthy living trees through mass colonization by conspecifics (Paine *et al.*, 1997). Reproductive female beetles initiate attacks on host trees by boring entrance holes through the rough outer bark of southern pines, creating a nuptial chamber (Fig. 13.2) and releasing a pheromone

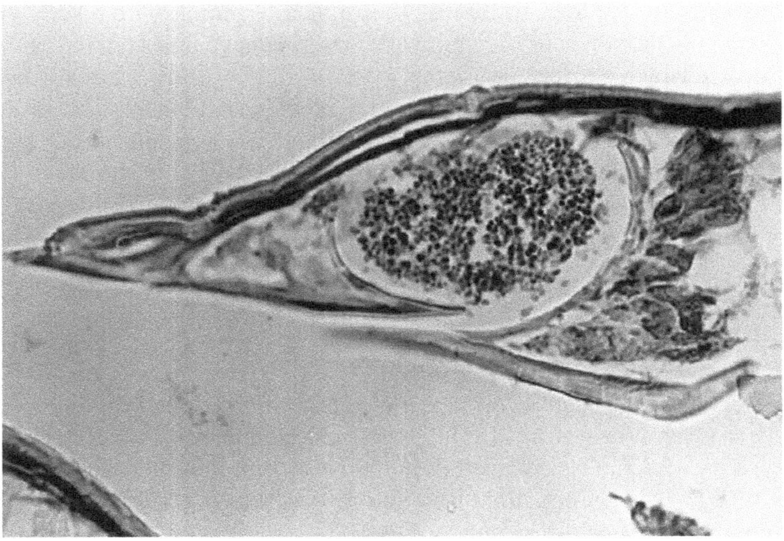

Fig. 13.1. Southern pine beetle mycangium. Light micrograph of cross section of mycangium with fungal spores contained within.

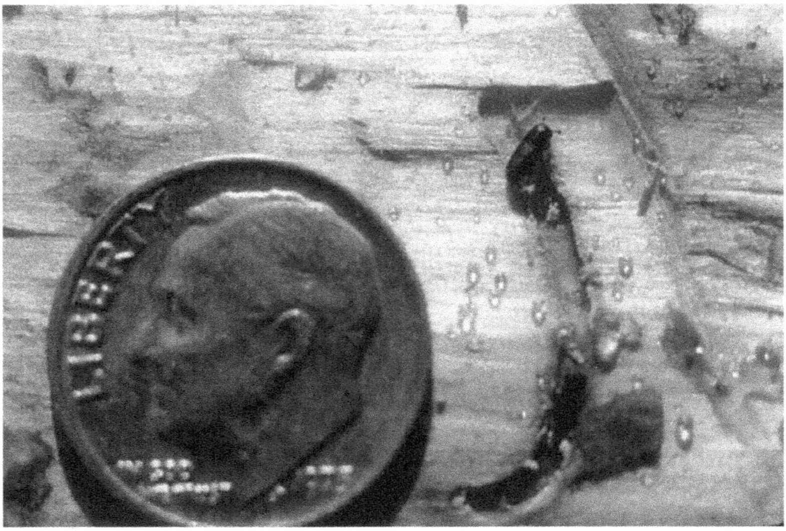

Fig. 13.2. Southern pine beetle adults in pine phloem. The outer bark has been stripped away to reveal the male and female near the nuptial chamber, the female is beginning to create an ovipositional gallery.

to attract more beetles to the tree. The host tree attempts to repel the attack primarily through the release of preformed (constitutive) resin (Hodges *et al.*, 1979; Lewinsohn *et al.*, 1991a,b; Nebeker *et al.*, 1993; Ruel *et al.*, 1998). If enough SPB attack the tree in this manner, the tree's resin system is overcome, the beetles are able to complete development and the tree dies (essentially from disruption of water flow within the vascular system) (Fig. 13.3). Once the female beetle has mated, she begins chewing ovipositional (egg) galleries within the inner bark and phloem of the tree (Thatcher, 1960; Payne, 1983). As she does so, the female SPB inoculates several fungi into the phloem tissue (Bramble and Holst, 1940). Although many fungi have been associated with galleries of SPB in pine phloem, three have been the focus of most SPB–fungal research, and appear to have the most significant impacts on the SPB life cycle: *Ophiostoma minus* (Hedgc.) H. and P. Sydow, *Ceratocystiopsis ranaculosus* Perry and Bridges and *Entomocorticium* sp. A (an undescribed basidiomycete, formerly referred to in the literature as isolate SJB122).

Ophiostoma minus, the causal agent of the 'blue stain', often found in the xylem and phloem of SPB-infested wood is an ascomycetous fungus (Fig. 13.4a) carried phoretically on the SPB exoskeleton (Rumbold, 1931; Bridges and Moser, 1983) and by phoretic mites (Bridges and Moser, 1983), which we will discuss in detail below. Early research into the SPB–fungi system focused on the putative role of *O. minus* as a tree-killing pathogen (Nelson, 1934; Caird, 1935; Bramble and Holst, 1940; Mathre, 1964; Basham, 1970). However, the fungus is apparently not necessary for tree death to occur (Hetrick, 1949; Bridges, 1985; Bridges *et al.*, 1985). Although artificial inoculations of southern pines with *O. minus* do cause resinosis and tissue damage (Fig. 13.4b), they do not result in mortality of mature trees (Nelson, 1934; Cook *et al.*, 1986; Cook and Hain, 1987; Parmeter *et al.*, 1992; Ross *et al.*, 1992; Nevill *et al.*, 1995; Popp *et al.*, 1995). It seems probable that *O. minus*, in concert with SPB tunnelling, hastens tree death (Paine *et al.*, 1997). The benefits of this relationship to the fungus are clearer. Bark beetles and their arthropod associates serve as the only effective means by which stain fungi gain access to new host tissue (Dowding, 1969). Thus, at the early stages of attack, the SPB–*O. minus* relationship may be categorized as mutualistic, although the frequency with which these organisms are associated does not necessarily imply this (Harrington, 1993). Subsequent research has focused on the impacts of *O. minus* on SPB larval development. As SPB eggs hatch within the niches the female has created in the pine phloem, the fungi she inoculated begin growing and colonizing the tissue as well. Within this community of organisms, patches of *O. minus* develop (Fig. 13.5). When these areas of heavy colonization by the blue stain fungus overlap areas within which the developing larvae are feeding, the SPB almost always suffers. Although much of the evidence has been circumstantial, higher levels of phloem colonization with *O. minus* are correlated with reduced developmental success – inhibited egg production, slower larval growth and development, even larval mortality (Fig. 13.6) (Barras, 1970; Franklin, 1970). In addition, overall levels of *O. minus* within SPB infestations have been negatively correlated with SPB population increase (Bridges, 1985). The relationship here seems simple. The more blue stain that is present, the less SPB reproductive success will occur (Lombardero *et al.*, 2000). At the time of larval development, *O. minus* appears to be a competitor and antagonist of SPB (Barras, 1970). The mechanism of this antagonism, however, has remained unclear. Some have speculated that *O. minus* leaves the phloem nutrient impoverished and deprives the developing larvae of necessary sustenance (Hodges *et al.*, 1968; Barras and Hodges, 1969; Barras, 1970). As such, it has also been suggested that the beneficial roles of the two other major fungal associates of SPB consist largely of outgrowing or outcompeting *O. minus* and keeping this blue stain fungus out of SPB larval galleries (Bridges and Perry, 1985).

The antagonism of SPB larvae by *O. minus*, which at first seemed contradictory to the pattern seen between *O. minus* and attacking SPB adults, may be partially explained when the interactions of SPB with its two other significant fungal associates are examined. Each female SPB possesses a prothoracic structure specialized for transporting fungi (Fig. 13.7). This mycangium

Fig. 13.3. A pine tree mass attacked by southern pine beetle. Each of the numerous pitch tubes are an attempt by the tree to flood the beetles out of the tree through heavy resin flow.

consists of paired invaginations of the exoskeleton each of which has one pore-like ventral opening and contains two types of secretory cells (Happ *et al.*, 1971; Barras and Perry, 1972). Within each side of the mycangium, the female SPB is able to maintain a pure culture of either *C. ranaculosus* (a hyaline ascomycete) (Fig. 13.8a) (Barras and Taylor, 1973) or *Entomocorticium* sp. A,

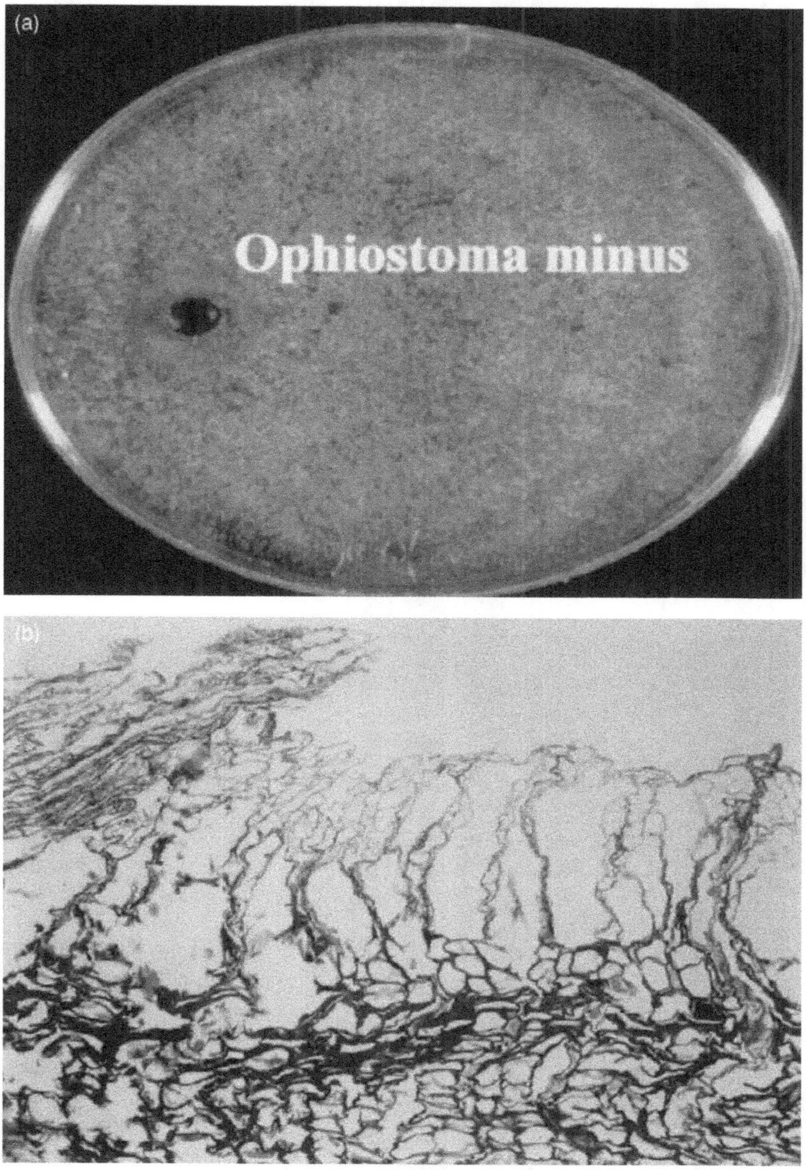

Fig. 13.4. *Ophiostoma minus.* (a) Culture grown on malt extract agar. (b) Tissue damage from inoculation of *Pinus taeda* with *O. minus.* Note heavy accumulation of tannins and related defence compounds in cambial tissue.

Fig. 13.5. Areas of 'blue stain', within southern pine beetle-infested pine, due to infection with *O. minus*.

Fig. 13.6. Southern pine beetle larval galleries within pine logs. Both logs infested with surface-sterilized beetles, log on right was inoculated with *O. minus*, log on left was not inoculated. Larval development in *O. minus*-infected logs was heavily reduced compared with uninfected logs.

formerly referred to in the literature as SJB122 (Fig. 13.9a) (Barras and Perry, 1972; Happ *et al.*, 1976). This slow-growing fungus is an amber-coloured basidiomycete whose sexual stage remains undescribed, but which appears to belong in this genus *Entomocorticium* (Hsiau, 1996). Each female may carry either one (rarely both) of the two fungi, or no fungi, in either of the two mycangial pouches (Bridges, 1985). Although it seems likely that the majority of inoculation of mycangial fungi into pine phloem occurs later (Barras, 1975) perhaps during oviposition, the relative virulence of these two fungi in healthy trees has also been investigated. Inoculations of both *C. ranaculosus* and *Entomocorticium* sp. A invariably result in smaller amounts of tree damage (e.g. resinous lesions (Figs 13.8b and 13.9b)) than do inoculations with *O. minus* (Cook and Hain, 1985; Paine *et al.*, 1997). However, both mycangial fungi do cause reactions, especially at the tissue and cellular level, that differ from those seen in response to mere mechanical wounding (Figs 13.8b, 13.9b, 13.10). SPB mycangial fungi do not appear to be highly virulent in their pine hosts nor do they seem to assist in any meaningful way in tree killing. It seems more likely that the proper window in time to evaluate the role of the mycangial fungi in the SPB life cycle is post-mass attack. Once the tree's resistance has been overcome, as the female SPB deposits her eggs within the pine phloem, she may inoculate the area immediately surrounding the eggs with the contents of her mycangium. As the eggs hatch the early instar larvae begin feeding, constructing fine, sinuous galleries as they go (Payne, 1983). Eventually, the larvae cease moving forward and begin enlarging their feeding area

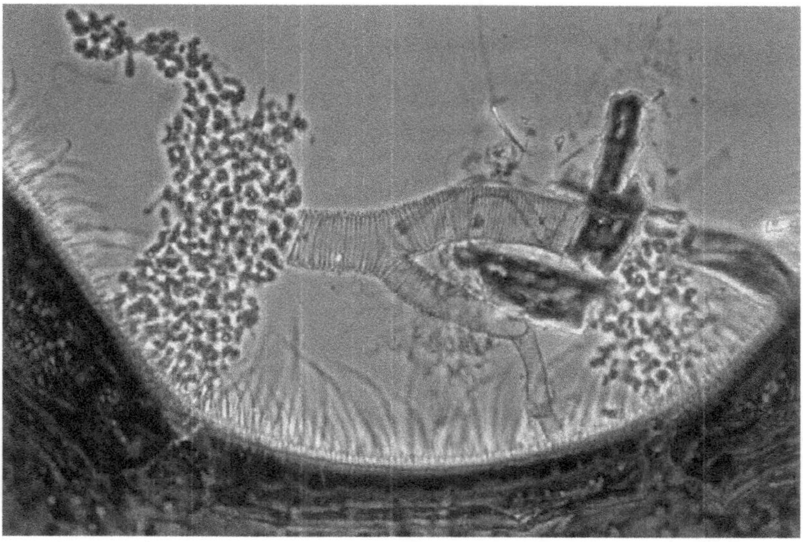

Fig. 13.7. Mycangium dissected from a female southern pine beetle. The head (above) and prothoracic legs (below) have been removed. Two streams of yeast-like spores of the fungi contained within the mycangium can be seen streaming from the pore-like openings of the structure.

to an obovate shape. It is within these 'feeding chambers' that one can find luxuriant growth of either of the two mycangial fungi (Fig. 13.11). It is assumed that the mid- to late instar larvae feed on fungal hyphae and spores, although due, in part, to difficulties in artificially rearing SPB, it has never been explicitly demonstrated. It appears extremely likely that larval SPB get the

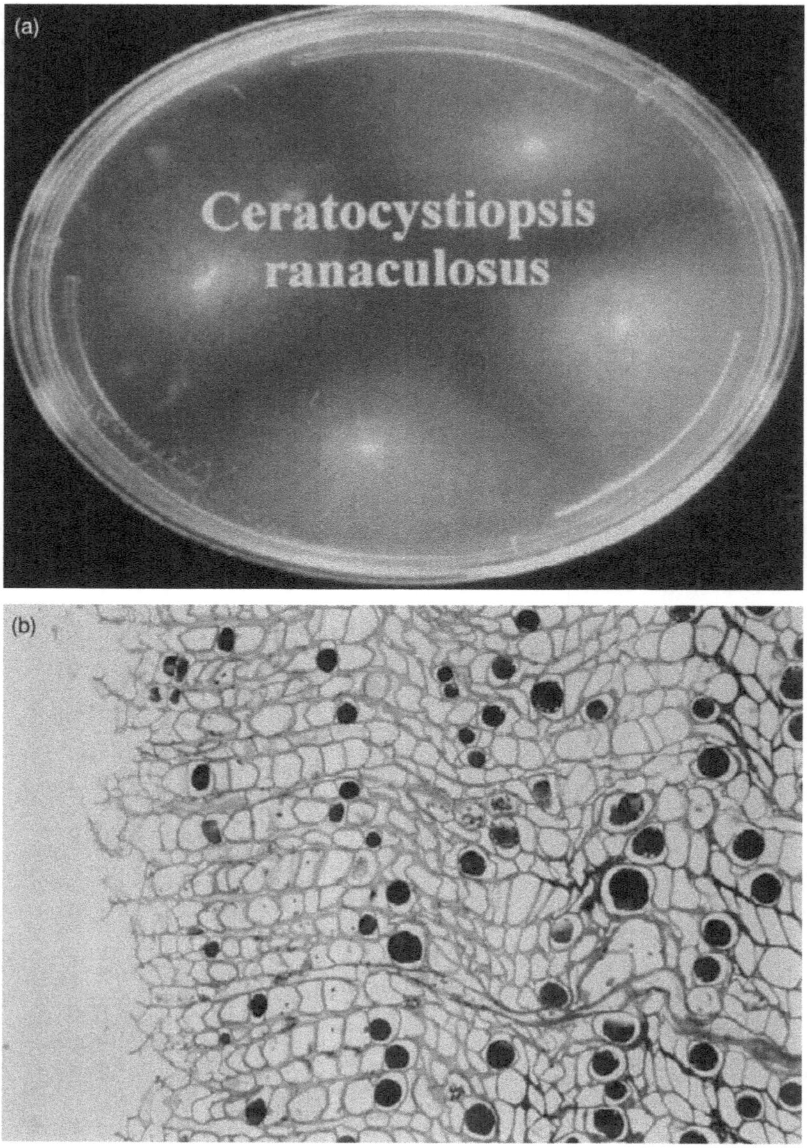

Fig. 13.8. *Ceratocystiopsis ranaculosus.* (a) Culture grown on malt extract agar. (b) Tissue damage from inoculation of *Pinus taeda* with *C. ranaculosus.* Note only moderate accumulation of tannins and related defence compounds in cambial tissue.

majority of their nutrition from the fungal growth within their feeding chambers rather than from the phloem itself. The mycangial fungi may, in fact, provide their most substantial benefits to SPB by concentrating dietary N for larvae (Fig. 13.12) (Ayres *et al.*, 2000). For the fungi, again, the advantages of association with SPB are clear. The fungi obtain a selective medium within which to grow as they are borne, protected and pure, to the next available

Fig. 13.9. *Entomocorticium* sp. A (a) Culture grown on malt extract agar. (b) Tissue damage from inoculation of *Pinus taeda* with *Entomocorticium* sp. A. Note only moderate accumulation of tannins and related defence compounds in cambial tissue.

resource (Happ *et al.*, 1971). The benefits to the beetle from these fungi appear obvious as well. Beetles containing *Entomocorticium* sp. A are more fecund, heavier and have higher lipid contents than those containing *C. ranaculosus*. In turn, beetles containing *C. ranaculosus* tend to be more fit than those whose mycangia contain no fungi (Bridges, 1985; Goldhammer *et al.*, 1990;

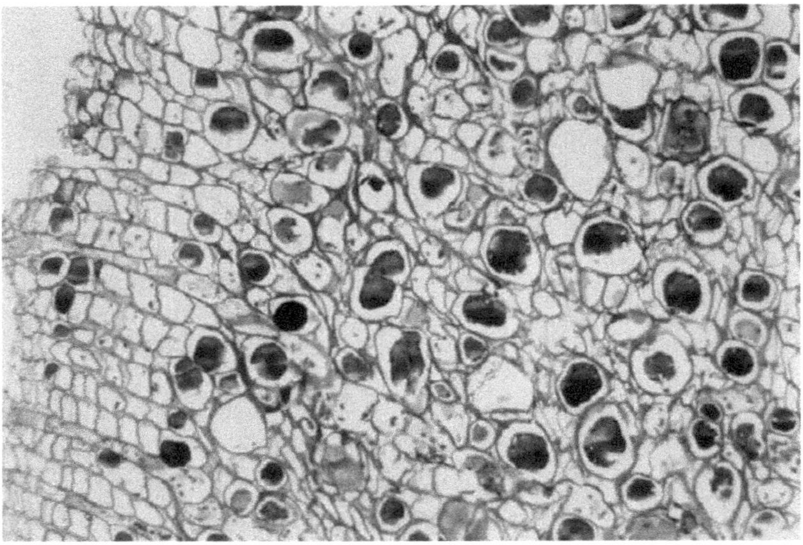

Fig. 13.10. Tissue damage due to mechanical wounding of *Pinus taeda*. Note lack of tannin accumulation and related defence compounds, and presence of callus growth, in cambial tissue.

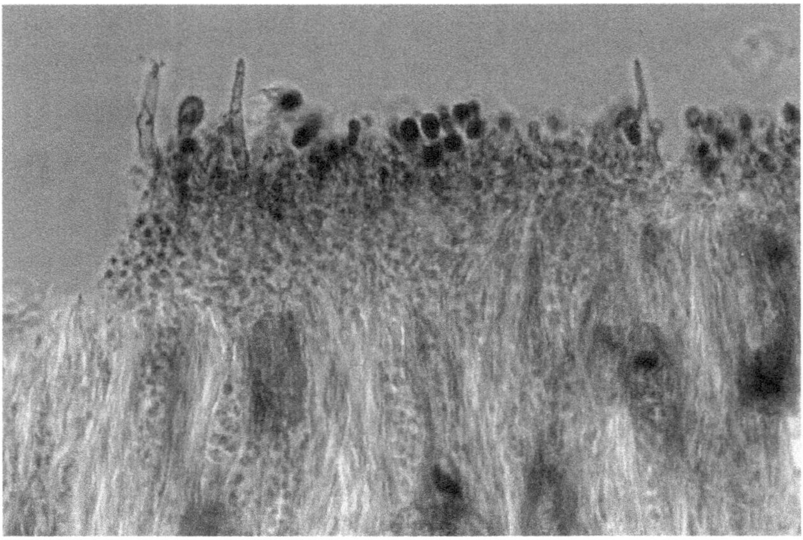

Fig. 13.11. Growth of mycangial fungi in southern pine beetle pupal chamber. Note sporulation.

Coppedge *et al.*, 1995). Thus, the two mycangial fungi can be considered to be nutritional mutualists of SPB.

Attempts to adequately describe the complexity of SPB–fungal ecology must, in addition, involve consideration of the mites associated with both the beetle and the fungi. The SPB is associated with, and may transport from tree to tree, over 57 species of mites (Moser and Roton, 1971; Moser *et al.*, 1971, 1974). The SPB-associated acarofauna includes parasitic, predatory, fungivorous and omnivorous species. Most species within this complex are truly phoretic, in which the mite is transported on the external surface of the beetle and does not undergo feeding or ontogenesis during this period of transport (Lindquist, 1969; Smiley and Moser, 1974). In particular, phoretic mites within the genus *Tarsonemus* have been the focus of most of the limited amount of research conducted in bark beetle–mite interactions (Moser and Roton, 1971; Smiley and Moser, 1974; Moser, 1976; Bridges and Moser, 1983; Moser and Bridges, 1986). We have concentrated on three mite species, *Tarsonemus ips* Lindquist, *Tarsonemus krantzii* Smiley and Moser and *Tarsonemus fusarii* Cooreman. All three of these mites are common SPB associates (though *T. fusarii* is less common and seemingly more of a generalist than the other two species). *Tarsonemus ips*, *T. krantzii* and *T. fusarii* are all phoretic on SPB, obtaining transport to new, suitable host material with no – directly – discernible deleterious effects on the beetle. However, all three mites have shown at least the potential to impact the SPB–fungus–tree interaction.

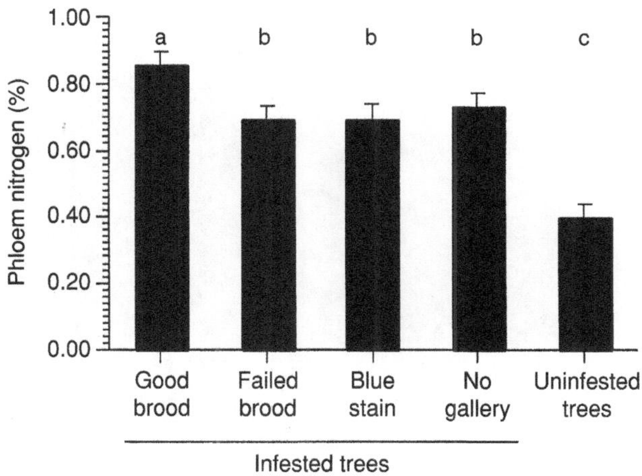

Fig. 13.12. Phloem N (%) in southern pine beetle infested phloem. Good brood = lack of blue stain, growth of mycangial fungi, abundant larval feeding galleries and pupal chambers; Failed brood = poor larval feeding and development, and lack of pupal chambers; Blue stain = abundant growth of *Ophiostoma minus* in larval gallery system, poor larval development; No gallery = no larval galleries present in area of sampling; all compared to 'uninfested trees' which contained no southern pine beetles. Bars (mean + SE) followed by different letters are significantly different at $P < 0.05$ level.

Two of these three mites possess sporothecae, which are specialized, flap-like structures of the integument (Fig. 13.13). In *T. ips* and *T. krantzii*, these sporothecae have been found, relatively frequently, to transport ascospores of *O. minus* (Bridges and Moser, 1983; Moser, 1985) and *C. ranaculosus* (Moser *et al.*, 1995). Despite the possibilities raised by these circumstances, nothing more was known about the relationships between the mites and their associated fungi, nor about the implications of these interactions to the beetle–fungus relationship.

Fig. 13.13. (a) *Tarsonemus* sp. mites between legs of beetle, carrying (b) crescent-shaped spores of *Ophiostoma minus*, and (c) tadpole-shaped spores of *Cerato-cystiopsis ranaculosus* within sporothecae (laterally located on the mite body).

Ecological Interactions in the SPB Community

We have attempted to unravel the complex ecological interactions among tree-killing bark beetles, fungi and mites using SPB as our study organism. Taking a reductionist approach, we have considered the manner in which SPB-associated fungi compete with one another and thus facilitate or interfere with the success of SPB. We have also considered the role of mites as indirect facilitating, and/or interfering, agents in fungus–insect–tree host interactions.

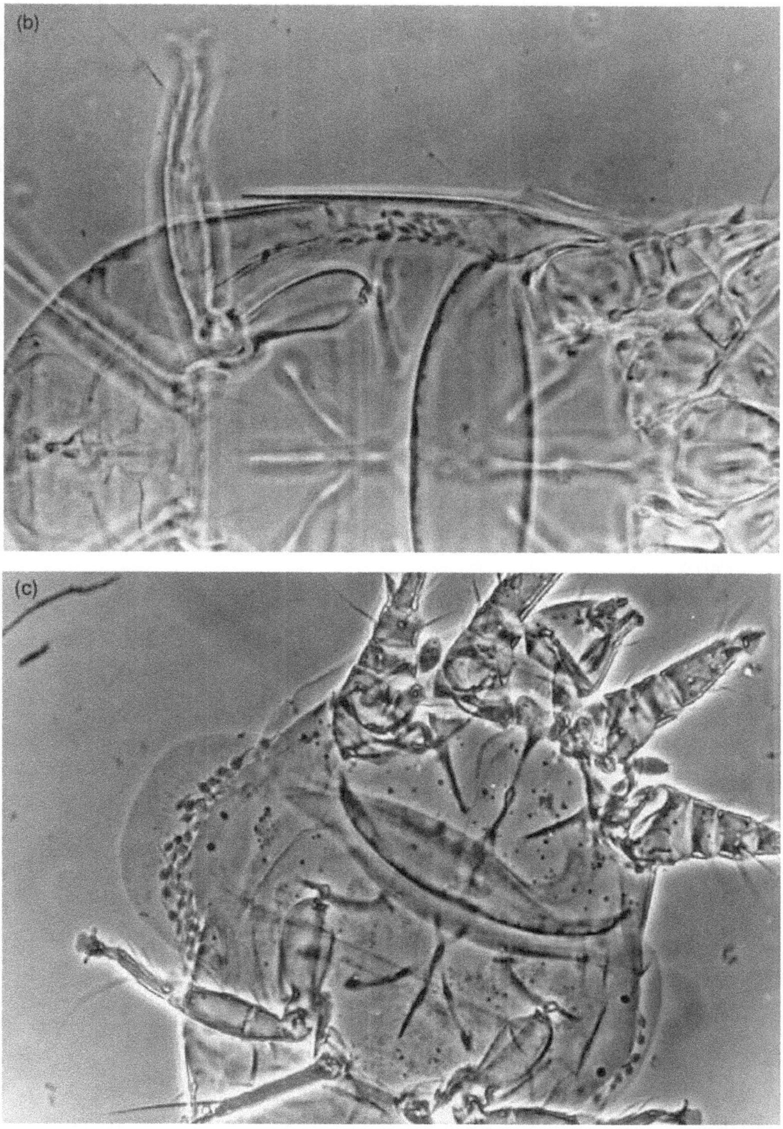

Fig. 13.13 (cont.)

Fungal competition

The three major SPB-associated fungi, *O. minus*, *Entomocorticium* sp. A and *C. ranaculosus*, compete for the rare and ephemeral resource of uncolonized pine phloem (Klepzig and Wilkens, 1997). It is likely that, in doing so, these fungi follow the previously stated model of primary resource capture, followed by direct interaction, which can lead to defence, and/or secondary resource capture. Likewise, the degree to which these fungi differentially compete can be quantified. The de Wit replacement series has been used extensively to study plant competition, and is being increasingly accepted as an analytical tool for microbial competition (Adee *et al.*, 1990; Snaydon, 1991; Wilson and Lindow, 1994; Klepzig and Wilkens, 1997; Klepzig, 1998) but see a cautionary note in Newton *et al.* (1998). In using this technique with microbes, varying proportions of inoculum of potentially competing microbes are introduced on to a substrate. In the case of competing fungal hyphae, this may consist of inoculating substrate (e.g. agar medium, pine billets) with varying numbers of agar discs colonized with hyphae of one fungal species. The initial inoculum of one species is increased with each replicate as the initial inoculum of the other species is decreased. The population size (in the case of fungal hyphae, the area colonized) is determined at the end of the experiment as a function of initial population size (in this case, the percentage of each species in the original population). If no differential competition is occurring between the two species, it is expected that there will be a close to one-to-one linear relationship between proportion of the fungus in the initial inoculum and its representation (area colonized) in the final population (Fig. 13.14) (Wilson and Lindow,

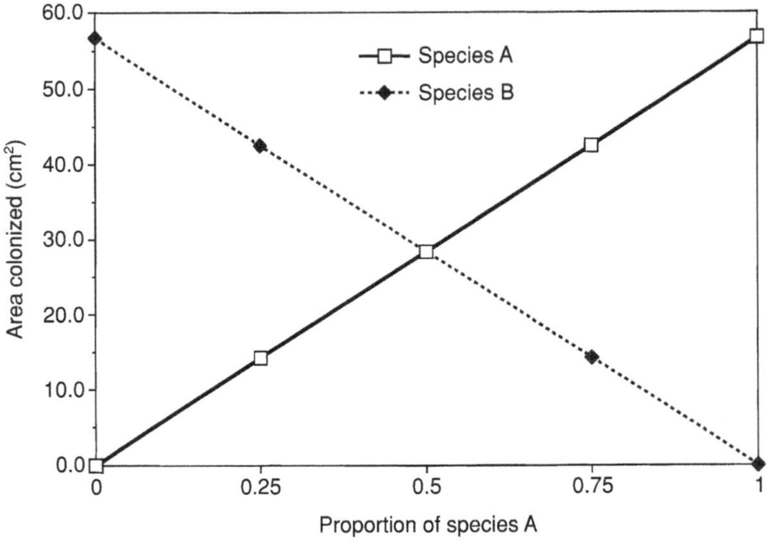

Fig. 13.14. DeWit replacement series diagram which would theoretically result if there was not differential competition between two co-occurring species.

1994). Differential competition between two fungi is indicated when there is a significant positive deviation from linearity for one and a significant negative deviation from linearity for the other. To determine this, the areas colonized by each fungus at the end of the experiment are recorded, and the means are calculated and log transformed. An ANOVA is performed on the transformed means to test for deviations from linearity in the relationships between final area colonized and initial inoculum proportion, for each fungal species. Specifically, pairwise competitions between *O. minus*, *Entomocorticium* sp. A and *C. ranaculosus* can be conducted to determine the degree to which differential competition occurs among these frequently co-occurring fungi vectored by the same beetle. From laboratory experiments, it is absolutely clear that differential competition occurs amongst these three fungi (Klepzig and Wilkens, 1997). In all three pairwise comparisons, there were significant deviations from linearity in the relationships between initial and final population representation in the competing fungi (Fig. 13.15). The clearly superior competitor, at least on the artificial media used, was *O. minus* whose rapid growth rate and aggressive resource capture tactics overwhelmed the two mycangial fungi at even the lowest levels of *O. minus* inoculum (Fig. 13.16). The mycangial fungi were rapidly outcompeted by *O. minus* for the available substrate. *Entomocorticium* sp. A and *C. ranaculosus*, however, were very similar in their relative competitive abilities, and the graph of their de Wit replacement series reflected this (Fig. 13.15c). There is even the appearance of the classic 'X'-shaped pattern in the data, which would (but for the skewing in the favour of *C. ranaculosus*) suggest a lack of differential competition.

Beyond the determination of the existence of differential competition with SPB-associated fungi, is the question of the outcomes of competition among these fungi. Which, for example, of the SPB fungi is best able to hold on to colonized substrate in the face of a concerted secondary resource capture effort by another fungal species? We have measured the relative primary and secondary resource capture capabilities of these three fungi. When the three SPB-associated fungi are forced to compete one-on-one on both artificial medium (malt extract agar) and natural substrate (loblolly pine billets), *O. minus* invariably comes out the victor in primary resource capture. Due again, in no small part, to its relatively rapid growth rate, *O. minus* can quickly colonize and gain control of substantially more of the available territory (uncolonized agar as well as pine phloem) than can either of the two mycangial fungi (Fig. 13.17). As is the case in considering differential competition, the two mycangial fungi are approximately equal competitors for the capture of primary resource. However, due to its higher growth rate, *C. ranaculosus* is significantly able to outcompete *Entomocorticium* sp. A.

Once the primary resource capture phase of the one-on-one competition is over, however, and the direct confrontations begin, the SPB-associated fungi differ in their competitive abilities in interesting ways. When *O. minus* grows into the same area of substrate as *C. ranaculosus*, aerial hyphae begin developing at the colony margins of both species. Within a short while (about 4 days), however, the heavily melanized hyphae of *O. minus* have grown over the margins of the hyaline *C. ranaculosus* colonies and begun the process of secondary resource capture (Fig. 13.17b). By the 11th day of competition between these two fungi, *C. ranaculosus* colonies have most often been completely overgrown by *O. minus*. The competition between the fast-growing *O. minus* and the

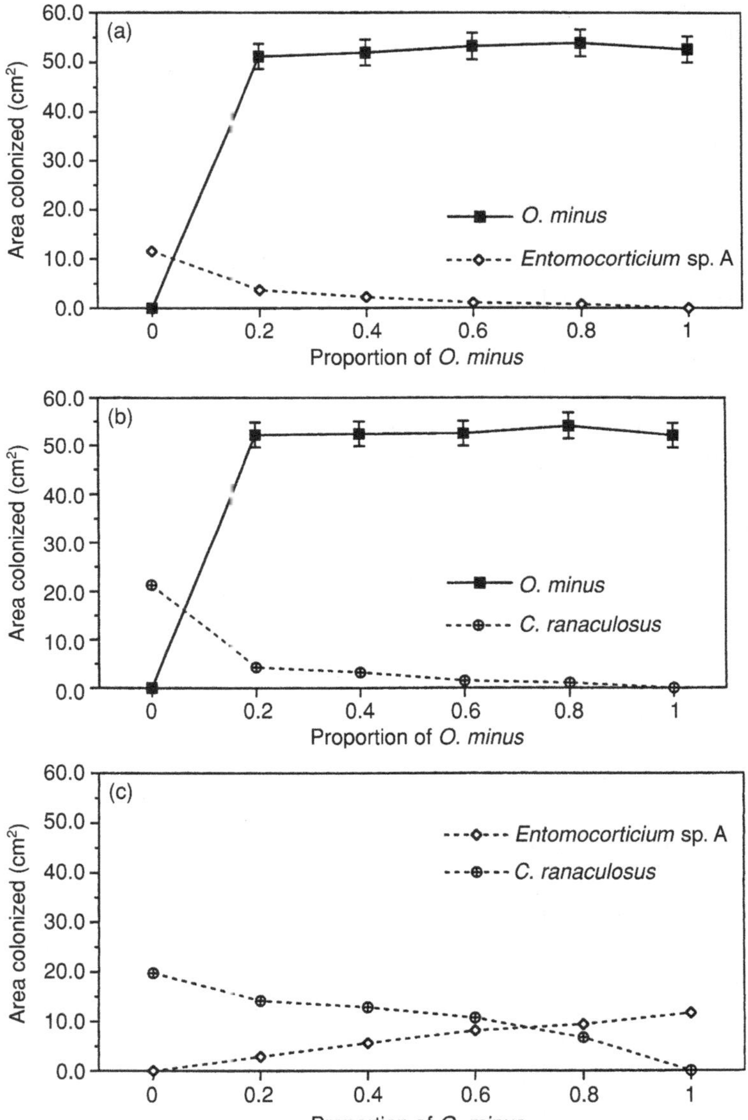

Fig. 13.15. DeWit replacement series diagrams resulting from competition between: (a) *Ophiostoma minus* and *Entomocorticium* sp. A. (b) *O. minus* and *Ceratocystiopsis ranaculosus*; and (c) *C. ranaculosus* and *Entomocorticium* sp. A. Standard errors are given about each mean.

slow-growing, amber-coloured, floccose basidiomycete, *Entomocorticium* sp. A unfolds in a much different fashion. Although, due to the slow growth rate of *Entomocorticium* sp. A, *O. minus* is able to capture a great deal of uncolonized substrate before it reaches the basidiomycete, the direct interaction of these

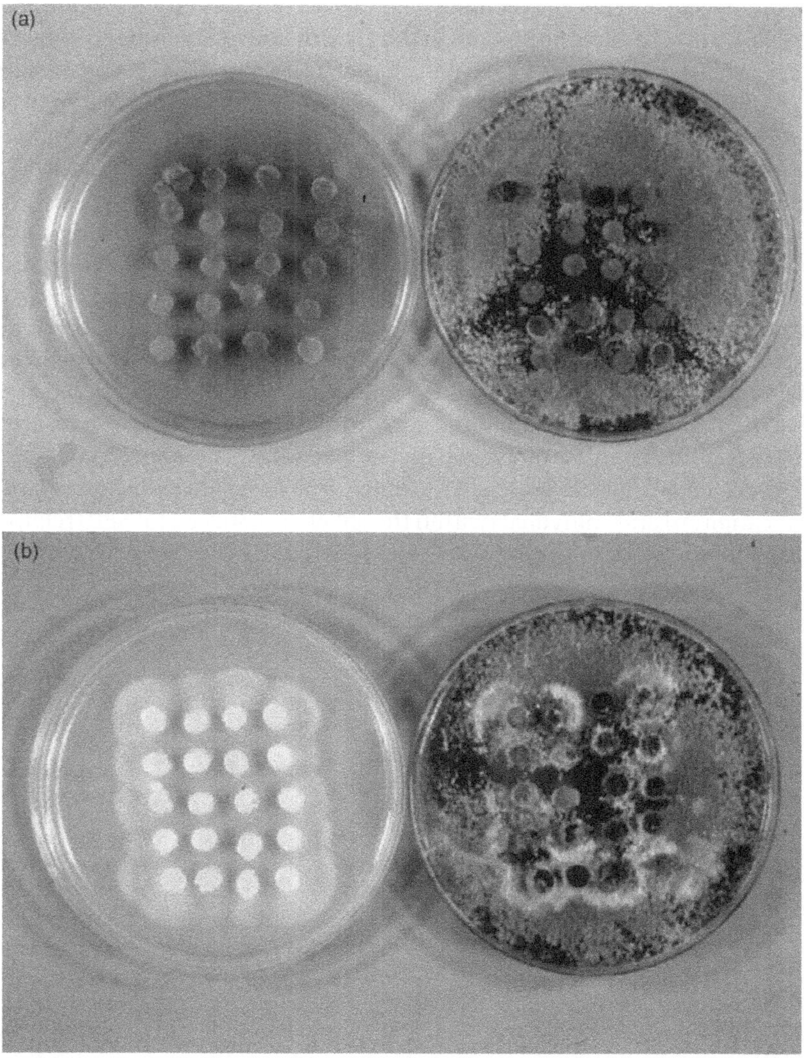

Fig. 13.16. Differential competition between fungi associated with southern pine beetle: 100% (right) and 80% (left) of mycangial fungus in the initial inoculum. (a) *Ophiostoma minus* and *Entomocorticium* sp. A. Note that although *O. minus* is outcompeting *Entomocorticium* sp. A, the original inoculum discs of the mycangial fungus are still uncolonized by *O. minus*; (b) *O. minus* and *Ceratocystiopsis ranaculosus*. Note that although *O. minus* is outcompeting *C. ranaculosus*, the original inoculum discs of this mycangial fungus have been colonized by *O. minus*.

two fungi slows *O. minus* drastically. Very slightly before the growing hyphae of *O. minus* reach the *Entomocorticium* sp. A colony margins, they slow in growth rate. There is little to no development of the aerial hyphae seen in the *O. minus–C. ranaculosus* interaction. The *O. minus* colony, if it grows further, grows around the *Entomocorticium* sp. A colony, never growing over the basidiomycete and never accomplishing any secondary resource capture (Fig. 13.17a). This dramatic limitation on the growth and further spread of *O. minus* suggests either very close range diffusion of antibiotics from *Entomocorticium* sp. A to *O. minus*, or localized nutrient depletion by *Entomocorticium* sp. A such that *O. minus* cannot develop further in substrate which has been colonized by *Entomocorticium* sp. A. These same patterns of competitive interactions also hold true within loblolly pine billets.

Environmental (abiotic) factors may also alter the intensity and nature of competitive interactions (Callaway and Walker, 1997). Temperature drastically affects growth rates in all three SPB-associated fungi (Fig. 13.18). Of particular note are the differences in the manner in which the three fungi respond to varying temperatures. *Ophiostoma minus* seems particularly adaptable to a range of temperatures; its range of optimal temperatures for growth is wider, and its minimum growth temperature lower, than are the same variables for either of the two mycangial fungi. This may be due, in part, to the protected manner in which the mycangial fungi are transported (within a mycangium) and cultivated (within the galleries of successful SPB) relative to

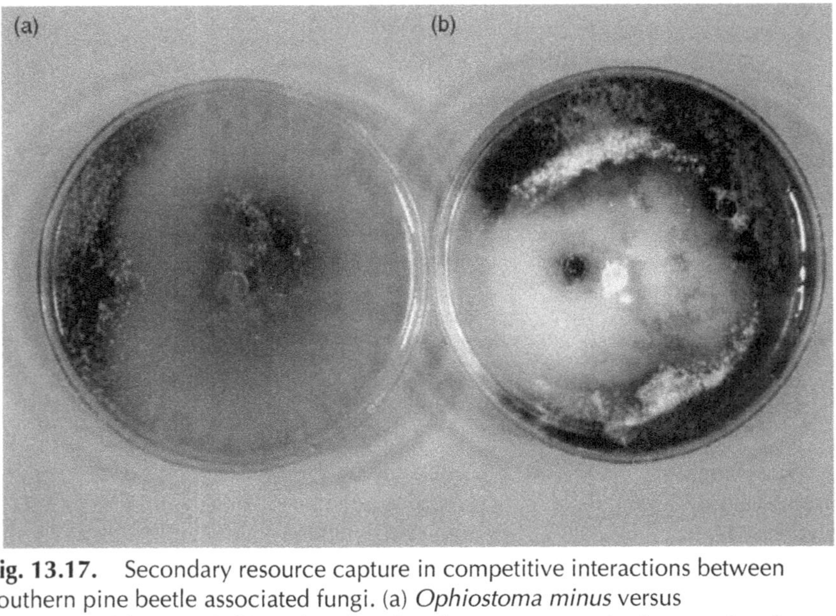

Fig. 13.17. Secondary resource capture in competitive interactions between southern pine beetle associated fungi. (a) *Ophiostoma minus* versus *Entomocorticium* sp. A. Note that *O. minus* has not captured substrate already colonized by *Entomocorticium* sp. A. (b) *O. minus* versus *Ceratocystiopsis ranaculosus*. Note that *O. minus* has captured substrate already colonized by *C. ranaculosus*.

O. minus (which is transported on beetle and mite exoskeletons, and inoculated by SPB attacking living trees). Nutrient levels within phloem may also impact growth and competitive interactions among SPB fungi.

Several implications for SPB and its pine host arise from the interactions described above. It is apparent that *O. minus* is best equipped to capitalize on the uncolonized phloem available in the early stages of SPB attack in pines. Not only does this aggressive fungus grow more rapidly than the two mycangial fungi, it is also more tolerant of pine allelochemicals than *Entomocorticium* sp. A (Bridges, 1987). This, of course, may be advantageous to the beetle and, especially if *O. minus* does assist in killing the tree, disadvantageous to the tree. As tree resistance is overcome, and the female beetles begin inoculating the mycangial fungi into the phloem, the aggressive saprophytic (for the tree is essentially dead at this point) characteristics of *O. minus* become a disadvantage for SPB. At this point, the female needs to establish colonies of either *Entomocorticium* sp. A or *C. ranaculosus* in the vicinity of the larvae, and far enough away from growth of *O. minus* for the fungi to become established and serve as a larval food source. At this point the differences between the two mycangial fungi come into focus. One of the fungi, *C. ranaculosus*, grows marginally faster than the other, but – once *O. minus* reaches it – does not seem capable of defending this territory enough to allow larval development (Klepzig and Wilkens, 1997). This fungus, especially when considered with its apparent relative inferiority as a larval nutritional substrate (Bridges, 1983; Goldhammer *et al.*, 1990; Coppedge *et al.*, 1995) would seem to be of less value as a symbiont than *Entomocorticium* sp. A. *Entomocorticium* sp. A, while slower growing than *C. ranaculosus*, is definitely capable of growing and providing nutrition for SPB larvae, even when surrounded by *O. minus*. The key to larval success, then, may be establishing a thriving culture of *Entomocorticium* sp. A soon enough, or far enough away, that it can grow without interference from

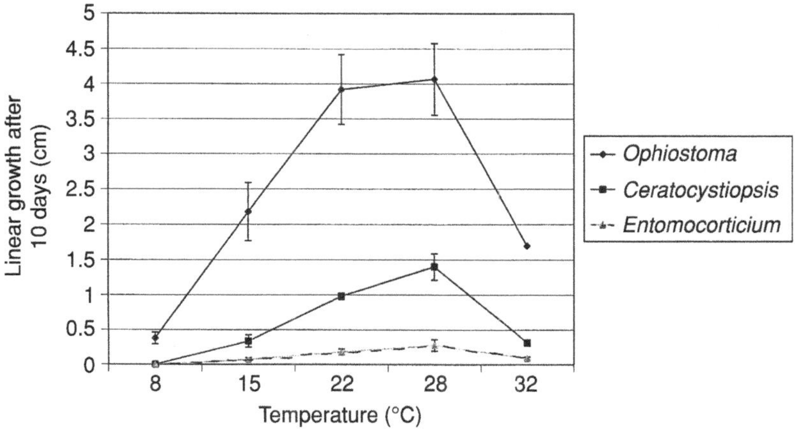

Fig. 13.18. Effects of temperature on linear growth of southern pine beetle-associated fungi (*Ophiostoma minus, Entomocorticium* sp. A and *Ceratocystiopsis ranaculosus*) growing on malt extract agar.

O. minus (recalling that when these two fungi compete for uncolonized substrate, *O. minus* wins). In this sense, as well as in the nutritional sense, *Entomocorticium* sp. A is apparently the superior of the two mycangial fungi.

Here we find the SPB system posing another conundrum. The question might be stated thus 'If *O. minus* is antagonistic to SPB larvae, and *C. ranaculosus* is of only moderate (or negative) value as a symbiont, why are these fungi so consistently associated with the beetle? Where is the selection pressure for maintaining fungal relationships of dubious value?' Recalling that 'the success of species in a community is affected not only by direct interactions between species, but also by indirect interactions among groups of species' (Miller, 1994, as cited in Callaway and Walker, 1997), the phoretic mites of SPB seem deserving of consideration.

Mite–fungus interactions

As described above, both *T. ips* and *T. krantzi* possess sporothecae, within which they carry spores of *O. minus* and/or *C. ranaculosus* (Bridges and Moser, 1983; Moser, 1985; Moser *et al.*, 1995). Neither of these mites have ever been found to transport *Entomocorticium* sp. A. Until recently, however, the nature of the relationship between these tarsonemid mites and the fungi they apparently vector into pine phloem, remained undescribed. When cultures of *T. ips*, *T. krantzi* and *T. fusarii* are initiated on pure cultures of the three major SPB fungal associates, reproduction occurs, but the results vary in a manner that helps explain the questions raised by the fungal competition research described above (Lombardero *et al.*, 2000). All three mites can successfully reproduce, and their offspring thrive (larval survival to first reproduction has been conservatively estimated at 90%), on colonies of *O. minus* (Table 13.1). Indeed,

Table 13.1. Demographic parameters for three species of *Tarsonemus* feeding on *Ophiostoma minus*.

	T. ips	*T. fusarii*	*T. krantzi*	*F* statistic (df)
Time to egg hatch (days)	2.20 ± 0.23	1.81 ± 0.13	2.70 ± 0.16	9.26** (2, 28)
Larval to adult (days)	3.90 ± 0.10	5.00 ± 0.19	4.87 ± 0.09	27.22*** (2, 47)
Age of 1st reproduction (days)	8.10 ± 0.10	8.81 ± 0.19	9.57 ± 0.09	53.61*** (2, 47)
Survival: egg to adult	>90%	>90%	>90%	
Adult longevity (days)	>28	>28	>28	
Fecundity	0.92 ± 0.11	0.87 ± 0.19	1.33 ± 0.13	3.36† (2, 14)
Population growth rate[a], *r*	0.133	0.128	0.149	
Mites per mite after 40 days	209	173	384	

[a]Based on life table analyses.
**$P < 0.01$; ***$P < 0.001$ (one-way ANOVA comparing species).
†$P = 0.06$.

colonies of all three mite species have positive growth rates when feeding upon new hyphal growth of the fungal species they transport (Table 13.2), *O. minus* and *C. ranaculosus*. However, none of the three mites had significant population growth when feeding on the one fungus they do not transport, *Entomocorticium* sp. A. When *T. fusarii* colonies were established on two other fungal species (which are commonly vectored by other bark beetles, but only occasionally associated with SPB), the colonies reproduced successfully on *Leptographium terebrantis* Barras and Perry but not on *Ophiostoma ips* (Rumbold) Nannf. Field observations showing that over ten times the number of tarsonemid mites are found within patches of *O. minus*-infested phloem vs. other areas, further the case for a symbiotic association of these two organisms (Lombardero *et al.*, 2000).

Ecological/economic implications

The web of complex relationships between mycangial fungi, phoretic fungi and phoretic mites associated with SPB, have significant implications to its life cycle and population dynamics of SPB. The possibility that *O. minus* assists SPB in killing tree hosts, or at least in overcoming tree resistance and/or conditioning the host tissue means that it may be vitally important that this fungus is present on the beetles or on their phoretic mites. Subsequently, the apparent dependence of developing larvae on vigorous growth of the mycangial fungi (especially *Entomocorticium* sp. A) demonstrates the importance of the presence of this fungus in the mycangium. If all three of the fungi are present within the network of SPB galleries, then the outcomes of fungal competition becomes extremely important. If *O. minus* is able to colonize the phloem around developing larvae, either because *Entomocorticium* sp. A has not yet become sufficiently established or because *C. ranaculosus* became established but was outcompeted by *O. minus*, larval development may be severely reduced. If *Entomocorticium* sp. A is outcompeted in the phloem by *C. ranaculosus*, the outcome may be similar, due to the relative inability of *C. ranaculosus* to exclude *O. minus* as well as its inferiority as a nutritional substrate. The success of the phoretic mites is linked similarly to the outcome of fungal competition for phloem. All three *Tarsonemus* species seem to be highly dependent upon the successful vectoring, inoculation and growth of the fungi they perform best on, *O. minus* and *C. ranaculosus*. The possibility also arises of exploiting the interdependencies between beetle and fungus as control or management options for SPB.

The negative effects of *O. minus* on SPB larval development could be seen as a positive, if the aim was to lower SPB population levels. However, augmentation of *O. minus* levels in the field might be counterproductive if it resulted in greater amounts or degrees of blue-stained wood, which is of lesser economic value both as lumber and pulp (Seifert, 1993). This has led to investigations into the use of a similar fungus, marketed under the trade name of Cartapip

Table 13.2. Population realized growth rates[a] (mean ± SE) for colonies of three *Tarsonemus* mite species feeding on five fungal species (*Ophiostoma minus, Ceratocystiopsis ranaculosus* and *Entomocorticium* sp. A are all associated with the focal bark beetle, *Dendroctonus frontalis*. *Leptographium terebrantis* and *O. ips* are associated with other bark beetles in the same forest).

	T. ips			*T. krantzi*			*T. fusarii*		
	r (mites per mite per day)	Colonies surviving (%)	*n*	*r* (mites per mite per day)	Colonies surviving (%)	*n*	*r* (mites per mite per day)	Colonies surviving (%)	*n*
O. minus				0.044 ± 0.014	47	15	0.045 ± 0.012	100	9
C. ranaculosus				0.022 ± 0.009	53	15	0.062 ± 0.004	100	7
E. sp. A	0.012 ± 0.012	16	6	0.002 ± 0.002	10	10	0.014 ± 0.015	80	5
L. terebrantis							0.044 ± 0.015	100	5
O. ips							–0.003 ± 0.004	60	5

[a] $r = \dfrac{\ln N_t - \ln(N_0)}{t}$.

(Clariant Corporation, Charlotte, North Carolina). Cartapip is a colourless strain of *Ophiostoma piliferum* (Fries) H. and P. Sydow which has been used to degrade pitch in wood chips (Blanchette *et al.*, 1992) and outcompete blue stain fungi. This white fungus differentially competes with all three SPB-associated fungi (Fig. 13.19) and outcompetes the mycangial fungi (and to a

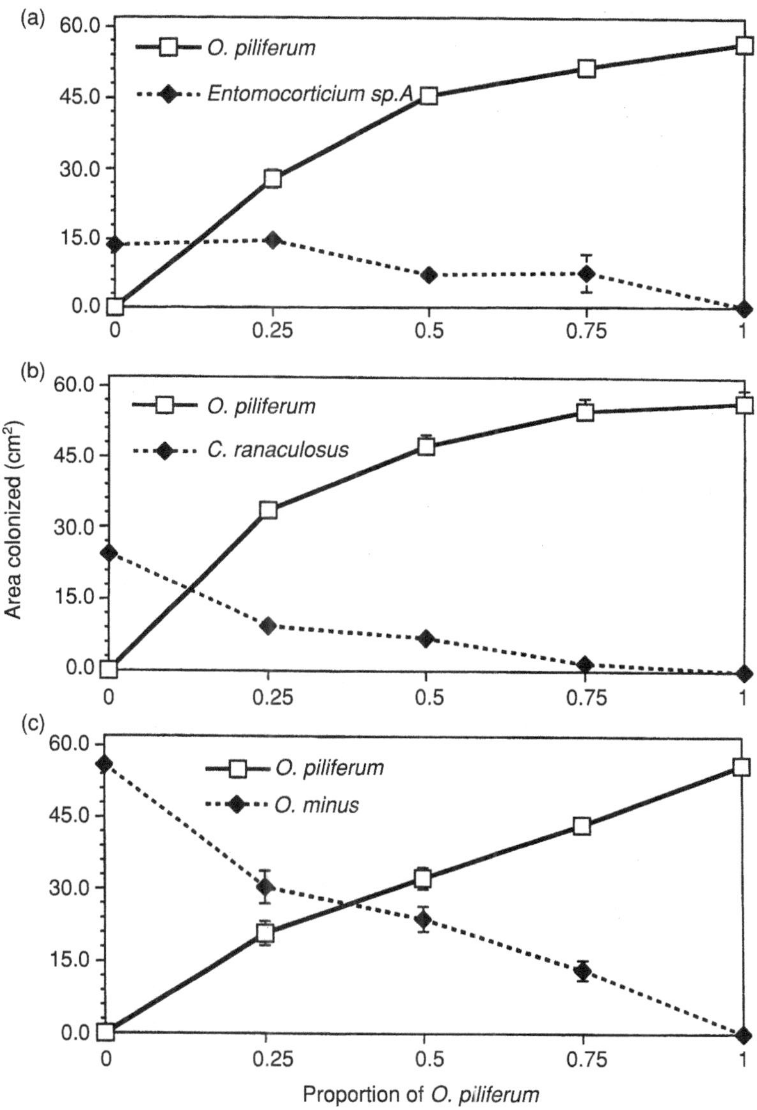

Fig. 13.19. DeWit replacement series diagrams resulting from competition between *Ophiostoma piliferum* (Cartapip) and: (a) *Entomocorticium* sp. A; (b) *Ceratocystiopsis ranaculosus*; and (c) *Ophiostoma minus*. Standard errors are given about each mean.

lesser degree, *O. minus*) in primary resource capture (Fig. 13.20) (Klepzig, 1998). While Cartapip is not able to capture already colonized substrate from *Entomocorticium* sp. A or *O. minus*, it does show some promise as a possible biocontrol agent of SPB, by virtue of its ability to interfere with the symbiotic relationships between SPB and its fungi.

Conclusions

Relationships among symbiotic organisms may change over time and ranges of resources. Other organisms may indirectly facilitate or interfere with these relationships. Interactions among bark beetles and their associated fungi and mites are complex examples of the manner in which symbioses change and are indirectly affected by other organisms. These complex relationships have been extensively studied in the southern pine beetle (SPB), a bark beetle that kills healthy living trees through mass colonization. The SPB is consistently associated with three main fungi. Two of these fungi (*Ceratocystiopsis ranaculosus* and *Entomocorticium* sp. A) are carried in a specialized structure (mycangium) in female SPB. The third fungus is carried phoretically on the exoskeleton. Both *O. minus* and *Entomocorticium* sp. A are also carried by phoretic mites of SPB. Due to the effects of these fungi on SPB larval development, their competitive interactions have significant implications. The two mycangial fungi provide nutrition to developing larvae, while the phoretic fungus interferes with larval

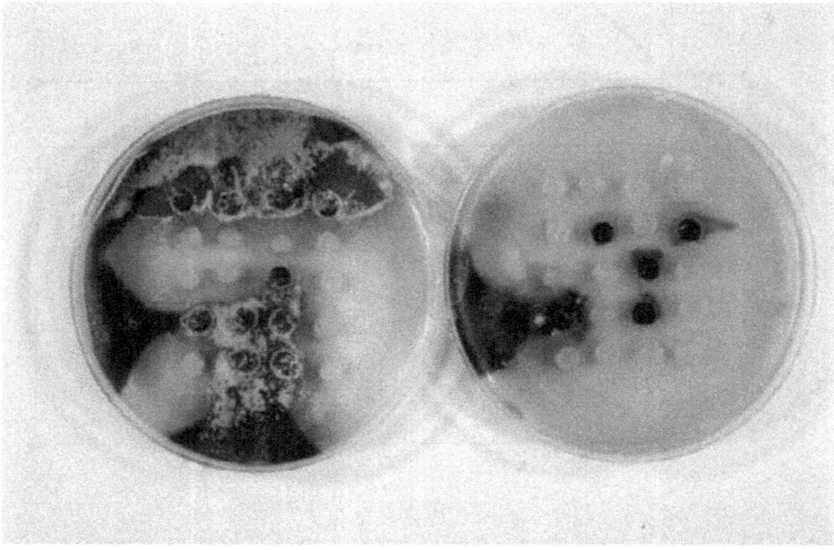

Fig. 13.20. Differential competition between *Ophiostoma piliferum* (Cartapip) and *Ophiostoma minus*. Note the mutual ability of each species to keep the other from colonizing the substrate it holds.

development. These interactions appear to be mediated by phoretic mites which have mutualistically symbiotic relationships with the SPB-associated fungi they vector. The multiple interdependencies in this system provide novel opportunities for control of, and further research on, this damaging forest pest complex.

References

Adee, S.R., Pfender, W.F. and Hartnett, D.C. (1990) Competition between *Pyrenophora tritici-repentis* and *Septoria nodorum* in the wheat leaf as measured with de Wit replacement series. *Phytopathology* 80, 1177–1182.

Ayres, M.P., Wilkens, R.T., Ruel, J.J. and Vallery, E. (2000) Fungal relationships and the nitrogen budget of phloem-feeding bark beetles (Coleoptera: Scolytidae). *Ecology* 81, 2198–2210.

Barras, S.J. (1970) Antagonism between *Dendroctonus frontalis* and the fungus *Ceratocystis minor*. *Annals of the Entomological Society of America* 63, 1187–1190.

Barras, S.J. (1975) Release of fungi from mycangia of southern pine beetles observed under a scanning electron microscope. *Zeitschrift für Angewandte Entomologie* 79, 173–176.

Barras, S.J. and Hodges, J.D. (1969) Carbohydrates of inner bark of *Pinus taeda* as affected by *Dendroctonus frontalis* and associated microorganisms. *Canadian Entomologist* 101, 489–483.

Barras S.J. and Perry, T.J. (1972) Fungal symbionts in the prothoracic mycangium of *Dendroctonus frontalis*. *Zeitschrift fur angewandte Entomologie* 71, 95–104.

Barras, S.J. and Taylor, J.J. (1973) Varietal *Ceratocystis minor* identified from mycangium of *Dendroctonus frontalis*. *Mycopathologia et Mycologia Applicata* 50, 203–305.

Basham, H.G. (1970) Wilt of loblolly pine inoculated with blue-stain fungi of the genus *Ceratocystis*. *Phytopathology* 60, 750–754.

Blanchette, R.A., Farrell, R.L., Burnes, T.A., Wendler, P.A., Zimmerman, W., Brush, T.S. and Snyder, R.A. (1992) Biological control of pitch in pulp and paper production by *Ophiostoma piliferum*. *Tappi Journal* 75, 102–106.

Bramble, W.C. and Holst, E.C. (1940) Fungi associated with *Dendroctonus frontalis* in killing shortleaf pines and their effect on conduction. *Phytopathology* 30, 881–899.

Bridges, J.R. (1983) Mycangial fungi of *Dendroctonus frontalis* and their relationship to beetle population trends. *Environmental Entomology* 12, 858–861.

Bridges, R.L. (1985) Relationship of symbiotic fungi to southern pine beetle population trends. In: Branham, S.J. and Thatcher, R.C. (eds) *Integrated Pest Management Research Symposium: the Proceedings*. USDA Forest Service, General Technical Report SO-56, Asheville, North Carolina, pp. 127–135.

Bridges, J.R. (1987) Effects of terpenoid compounds on growth of symbiotic fungi associated with the southern pine beetle. *Phytopathology* 77, 83–85.

Bridges, J.R. and Moser, J.C. (1983) Role of two phoretic mites in transmission of bluestain fungus, *Ceratocystis minor*. *Ecological Entomology* 8, 9–12.

Bridges, J.R. and Perry, T.J. (1985) Effects of mycangial fungi on gallery construction and distribution of bluestain in southern pine beetle-infested pine bolts. *Journal of Entomological Science* 20, 271–275.

Bridges, J.R., Nettleton, W.A. and Connor, M.D. (1985) Southern pine beetle (Coleoptera: Scolytidae) infestations without the bluestain fungus, *Ceratocystis minor*. *Journal of Economic Entomology* 78, 325–327.

Caird, R.W. (1935) Physiology of pines infested with bark beetles. *Botanical Gazette* 96, 709–733.

Callaway, R.M. and Walker, L.R. (1997) Competition and facilitation: a synthetic approach to interactions in plant communities. *Ecology* 78, 1958–1965.

Cook, S.P. and Hain, F.P. (1985) Qualitative examination of the hypersensitive response of loblolly pine, *Pinus taeda* L., inoculated with two fungal associates of the southern pine beetle, *Dendroctonus frontalis* Zimmermann (Coleoptera: Scolytidae). *Environmental Entomology* 14, 396–400.

Cook, S.P. and Hain, F.P. (1987) Four parameters of the wound response of loblolly and shortleaf pines to inoculations with the blue-staining fungus associated with the southern pine beetle. *Canadian Journal of Botany* 65, 2403–2409.

Cook, S.P., Hain, F.P. and Nappen, P.B. (1986) Seasonality of the hypersensitive response by loblolly and shortleaf pine to inoculation with a fungal associate of the southern pine beetle (Coleoptera: Scolytidae). *Journal of Entomological Science* 21, 283–285.

Coppedge, B.R., Stephen, F.M. and Felton, G.W. (1995) Variation in female southern pine beetle size and lipid content in relation to fungal associates. *Canadian Entomologist* 127, 145–154.

Dowding, P. (1969) The dispersal and survival of spores of fungi causing bluestain in pine. *Transactions of the British Mycological Society* 52, 125–137.

Drooz, A.T. (1985) *Insects of Eastern Forests*. USDA Forest Service Miscellaneous Publication 1426, Washington, DC.

Franklin, R.T. (1970) Observations on the blue stain–southern pine beetle relationship. *Journal of the Georgia Entomological Society* 5, 53–57.

Goldhammer, D.S., Stephen, F.M. and Paine, T.D. (1990) The effect of the fungi *Ceratocystis minor*, *Ceratocystis minor* var. *barassii* and SJB 122 on reproduction of the southern pine beetle, *Dendroctonus frontalis*. *Canadian Entomologist* 122, 407–418.

Happ, G.M., Happ, C.M. and Barras, S.J. (1971) Fine structure of the prothoracic mycangium, a chamber for the culture of symbiotic fungi, in the southern pine beetle, *Dendroctonus frontalis*. *Tissue and Cell* 3, 295–308.

Happ, G.M., Happ, C.M. and Barras, S.J. (1976) Bark beetle–fungal symbiosis. II. Fine structure of a basidiomycetous ectosymbiont of the southern pine beetle. *Canadian Journal of Botany* 54, 1049–1062.

Harrington, T.C. (1993) Diseases of conifers caused by species of *Ophiostoma* and *Leptographium*. In: Wingfield, M.J., Seifert, K.A. and Webber, J.F. (eds) *Ceratocystis and Ophiostoma: Taxonomy, Ecology and Pathogenicity*. APS Press, St Paul, Minnesota, pp. 161–172.

Hetrick, L.A. (1949) Some overlooked relationships of the southern pine beetle. *Journal of Economic Entomology* 42, 466–469.

Hodges, J.D., Barras, S.J. and Mauldin, J.K. (1968) Amino acids in inner bark of loblolly pine, as affected by the southern pine beetle and associated microorganisms. *Canadian Journal of Botany* 46, 1467–1472.

Hodges, J.D., Elam, W.W., Watson W.F. and Nebeker, T.E. (1979) Oleoresin characteristics and susceptibility of four southern pines to southern pine beetle attacks. *Canadian Entomologist* 111, 889–896.

Hsiau, O.T-W. (1996) The taxonomy and phylogeny of the mycangial fungi from *Dendroctonus brevicomis* and *Dendroctonus frontalis* (Coleoptera: Scolytidae). PhD thesis, Iowa State University, Ames, Iowa.

Klepzig, K.D. (1998) Competition between a biological control fungus, *Ophiostoma piliferum*, and symbionts of the southern pine beetle. *Mycologia* 90, 69–75.

Klepzig, K.D. and Wilkens, R.T. (1997) Competitive interactions among symbiotic fungi of the southern pine beetle. *Applied and Environmental Microbiology* 63, 621–627.

Lewinsohn, E., Gijzen, M., Savage, T.J. and Croteau, R. (1991a) Defense mechanisms of conifers: relationship of monoterpene cyclase activity to anatomical specialization and oleoresin monoterpene content. *Plant Physiology* 96, 38–43.

Lewinsohn, E., Gijzen, M., Savage, T.J. and Croteau, R. (1991b) Defense mechanisms of conifers: differences in constitutive and wound-induced monoterpene biosynthesis among species. *Plant Physiology* 96, 44–49.

Lindquist, E. (1969) New species of *Tarsonemus* (Acarina: Tarsonemidae) associated with bark beetles. *Canadian Entomologist* 101, 1291–1314.

Lombardero, M.J., Klepzig, K.D., Moser, J.C. and Ayres, M.P. (2000) Biology, demography and community interactions of *Tarsonemus* (Acarina: Tarsonemidae) mites phoretic on *Dendroctonus frontalis* (Coleoptera: Scolytidae). *Agricultural and Forest Entomology* (in press)

Malloch, D., and Blackwell, M. (1993) Dispersal biology of the Ophiostomatoid fungi. In: Wingfield, M.J., Seifert, K.A. and Webber, J.F. (eds) *Ceratocystis and Ophiostoma: Taxonomy, Ecology and Pathogenicity*. APS Press, St Paul, Minnesota, pp. 195–206.

Mathre, D.E. (1964) Pathogenicity of *Ceratocystis ips* and *Ceratocystis minor* to *Pinus ponderosa*. *Contributions to the Boyce Thompson Institute* 22, 363–388.

Moser, J.C. (1976) Surveying mites (Acarina) phoretic on the southern pine beetle (Coleoptera: Scolytidae) with sticky traps. *Canadian Entomologist* 108, 809–813.

Moser, J.C. (1985) Use of sporothecae by phoretic *Tarsonemus* mites to transport ascospores of coniferous bluestain fungi. *Transactions of the British Mycological Society* 84, 750–753.

Moser, J.C. and Bridges, J.R. (1986) *Tarsonemus* mites phoretic on the southern pine beetle: attachment sites and numbers of bluestain ascospores carried. *Proceedings of the Entomological Society of Washington* 88, 297–299.

Moser, J.C. and Roton, L.M. (1971) Mites associated with southern pine bark beetles in Allen Parish, Lousiana. *Canadian Entomologist* 103, 1775–1798.

Moser, J.C., Thatcher, R.C. and Pickard, L.S. (1971) Relative abundance of southern pine beetle associates in East Texas. *Annals of the Entomological Society of America* 64, 72–77.

Moser, J.C., Wilkinson, R.C. and Clark, E.W. (1974) Mites associated with *Dendronctonus frontalis* Zimmerman (Scolytidae: Coleoptera) in Central America and Mexico. *Turrialba* 24, 379–381.

Moser, J.C., Perry, T.J., Bridges, J.R. and Yin, H.-F. (1995) Ascospore dispersal of *Ceratocystiopsis ranaculosus*, a mycangial fungus of the southern pine beetle. *Mycologia* 87, 84–86.

Nebeker, T.E., Hodges, J.D. and Blanche, C.A. (1993) Host response to bark beetle and pathogen colonization. In: Schowalter, T.D. and Filip, G.M (eds) *Beetle–Pathogen Interactions in Conifer Forests*. Academic Press, London, pp. 157–173.

Nelson, R.M. (1934) Effect of bluestain fungi on southern pines attacked by bark beetles. *Phytopathologische Zeitschrift* 7, 327–426.

Nevill, R.J., Kelley, W.D., Hess, N.J. and Perry, T.J. (1995) Pathogenicity to loblolly pines of fungi recovered from trees attacked by southern pine beetles. *Southern Journal of Applied Forestry* 19, 78–83.

Newton, M.R., Kinkel, L.L. and Leonard, K.J. (1998) Constraints on the use of de Wit models to analyze competitive interactions. *Phytopathology* 88, 873–878.

Paine, T.D., Raffa, K.F. and Harrington, T.C. (1997) Interactions among Scolytid bark beetles, their associated fungi, and live host conifers. *Annual Review of Entomology* 42, 179–206.

Parmeter, J.R. Jr, Slaughter, G.W., Chen, M. and Wood, D.L. (1992) Rate and depth of sapwood occlusion following inoculation of pines with bluestain fungi. *Forest Science* 38, 34–44.

Payne, T.L. (1983) Behavior. In: *History, Status and Future Needs for Entomology Research in Southern Forests: Proceedings of the 10th Anniversary of the East Texas Forest Entomology Seminar.* Texas Agricultural Experiment Station, Texas A&M University System, College Station, Texas.

Popp, M.P., Johnson, J.D. and Lesney, M.S. (1995) Characterization of the induced response of slash pine to inoculation with bark beetle vectored fungi. *Tree Physiology* 15, 619–623.

Price, T.S., Doggett, C., Pye, J.M. and Holmes, T.P. (1992) *A History of Southern Pine Beetle Outbreaks in the Southeastern United States.* Georgia Forestry Commission, Macon, Georgia.

Raffa, K.F., Phillips, T.W. and Salom, S.M. (1993) Strategies and mechanisms of host colonization by bark beetles. In: Schowalter, T. and Filip, G. (eds) *Beetle–Pathogen Interactions in Conifer Forests.* Academic Press, San Diego, California, pp. 102–128.

Rayner, A.D.M. and Webber, J.F. (1984) Interspecific mycelial interactions - an overview. In: Jennings, D.H. and Rayner, A.D.M. (eds) *The Ecology and Physiology of the Fungal Mycelium.* Cambridge University Press, Cambridge, pp. 383–417.

Ross, D.W., Fenn, P. and Stephen, F.M. (1992) Growth of southern pine beetle associated fungi in relation to the induced wound response in loblolly pine. *Canadian Journal of Forest Research* 22, 1851–1859.

Ruel, J.J., Ayres, M.P. and Lorio, P.L. Jr (1998) Loblolly pine responds to mechanical wounding with increased resin flow. *Canadian Journal of Forest Research* 28, 596–602.

Rumbold, C.T. (1931) Two blue-stain fungi associated with bark beetle infestation of pines. *Journal of Agricultural Research* 43, 847–873.

Seifert, K.A. (1993) Sapstain of commercial lumber by species of *Ophiostoma* and *Ceratocystis*. In: Wingfield, M.J., Seifert, K.A. and Webber, J.F. (eds) *Ceratocystis and Ophiostoma: Taxonomy, Ecology and Pathogenicity.* APS Press, St Paul, Minnesota, pp. 141–152.

Smiley, R.T. and Moser, J.C. (1974) New Tarsonemids associated with bark beetles (Acarina: Tarsonemidae). *Annals of the Entomological Society of America* 67, 713–715.

Snaydon, R.W. (1991) Replacement or additive designs for competition studies? *Journal of Applied Ecology* 28, 930–946.

Thatcher, R.C. (1960) Bark beetles affecting southern pines: review of current knowledge. *USDA Forest Service Occasional Paper 180.* USDA Forest Service, Pineville, Louisiana.

Thatcher, R.C., Searcy, J.L., Coster, J.E. and Hertel, G.D. (1980) The southern pine beetle. *USDA Forest Service Science Education Administration. Technical Bulletin No. 1631.* USDA Forest Service, Pineville, Louisiana.

Wilson, M. and Lindow, S.E. (1994) Ecological similarity and coexistence of epiphytic ice-nucleating (Ice+) *Pseudomonas syringae* strains and a non-ice-nucleating (Ice–) biological control agent. *Applied and Environmental Microbiology* 60, 3128–3137.

Zook, D. (1998) A new symbiosis language. *Symbiosis News* 1, 1–3.

The Implications for Plant Health of Nematode–Fungal Interactions in the Root Zone

14

R.J. Hillocks

NRI-University of Greenwich, Central Avenue, Chatham Maritime, Chatham, Kent ME4 4TB, UK

Introduction

The soil is a microbiologically complex environment and within it the rhizosphere may be regarded as a niche with additional levels of complexity introduced into the system by intrusion of the plant root network. The root affects the physical dynamics of the soil at the root–soil interface in three main ways. It exudes nutrients, mainly in the form of nitrogen compounds and carbohydrate into the soil, it contributes organic matter through sloughing of epidermal tissue and it withdraws water from the system. The rhizosphere effect is defined by the impact of these changes, particularly root exudation, on microbial dynamics (Campbell and Greaves, 1990).

While it would be beyond the scope of this chapter to digress into a fuller description of the rhizosphere effect and microbial dynamics within the root zone, this brief mention serves as a reminder that microbial interactions in the soil are multiple with continuously varying permutations and opportunities for trophic interaction. The effects of rhizosphere associations on the plant may be beneficial, detrimental or neutral (Lynch, 1990). In considering only the nematode–fungal interactions we have already created a very simple model of only a small part of the biodiversity within the system.

Microbial Dynamics in the Soil

Nematodes interact with fungi either directly in the soil or indirectly, mediated through the host plant. Direct interactions are of two main types: fungi feeding on nematodes and nematodes feeding on fungi. Nematophagous fungi may be parasitic as in the case of *Verticillium chlamydosporum* (Kerry and Bourne, 1996), *Paecilomyces lilacinous* and *Nematophthora gynophila* or they may be predaceous as in the case of *Arthrobotrys* spp., *Dactylella* spp. and *Dactylaria* spp. Several of the parasitic and predaceous associations have been investigated for their potential in biological control of plant-parasitic nematodes (Stirling, 1991). The limitations of such biocontrol systems are based in the density-dependency of the effect these fungi have on nematode populations and the natural buffering capacity of the soil which tends to restore the natural microbial balance (see Fig. 14.1). The mycophagous nematodes are free-living in the soil, feeding largely on saprophytic fungi. They also feed on parasitic fungi where they are exposed on the host root surface and on mycorrhizal

Addition of organic material to field soil

leads to:

Increase in saprophytic fungi

and therefore to:

Increase in free-living nematodes feeding on saprophytic fungi

then:

Increase in nematode populations

leads to:

Increase in parasitic and predaceous fungi attacking nematodes

=control:

Decrease in plant-parasitic nematodes

but also:

Decrease in free-living nematodes

therefore:

Increase in saprophytic fungal populations

and:

Suppressive effect on predaceous fungi

consequently:

Decrease in efficacy of biocontrol of plant-parasitic nematodes

Fig. 14.1. Theoretical scheme to explain short-term benefits from addition of organic amendments to soil to manage plant-parasitic nematodes.

fungi, as in the case of *Aphelenchus avenae* feeding on vesicular-arbuscular mycorrhizae (VAMs) (Francl, 1993). The indirect interactions between nematodes and fungi occur between root-invading plant-parasitic nematodes and plant-pathogenic soil-borne fungi and also the root-colonizing rhizosphere bacteria. Further information on the direct nematode–fungal interactions, particularly with respect to the implications for biological control, can be found elsewhere (e.g. Kerry, 1990). With the exception of the special case of nematode–mycorrhizal interactions, only the indirect associations (i.e. those mediated by the host plant) with plant-pathogenic fungi will be considered further.

Mycorrhizal Interactions

Associations between nematodes and mycorrhizae are included in this review because although nematodes interact directly by grazing on VAMs, a host plant is required as a substrate for the fungus and the interaction has implications for plant health. The subject of nematode–mycorrhizal interactions has been more widely reviewed by Francl (1993).

Mycorrhizal fungi establish a symbiotic association with the plant, which allows the fungal partner access to carbon products of plant photosynthesis. To the benefit of the host plant, mycorrhizae in effect extend the rhizosphere (mycorhizosphere) and increase the uptake of soil nutrients, particularly phosphorus.

Mycophagous nematodes can potentially reduce the effectiveness of the mycorrhizal network in delivering nutrient to the plant through damage caused by their grazing activity. The evidence for this is drawn mainly from pot experiments and contradictory results seem almost inevitable in view of the problems of standardization of soil nutrient status and variability in populations of plant host, mycorrhizal fungus and nematode. Although species of *Aphelenchus, Aphelencoides, Bursaphelebchus* and *Ditylenchus* are all mycophagous, it is mainly the first two of these that have been shown to reproduce by feeding on mycorrhizal fungi and consequent reductions in growth have been manifested in decreased plant growth. Of 53 ectomycorrhizal species found by Riffle (1971) to support reproduction of *Aphelenchoides cibolensis*, 15 had their mycelial growth reduced by between 20 and 100%. *Aphelenchus avenae* inhibited sporulation of *Glomus* sp. on soybean, resulting in reduced plant growth (Salawu and Estey, 1979). In contrast, no such effect of *A. avenae* could be detected for either *Glomus etunicatum* or *Gigaspora margarita* on cotton (Hussey and Roncadori, 1981). *Tylenchulus semipenetrans* on *Citrus limon* prevented vesicle formation by *Glommus mosae. Radopholus similis* had a similar effect on *G. etunicatum* on the same host.

The presence of mycorrhizae in the root has also been reported to decrease nematode colonization and development. For instance, *G. fasciculatum* on cowpea decreased cyst production and reproduction of *Heterodera cajani* (Jain and Sethi, 1987).

In reports of nematode–mycorrhizal interactions, the effects on plant growth are often not reported. In the case of *G. intraradices* and *M. incognita* on cantaloupe (*Cucumis melo*), the growth-retarding effect of the nematode was decreased by 63% in mycorrhizal plants compared to the non-mycorrhizal ones (Heald *et al.*, 1989). *G. intraradices* has also been reported to increase the tolerance of cotton to *M. incognita* (Smith *et al.*, 1986).

Indirect Associations Between Nematodes and Plant Pathogens

Indirect nematode–fungal interactions or associations can, for convenience, be divided into two types: those in which the effect upon the host is localized or those in which the effect is systemic within the host plant. The localized effect may be a physical one, dependent on mechanical damage to the root caused by nematode invasion of the root or migration from it, or it may be a nutritional effect where sedentary endoparasitic nematodes create a nutrient sink effect to supply the developing female. These nutrients also provide a medium for enhanced growth of invading fungi and bacteria.

Systemic nematode–fungal interactions occur only in the case of sedentary endoparasitic nematodes and vascular pathogens. The evidence for such a systemic effect remains controversial and to some extent contradictory and will be discussed in more detail with particular reference to the associations between fusarium wilt and the root-knot nematodes in cotton (*Gossypium* spp.) and pigeonpea (*Cajanus cajan*). Before looking further at nematode effects on disease susceptibility, some taxonomic consideration must be given to the fungi and associated nematodes.

Which Plant-pathogenic Fungi Interact with Root-invading Nematodes?

The plant-pathogenic fungi involved in interactions with root-invading nematodes leading to increased plant disease can be broadly described as soil-borne plant pathogens. For the purposes of this review a soil-borne plant pathogen is defined as one which has the capacity for saprophytic survival in the soil and/or produces resting spores that remain viable in the soil for some time in the absence of a host.

The separation of soil-borne plant-pathogenic fungi into those which are soil invading or root-inhabiting and those which are soil-inhabiting (Garret, 1956) is a useful one (Table 14.1) in this context, as the distinction coincides with their mode of parasitism. The mode of parasitism in turn influences the type of nematode with which they are associated.

Table 14.1. Distinguishing characteristics of soil-inhabiting and soil-invading fungi.

Soil-invading (or root-inhabiting)	Soil-inhabiting
1. Suppressed by saprophytes on and in sterilized soil.	Often dominant on agar plates.
2. Do not grow through unsterilized soil in absence of host tissue.	May grow freely through unsterilized soil.
3. Typically invade only living root tissues.	Able to colonize moribund plant tissue as competitive saprophytes, but often unable to invade undamaged living tissue.
4. Distribution in soil depends on that of the host.	Distribution in soil is general.
5. Tend to have limited host range.	Usually have wide host ranges.

The soil-inhabiting pathogens tend to be non-specialized, causing damping-off and root-rot diseases (e.g. *Rhizoctonia solani, Pythium* spp.) while others cause root and stalk diseases (e.g. *Macrophomina phaseolina, Fusarium* spp.). The soil-invading pathogens tend to be more specialized and host-specific, causing vascular wilt diseases (e.g. *Fusarium oxysporum* f. sp. *vasinfectum, Verticillium dahliae*) or, they are ectotrophic root colonizers (e.g. *Armillaria mellea, Gauemannomyces gramminis*) or obligate parasites (e.g. *Plasmodiophora brassicae, Synchitrium endobioticum*), causing various types of root-rot diseases.

Which Plant-parasitic Nematodes Interact with Plant-pathogenic Fungi?

Plant-parasitic nematodes can be classified according to their feeding mode and this is a useful distinction as it has a significant bearing on the type of plant pathogens with which they interact.

There are four main categories of root-invading plant-parasitic nematodes:

1. Ectoparasitic. The nematode does not enter the plant tissue but feeds from outside the root using the stylet to puncture plant cells. Different nematodes feed on epidermal, mesodermal and endodermal cells and have appropriate stylet lengths, e.g. short stylet, *Tylenchorhynchus*; e.g. medium stylet, *Rotylenchus*; e.g. long stylet, *Criconemella*.

2. Endoparasitic/migratory. The nematode retains mobility and moves through the host tissues as it feeds. They can leave the root if the tissue becomes moribund and re-enter at another point or another root, e.g. *Pratylenchus, Hirschmanniella*.

3. Endoparasitic/sedentary. The nematode loses its mobility after arriving at the feeding site within the root and the female matures and produces eggs at the fixed feeding site. The egg sac may be extruded to the outside of the root, e.g. *Meloidogyne*, *Heterodera*.
4. Semi-endoparasitic. Similar in feeding habit to the sedentary endoparasites except that only the anterior portion of the nematode body enters the root, the posterior remains in the soil allowing the egg sac to be produced in the rhizosphere soil, e.g. *Rotylenchulus*, *Tylenchulus*.

The damage caused by the more surface-feeding ectoparasitic nematodes tends to be too superficial to cause increased fungal infection but some of those which feed on the cortex can provide entry sites for pathogens. Some of these nematodes such as *Xiphenema* spp. and *Longidorus* spp. have long feeding stylets able to pierce the vascular tissues and are often virus vectors. It is mainly the migratory endoparasites and sedentary endoparasites and semi-endoparasites that have been implicated in disease complexes. A disease complex is defined as a microbial association resulting in a higher disease incidence or severity than can be explained by the additive effects on the host of the two organisms.

The migratory endoparasites that feed on the root cortex are more often associated with the non-systemic diseases caused by non-specialized fungal pathogens, such as the root-rot and damping-off diseases. The sedentary endoparasites have a much more profound effect on the metabolism of their host and are therefore the nematodes more often associated with systemic diseases such as the vascular wilts.

Nematode Associations with Non-specialized Root-rot Pathogens

The spatial association of a nematode and root pathogen does not in itself indicate that there is an interaction. For instance, both the lesion nematode *Pratylenchus thorneii* and root-rot caused by *Bipolaris sorokiniana* and *Fusarium graminearumi*, have been implicated in wheat yield decline in Australia (Doyle *et al.*, 1987). No evidence has been found to show that root-rot is more severe in the presence of the nematode than in its absence. Associations between plant-parasitic nematodes and fungal pathogens in the rhizosphere, therefore, may be passive or interactive. The interactive associations may in turn be additive or synergistic. In the synergistic interaction there must be an effect on the host that is greater in magnitude than would be expected from the additive effects of the two parasites. Most of the research to demonstrate nematode–fungal interactions has been based on pot experiments. As Evans and Haydock (1993) have pointed out, synergistic interactions are difficult to distinguish from additive ones and many papers report synergistic effects without proof that they are significantly greater than additive.

Meloidogyne incognita has been shown to increase infection by *Rhizoctonia solani* and *Pythium aphanidermatum* on a wide range of crops such as chilli (Hasan, 1985), groundnut (Garcia and Mitchell, 1975), maize (Sumner *et al.*, 1985) and okra (Golden and van Gundy, 1975). On cowpea, both *M. incognita* and *Rotylenchulus reniformis* increased root infection by *R. solani* and caused greater reduction in plant growth than either parasite acting alone (Khan and Husain, 1989). The root-knot nematodes (RKN) are by far the most important and widely distributed plant-parasitic nematodes and are also those most often reported in disease complexes with root-rot fungi. However, a number of other nematodes increase root disease, the most common being *Rotylenchulus reniformis* and *Pratylenchus* spp. with *Fusarium, Rhizoctonia, Macrophomina, Pythium* and *Phytophthora* the most common fungal genera with which they have been associated (e.g. Ignaki and Powell, 1969; Kotcon *et al.*, 1985; Jordaan *et al.*, 1987; LaMondia and Martin, 1989).

Nematode Associations with Vascular Wilt Pathogens

Results from pot experiments that claim to show synergistic effects between nematodes and fungal pathogens should be interpreted with caution and provide only an indication that this might occur also in the field. Studies that show interactive effects and have a statistically significant effect on infection and even on symptom development are not of much practical value if there is no synergistic effect on yield.

The most common nematode–fungal disease complexes and the ones that have been most often studied are those between RKN and the vascular wilt fusaria. This is perhaps because similar disease complexes affect a number of commercially important crops. One of these is cotton, and, in this case, there have been a number of field studies undertaken. Much of the early evidence for the interaction between *M. incognita* and *Fusarium oxysporum* f. sp. *vasinfectum* was derived from field experiments in which there was a decrease in disease following nematicide application to the soil. The main criticism of this approach is that the direct effects of soil fumigation on the fungal pathogen were not taken into account. Furthermore, soil fumigants can cause yield increments in the absence of any nematode damage, due simply to the mineralizing effect they have on the soil.

In Tanzania, fusarium wilt affects cotton grown in an area of some 70,000 km^2 around the southern shore of Lake Victoria. Sandy soils predominate in the area and where they have been cultivated for long periods, large populations of RKN may be found. A field experiment was designed to quantify the effect of RKN on disease incidence. Populations of the nematode and of the fungal pathogen were increased by augmenting the naturally occurring soil populations with inocula. The soil was treated with nematicide to decrease the nematode population. Changes in the nematode population had a much greater effect on wilt incidence than changes in the *Fusarium* population. Higher nematode population in both the wilt-susceptible and wilt-resistant cotton varieties increased wilt incidence but final wilt incidence in the wilt-resistant one did not reach that of the wilt-susceptible ones in the presence of the nematode. Therefore, susceptibility to infection had been enhanced by the nematode but resistance was not totally nullified (Hillocks and Bridge, 1992).

Working with the same disease complex in the USA using microplots, no interaction could be demonstrated at high levels of *Fusarium* and low levels of *Meloidogyne* but a significant interaction occurred at intermediate populations of *Fusarium* and high nematode population (Starr *et al.*, 1989). When the effects on yield were examined, it was evident that the nematode had the larger impact on yield reduction. This observation has significant crop protection implications because the nematode is often more prevalent than the pathogen.

In breeding for resistance to the wilt complex, one approach advocated is to introduce resistance to RKN into the breeding material or select for nematode resistance in cotton lines that already have some resistance to fusarium wilt. A high level of resistance to RKN can be more effective in reducing losses to the wilt complex than moderate levels of resistance to both organisms (Hyer *et al.*, 1979). However, selection based on survival or low disease scores in sick plots infested with both RKN and the fusarium wilt fungus does not necessarily produce plants with high levels of nematode resistance. More intensive selection for nematode resistance is required.

Curiously, there are far fewer reports of any interaction between RKN and verticillium wilt. There are, however, several reports of interactions between this disease and *Pratylenchus* spp. (Mountain and McKeen, 1965; Faulkner *et al.*, 1970; Conroy and Green, 1974), but fusarium wilt is rarely associated with these nematodes. Verticillium and fusarium wilts are very similar diseases in terms of the pathogen and mode of infection and it seems questionable that there are major differences in the way they interact with nematodes. One possible explanation is that fusarium wilt diseases are more common on acid, sandy soils, which are also those more favourable to RKN, whereas verticillium wilt diseases occur on neutral to alkaline soils with more clay content. The limited evidence from pot experiments is that on cotton, RKN can enhance infection by *Verticillium dahliae* (Khoury and Alcorn, 1973; Katsandonis, 1999). As Sikora and Carter (1987) pointed out, if you put large enough numbers of nematodes into the pot, you are almost bound to get some sort of effect on disease. On the available evidence, RKN seems to enhance infection by *Verticillium* less than by *Fusarium* and if this is the case then, comparative studies of these two interactions may improve our understanding of the mechanisms of resistance to the two vascular wilt pathogens.

Mechanisms of Nematode-enhanced Disease

Localized effects

The localized effects of nematodes on the host plant that affect its susceptibility to infection are of three types: predisposition, rhizosphere modification and modification of host metabolism. Enhancement of diseases caused by non-specialized, root-inhabiting fungi such as *Rhizoctonia solani* is often due to the first of these effects: predisposition due to mechanical wounding which creates entry sites for pathogens that are unable to penetrate undamaged living tissue. Wounding may be of the minor type caused by the feeding of ectoparasitic nematodes or more severe as might be caused by migratory nematodes such as *Pratylenchus* spp. Severe damage is caused also by rupturing of the cortex and epidermis that takes place when the egg sac of a mature *Meloidogyne* female is extruded into the soil. Root wounding may have the additional effect of decreasing the efficiency of water uptake and decreasing the ability of the plant to withstand the effects of infection, particularly to vascular wilt diseases.

The second type of indirect effect of plant parasitic nematodes on soil-borne fungi is local in the sense that it is confined to the rhizosphere. Nematode invasion of the root enhances root exudation and the supply of exogenous nutrient in root exudates provides the energy to overcome fungistasis in the soil, allowing resting spores of plant-parasitic fungi to germinate. Growth of the germ tube towards the host then follows the nutrient gradient. For a given inoculum density, increased root exudation should increase the proportion of fungal propagules that germinate close enough to the host root for the germ tube to reach it and penetrate, before the energy stored in the spore is consumed.

The third type of local effect involves modification of host metabolism to create a nutrient sink effect. This provides the nutrients on which sedentary endoparasitic nematodes feed and develop to maturity. Prolific growth of vascular wilt fungi has been observed in these areas of nutrient accumulation (Minton and Minton, 1966; Melendez and Powell, 1967). The nutrient sink effect is well documented for the RKN that disrupt the normal xylem tissue, causing hypertrophy and hyperplasia that results in the formation of giant cells. These cells are highly differentiated with a structure determined by their function of providing nutrient for the developing nematode (Bird, 1974; Jones, 1981). The interior surface of the giant cell wall is invaginated to increase the surface area for the inward secretion of nutrient solution consisting largely of amino acids and sugars (Melendez and Powell, 1967).

Systemic effects

The effects of the sedentary endoparasitic nematodes on host metabolism have a systemic as well as a localized dimension. Invasion of the root by *Meloidogyne* spp. can cause changes in host metabolism that interfere with normal defence mechanisms against vascular invasion by plant pathogens in tissues some distance above the feeding site of the nematode. However, the available evidence suggests that such a systemic effect of RKN on host defence mechanisms may not be common to all disease complexes involving *Meloidogyne* spp. and fusarium wilt fungi. There is strong evidence for systemic effects in both cotton and pigeonpea but no such effect could be demonstrated for fusarium wilt in tobacco, for instance.

The first part of the evidence that the effect of RKN on vascular infection by *Fusarium oxysporum* is something more profound than root wounding is theoretical. Nematodes that cause extensive damage to the cortex such as the burrowing nematode (*Radopholus similis*) on banana and *Hoplolaimus galeatus*, a migratory endoparasite on cotton, are not known to increase infection by the respective wilt pathogens *F. oxysporum* f. sp. *cubense* and *F. oxysporum* f. sp. *vasinfectum*. Also, despite being similar in the damage they cause to the host root system, species of *Meloidogyne* vary in the extent to which they enhance vascular infection. In fusarium wilt of chrysanthemum for instance, *M. javanica* enhances infection to a greater extent than either *M. hapla* or *M. incognita*. The second part of the evidence is empirical and is derived from numerous experiments conducted with a range of hosts and their respective vascular wilt pathogens. The evidence in favour of a systemic effect is perhaps strongest in the case of the association between *Meloidogyne incognita* and *Fusarium oxysporum* f. sp. *lycopersici* on tomato. The stems of tomato plants were layered into separate pots to encourage the production of adventitious roots at one of the nodes. Once the roots were established in the pot, the soil was inoculated with nematodes and/or fungus. The treatments consisted of separation of nematode and fungal inoculum by placing the nematode on the main root and fungus on the adventitious roots, or combining both inocula at either the main root or the adventitious root. Disease incidence/severity was increased in both treatments compared to inoculation with the fungus alone (Sidhu and Webster, 1977).

Similar split root experiments have been conducted with tobacco and cotton with very different results. In tobacco plants, although RKN increased wilt incidence, this only occurred when nematode and fungus were inoculated on to the roots together. A similar result was obtained with cotton plants inoculated with *F. oxysporum* f. sp. *vasinfectum* and *M. incognita* (Hillocks, 1986). However, in the experiments with cotton, disease severity was increased when inoculating fungal spores directly into the stem xylem thereby spatially separating nematode and fungus. It would appear from this result that nematode enhancement of disease has both a localized and systemic component in cotton plants. The effect on disease incidence is a localized one and the effect on disease severity, following xylem invasion, is a systemic one. If only disease incidence is measured then systemic effects might be overlooked.

The device of spatially separating nematode and fungal inoculum by using stem-puncture inoculation with the fungus has also been used for studies on the interaction between RKN and fusarium wilt (*F. udum*) on pigeonpea. That RKN increases wilt incidence on pigeonpea is well documented from field studies (Sharma and Nene, 1990). The cultivar ICP 9145 was selected at ICRISAT in India for resistance to fusarium wilt and became widely grown in Malawi where fusarium wilt was a major problem. In pot experiments with ICP 9145, it was shown that RKN increased disease severity and furthermore, since the plants were inoculated with the fungus by stem puncture, the effect was systemic (Fig. 14.2). The primary mechanism of resistance to wilt in ICP 9145 was shown to be the ability for more rapid accumulation of the phytoalexin, cajanol, than occurred in more susceptible cultivars (Marley and Hillocks, 1993). The importance of cajanol in the resistance mechanism was confirmed when it was found that once the nematode was established within the root, cajanol accumulation in the xylem following stem inoculation with the wilt pathogen was retarded compared to similar plants free of nematodes (Fig. 14.3) (Marley and Hillocks, 1994). Care should be taken in drawing general conclusions from work on specific interactions with single strains of nematode and fungal pathogen. There was some indication from the work on pigeonpea that rapid cajanol accumulation may not be the mechanism of wilt resistance in all cultivars and that RKN may not have a systemic effect on resistance in all cultivars (Marley, 1992).

Since the work on pigeonpea, little progress has been made towards improving our understanding of the mechanisms involved in the systemic effect of RKN on vascular wilt diseases. There is some suggestion that RKN in

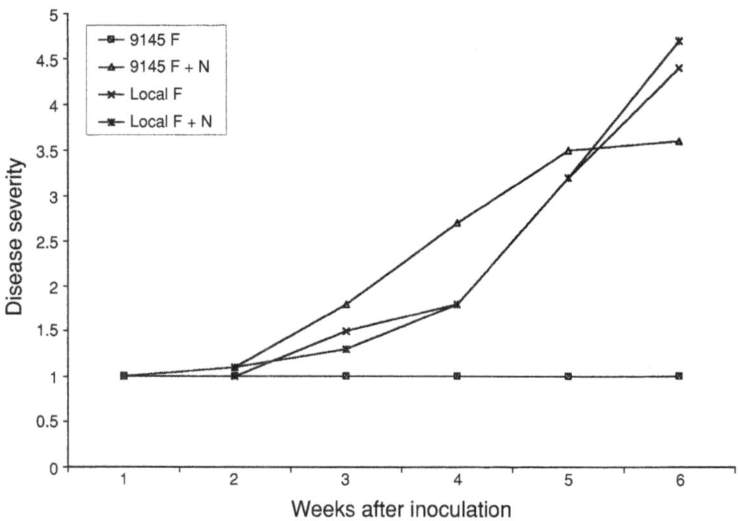

Fig. 14.2. Effect of root-knot nematode (RKN) on fusarium wilt severity in wilt-resistant pigeonpea cultivar ICP 9145 and wilt-susceptible cultivar 'Local' 1–6 weeks after stem-puncture inoculation with *Fusarium udum*. F = inoculated with *Fusarium* alone; F + N = inoculated with *Fusarium* and RKN.

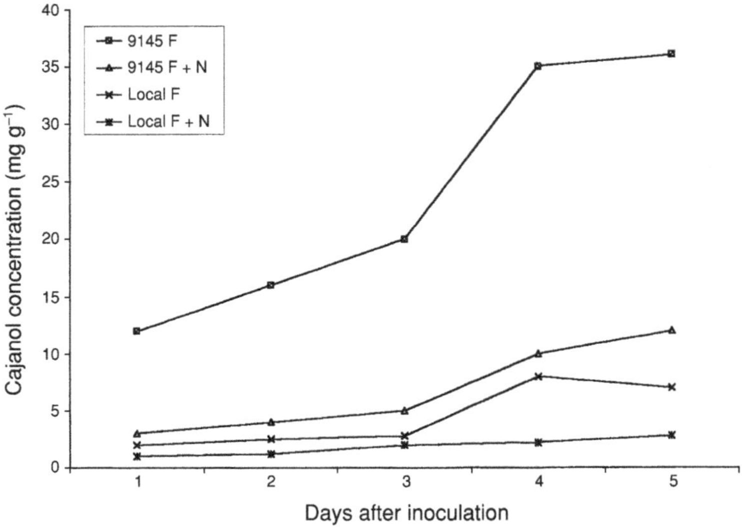

Fig. 14.3. Changing concentration of the phytoalexin cajanol in the stele tissue of a wilt-resistant (ICP 9145) and wilt-susceptible ('Local') pigeonpea cultivar 1–5 days after stem-puncture inoculation with *Fusarium udum* alone (F) or in the presence of RKN (F + N).

cotton roots decreases sugar levels and increases peroxidase activity in the xylem fluid. Decreased sugar levels could stimulate spore germination of *Fusarium* and increased peroxidase activity may weaken host resistance to vascular invasion by detoxifying terpenoid phytoalexins (Katsandonis, 1999).

Pot experiments on nematode–fungal interactions are difficult to carry out because of the difficulty of controlling inoculum dose and regulating growth conditions, particularly temperature, at the optimum for the interactive effects, which is not necessarily the same as the optimum for infection and growth of the parasites. Much remains to be done, therefore, to identify the translocated factor(s) involved in nematode-enhanced infection by vascular wilt pathogens.

Conclusions

In this chapter I have only been able to give a brief overview of the subject of nematode–fungal interactions and only a few of the reported examples have been cited. Although there is a large body of literature on the subject, there is still considerable controversy surrounding issues such as the proof of synergistic effects, the extrapolation of results from pot experiments to the field, and whether or not resistance to fungal disease can be 'broken' by nematodes. Despite the large amount of conflicting research, there remains much still to learn about the dynamics of these associations and about resistance to disease by investigation of the effects that nematodes (particularly root-knot nematodes) have on the expression of disease resistance in the host plant. The main areas which need addressing and on which further research would be welcome are:

1. More field studies should be conducted or results from pot experiments corroborated by field data.
2. More emphasis should be placed on measuring effects on plant growth or crop yield, rather than just infection levels and parasite populations.
3. Studies on the mechanism of systemic effects of root-knot nematodes on vascular wilt resistance are required to determine the mechanism by which the nematode is able to interfere with host resistance mechanisms to vascular invasion.
4. If pot experiments alone are to be conducted then they should be carried out under conditions that keep as closely to the field situation as possible, particularly with respect to population levels of the parasites. Studies that claim interactions where nematode populations have been used that are greatly in excess of those normally found in the field are of limited value.

References

Bird, A.F. (1974) Plant response to root-knot nematode. *Annual Review of Phytopathology* 12, 69–85.

Campbell, R. and Greaves, M.P. (1990) Anatomy and community structure of the rhizosphere. In: Lynch, J.M. (ed.) *The Rhizosphere*. John Wiley & Sons, Chichester, pp. 11–34.

Conroy, J. and Green, R.J. (1974) Interactions of the root-knot nematode *Meloidogyne incognita* and the stubby root nematode *Trichodorus christiei* with *Verticillium albo-atrum* on tomato at controlled inoculum densities. *Phytopathology* 64, 1118–1121.

Doyle, A.D., McLeod, R.W., Wong, P.T.W., Hetherington, S.E. and Southwell, R.J. (1987) Evidence for the involvement of the root lesion nematode *Pratylenchus thornei* in wheat yield decline in northern New South Wales. *Australian Journal of Experimental Agriculture* 27, 563–570.

Evans, K. and Haydock, P.P.J. (1993) Interactions of nematodes with root-rot fungi. In: Khan, M.W. (ed.) *Nematode Interactions*. Chapman & Hall, London, pp. 104–133.

Faulkner, L.R., Bolander, W.J. and Skotland, C.B. (1970) Interaction of *Verticillium dahliae* and *Pratylenchus minyus* in verticillium wilt of peppermint: influence of the nematode as determined by a double root technique. *Phytopathology* 60, 100–103.

Francl, L.J. (1993) Interactions of nematodes with mycorrhizae and mycorrhizal fungi. In: Khan, M.W. (ed.) *Nematode Interactions*. Chapman & Hall, London, pp. 203–216.

Garcia, R. and Mitchell, D.J. (1975) Synergistic interactions of *Pythium myriotylum* with *Fusarium solani* and *Meloidogyne arenaria* in pod rot of peanut. *Phytopathology* 65, 832–833.

Garret, S.D. (1956) *Biology of Root-infecting Fungi*. Cambridge University Press, Cambridge.

Golden, J.K. and van Gundy, S.D. (1975) A disease complex of okra and tomato involving the nematode, *Meloidogyne incognita* and the soil-inhabiting fungus, *Rhizoctonia solani*. *Phytopathology* 65, 265–273.

Hasan, A. (1985) Breaking resistance in chili to root-knot nematode by fungal pathogens. *Nematologica* 31, 210–217.

Heald, C.M., Bruton, B.D. and Davis, R.M. (1989) Influence of *Glomus intraradices* and soil phosphorus on *Meloidogyne incognita* infecting *Cucumis melo*. *Journal of Nematology* 21, 69–73.

Hillocks, R.J. (1986) Localised and systemic effects of root-knot nematode on incidence and severity of fusarium wilt in cotton. *Nematologica* 32, 202–208.

Hillocks, R.J. and Bridge, J. (1992) The role of nematodes in fusarium wilt of cotton in Tanzania. *Afro-Asian Journal of Nematology* 2, 35–40.

Hussey, R.S. and Roncadori, R.W. (1981) Influence of *Aphelenchus avenae* on vesicular-arbuscular endomycorrhizal growth response in cotton. *Journal of Nematology* 13, 48–52.

Hyer, A.H., Jorgenson, E.C., Garber, R.H. and Smith, S. (1979) Resistance to root-knot nematode in control of root-knot nematode–Fusarium wilt disease complex in cotton. *Crop Science* 19, 898–901.

Ignaki, H. and Powell, N.T. (1969) Influence of the root lesion nematode on black shank symptom development in flue-cured tobacco. *Phytopathology* 59, 1350–1355.

Jain, R.K. and Sethi, C.L. (1987) Pathogenicity of *Heterodera cajani* on cowpea as influenced by the presence of VAM fungi, *Glomus fasciculatumi* or *G. epigaeus*. *Indian Journal of Nematology* 17, 165–170.

Jones, M.G.K. (1981) The development and function of plant cells modified by endoparasitic nematodes. In: Zuckerman, B.M. and Rohde, R.A. (eds) *Plant Parasitic Nematodes*, Vol. III. Academic Press, New York, pp. 255–279.

Jordaan, E.M., Loots, G.C., Jooste, W.J. and de Waele, D. (1987) Effects of root-lesion nematodes (*Pratylenchus brachyurus* Godfrey and *P. zeae* Graham) and *Fusarium moniliforme* Sheldon alone or in combination on maize. *Nematologica* 33, 213–319.

Katsandonis, D. (1999) The nematode enhanced susceptibility to fusarium and Verticillium wilt of cotton. PhD thesis, University of Reading, UK.

Kerry, B.R. (1990) An assessment of progress towards microbial control of plant parasitic nematodes. *Annals of Applied Nematology* 22, 261–281.

Kerry, B.R. and Bourne, J.M. (1996) The importance of rhizosphere interactions in the biological control of plant parasitic nematodes – a case study using *Verticillium chlamydosporium*. *Pesticide Science* 47, 69–75.

Khan, T.A. and Husain, S.I. (1989) Relative resistance of six cowpea cultivars as affected by concomitance of two nematodes and a fungus. *Nematologica Mediterranea* 17, 39–41.

Khoury, F.Y. and Alcorn, M.A. (1973) Effects of *Meloidogyne incognita acrita* on the susceptibility of cotton plants to *Verticillium albo-atrum*. *Phytopathology* 63, 485–490.

Kotcon, J.B., Rouse, D.I. and Mitchell, J.E. (1985) Interactions of *Verticillium dahliae*, *Colletotrichum coccodes*, *Rhizoctonia solani* and *Pratylenchus penetrans* in the early dying syndrome of Russet Burbank potatoes. *Phytopathology* 75, 68–73.

LaMondia, J.A. and Martin, S.B. (1989) The influence of *Pratylenchus penetrans* and temperature on black root rot of strawberry by binucleate *Rhizoctonia* spp. *Plant Disease* 73, 107–110.

Lynch, J.M. (1990) Introduction: some consequences of microbial rhizosphere competence for plant and soil. In: Lynch, J.M. (ed.) *The Rhizosphere*. John Wiley & Sons, Chichester, pp. 1–10.

Marley, P.S. (1992) Resistance mechanisms to fusarium wilt in pigeonpea and the effect of interactions with root-knot nematodes. PhD thesis, University of Reading, UK.

Marley, P.S. and Hillocks, R.J. (1993) The role of phytoalexins in resistance to fusarium wilt in pigeonpea (*Cajanus cajan*). *Plant Pathology* 42, 212–218.

Marley, P.S. and Hillocks, R.J. (1994) Effect of root-knot nematodes on cajanol accumulation in the vascular tissues of pigeonpea after stem inoculation with *Fusarium udum*. *Plant Pathology* 43, 172–176.

Minton, N.A. and Minton, E.B. (1966) Effect of root-knot and sting nematodes on expression of fusarium wilt of cotton in three soils. *Phytopathology* 56, 319–322.

Melendez, P.L. and Powell, N.T. (1967) Histological aspects of the fusarium wilt–root-knot complex in flue-cured tobacco. *Phytopathology* 57, 286–291.

Mountain, W.B. and McKeen, C.D. (1965) Effects of transplant injury and nematodes on incidence of *Verticillium* wilt of eggplant. *Canadian Journal of Botany* 43, 619–624.

Riffle, J.W. (1971) Effects of nematodes on root-inhabiting fungi. In: Hacskaylo, E. (ed.) *Mycorrhizae; Proceedings of the First North American Conference on Mycorrhizae.* Miscellaneous Publication 1189. US Department of Agriculture, Washington, DC, pp. 97–113.

Salawu, E.O. and Estey, R.H. (1979) Observations on the relationship between a vesicular-arbuscular fungus, a fungivorous nematode and the growth of soybeans. *Phytoprotection* 60, 99–102.

Sharma, S.B. and Nene, Y.L. (1990) Effect of *Fusarium udum* alone and in combination with *Rotylenchulus reniformis* or *Meloidogyne* spp. on wilt incidence, growth of pigeonpea and multiplication of nematodes. *International Journal of Tropical Plant Diseases* 8, 95–101.

Sidhu, G. and Webster, J.M. (1977) Predisposition of tomato to the wilt fungus *Fusarium oxysporum* f. sp. *lycopersici* by the root-knot nematode (*Meloidogyne incognita*). *Nematologica* 23, 436–442.

Sikora, R.A. and Carter, W.W. (1987) Nematode interactions with fungal and bacterial plant pathogens: fact or fantasy. In: Veech, J.A. and Dickson, D.W. (eds) *Vistas on Nematology*. Society of Nematologists, Hyattsville, Maryland, pp. 307–312.

Smith, G.S., Hussey, R.S. and Roncadori, R.W. (1986) Penetration and postinfectional development of *Meloidogyne incognita* on cotton as affected by *Glomus intraradices* and phosphorus. *Journal of Nematology* 18, 429–435.

Starr, J.L., Jeger, M.J., Martyn, R.D. and Schilling, K. (1989) Effects of *Meloidogyne incognita* and *Fusarium oxysporum* f. sp. *vasinfectum* on plant mortality and yield of cotton. *Phytopathology* 79, 640–646.

Stirling, G.R. (1991) *Biological Control of Plant Parasitic Nematodes*. CAB International, Wallingford, UK.

Sumner, D.R., Dowler, C.C., Johnson, A.W., Chalfant, R.B., Glaze, N.C., Phatak, S.C. and Epperson, J.E. (1985) Effect of root diseases and nematodes on yield of corn in an irrigated multiple-cropping system with pest management. *Plant Disease* 69, 382–387.

Interactions of Plants, Soil Pathogens and Their Antagonists in Natural Ecosystems

15

W.H. Van der Putten

Multitrophic Interactions Department, Netherlands Institute of Ecology, NIOO-CTO, PO Box 40, 6666 ZG Heteren, The Netherlands

Introduction

Interest in the role of plant pathogens in natural ecosystems has increased strongly in the past two decades. Most of this research has focused on the role of above-ground pathogens in relation to the population biology and natural selection of plants (Burdon, 1987, 1993; Clay and Kover, 1996). Only a minority of the studies has concerned pathogens in natural soils, although this number is now increasing. The apparent bias may be due to the 'out-of-sight-out-of-mind-principle'.

In natural ecosystems, effects of soil pathogens may be incipient and difficult to identify and, subsequently, to prove. There are many factors involved in the spatio-temporal dynamics of natural vegetation, and effects of soil pathogens are likely to be overlooked because of invisibility. Failures of seedling survival may not be readily recognized as a result of soil pathogens, unless these are studied in detail both in the field and in controlled experimental conditions (Augspurger, 1990; Packer and Clay, 2000). The same holds for studies on the contribution of soil pathogens to directional succession (Van der Putten *et al.*, 1993) and cyclic succession or the maintenance of plant species richness in old field grasslands (Bever, 1984). Pathogenic soil fungi may be asymptomatic and interacting with mycorrhizal fungi; effects of soil fungicide treatments may be neutral, but this can be due to reduction of both the pathogen and the mutualist (Newsham *et al.*, 1994).

Most reports on soil pathogens in natural ecosystems are on pathogenic soil fungi. There are reports on plant-parasitic nematodes, mostly considering their role as herbivores (reviewed by Stanton, 1988; Mortimer *et al.*, 1999).

There have been only a few attempts to examine the role of plant-parasitic nematodes in interactions with pathogenic soil fungi (De Rooij-Van der Goes, 1995). Reports on soil bacterial diseases or virus transmission by soil organisms are rare, if indeed any exist at all (but see Szabo, 1999, for a case of pathogenic bacteria). Soil pathogens can be highly aggressive, especially facultative saprotrophic fungi, such as *Pythium* spp. and *Phytophthora* spp. As these pathogens may easily kill whole seedling cohorts, there is little, if any, opportunity for developing resistance (Jarosz and Davelos, 1995). Similar conclusions were drawn for other soil pathogens, such as *Actinomycetes* (Szabo, 1999) involved in replant diseases of *Malus* spp. (apple). There is hardly anything known on aggressiveness or virulence of plant-parasitic nematodes in natural ecosystems, although first results from coastal foredunes show that specialists may be less aggressive than generalists (Van der Stoel *et al.*, unpublished).

Several models have been developed to visualize how soil pathogens may be involved in plant–soil feedback in plant communities with cyclic succession (Bever *et al.*, 1997), directional succession (Van der Putten and Van der Stoel, 1998) and in the population dynamics of annual plants (Thrall *et al.*, 1997). However, biotic interactions of soil pathogens with other soil inhabitants have not yet been included as a factor, although natural antagonisms have been demonstrated to occur (Carey *et al.*, 1992; Newsham *et al.*, 1995a; Little and Maun, 1996; De Boer *et al.*, 1998a,b; Holah and Alexander, 1999, and reviews by Clay, 1991; Ingham and Molina, 1991; Roncadori, 1997).

This chapter will provide an overview of examples of soil pathogens in natural ecosystems. Soil pathogens have been studied in a fairly wide range of natural ecosystems from tropical rainforest to temperate forests, and from temperate grasslands and old fields to subtropical savannas, but the number of reports for each case is limited (Table 15.1). One of the most intensively studied plant–soil pathogen systems, that of coastal foredunes, will be elaborated to discuss possible interactions of plants, soil pathogens, and their possible natural antagonists. Conclusions on soil pathogens and antagonists in natural ecosystems will be related to the potential for avoiding expression of soil pathogens in production ecosystems. Finally, some grey areas are defined and suggestions will be given for future research on plant–soil pathogen–antagonist interactions in natural ecosystems.

Cases of Soil Pathogens in Natural Ecosystems

Seedling establishment and tree species diversity in forests

In tropical rainforests, Augspurger (1983) and Augspurger and Kelly (1984) have studied effects of soil pathogens (damping-off) on seedling establishment of *Platypodium elegans*, a wind-dispersed tree species. Close to the tree, seedlings were more likely to die than further away. This is in line with the Janzen–Connell hypothesis (Janzen, 1970; Connell, 1978) that associates the heterogeneity of tree species in tropical forest to the presence of (above-ground) specialist herbivores. Offspring of a tree species can only establish at a distance far enough from the parent (or other trees of the same species) where specialist herbivores are absent. Packer and Clay (2000) confirmed the existence of Janzen–Connell processes in temperate forest near Bloomington, Indiana, with wild cherry trees (*Prunus serotina*) as a model. *Pythium* spp. in the root zone of *Prunus serotina* trees cause seedling mortality near the parent tree. While no large saplings of *P. serotina* were found underneath *P. serotina* trees, juveniles of other tree species were able to establish (Packer and Clay, 2000). Therefore, specific soil pathogens may contribute to the generation of (small-scale) mosaics in tree species composition in both tropical and temperate forests.

Decline and succession of tree species in forests

Phytophthora cinnamomi is native in Europe where, for example in southern France, it is involved in the natural decline of oak trees (Robin *et al.*, 1998). Isolates of *P. cinnamomi* collected from trunks and soil of sites with *Quercus* species (cork and helm oaks) were aggressive to these tree species. There is some degree of specificity, as *Castanea sativa* was less and *Quercus robur* was more susceptible to these *P. cinnamomi* isolates (Robin *et al.*, 1998). In central Europe, other *Phytophthora* species are reported to be involved in the decline of oak stands (*Quercus robur* and *Q. petraea*) (Jung *et al.*, 1996). It has been assumed that increased nitrogen deposition and climate change may predispose for root damage of tree species (Jung *et al.*, 1996).

In southern Britain, *Phytophthora* species are involved in the die-back of common alder (*Alnus glutinosa*) (Gibbs *et al.*, 1999). Currently, a European-wide concerted action is making a full study of alder die-back in Europe in relation to *Phytophthora* species and other pathogenic fungi (J. Gibbs and C. Van Dijk, personal communication).

In western Oregon, Douglas fir (*Pseudotsuga menziesii*) and true firs may experience extensive root-rot caused by the native fungal pathogen *Phellinus weirii*. The fungus spreads from one tree to another by root contacts. The subsequent death of the trees may lead to enhanced cover of herbaceous species and to species diversity, albeit that the specific patterns depend on the tree species and the site involved (Holah *et al.*, 1993). Douglas fir is an early dominant of the successional sere and the fungus, by killing off its natural host, may enhance forest succession. This was demonstrated for one series of sites (Cascade sites), where western hemlock (*Tsuga heterophylla*) is the only late-successional tree species and succession in pathogen-affected forests was accelerated. However, at another series of sites (Coast range sites) there was more enhancement of shrub growth than of the late-successional tree species (Holah *et al.*, 1997). Therefore, the native pathogen may cause a change in the vegetation after killing off its natural host and it may, depending on the environmental conditions, speed up forest succession (Holah *et al.*, 1997).

Table 15.1. Overview of cases of natural soil pathogens, ecosystems and countries where identified, processes that are influenced, known cases of antagonists, and authors (or review in one case). (There are no reports from boreal forests or sub-arctic areas; aquatic vegetation has not been included. Cases where it was not sure if the plant species was natural or where the soil pathogen was collected from natural plant populations have been excluded.)

Ecosystem	Process	Plant species	Pathogen	Potential antagonists	Country	Author
Tropical rainforest	Forest composition	*Platypodium elegans*	*Pythium* sp.		Panama	Augspurger, 1983; Augspurger and Kelly, 1984
Temperate forest	Forest composition	*Prunus serotina*	*Pythium* spp.		USA (Indiana)	Packer and Clay, 2000
	Composition and succession	*Pseudotsuga menziesii*	*Phellinus weirii*		USA (Oregon)	Holah *et al.*, 1993, 1997
	Tree decline	*Alnus glutinosa*	*Phytophthora* spp.		UK/Europe	Gibbs *et al.*, 1999
	Tree decline	*Quercus* spp.	*Phytophthora* spp.		France, Germany and Central Europe	Robin *et al.*, 1998; Jung *et al.*, 1996
Coastal dunes	Grass decline and succession	*Ammophila arenaria, A. breviligulata*	Pathogenic fungi and parasitic nematodes	Mycoparasitism, nematode antagonists, arbuscular mycorrhizal fungi	The Netherlands, USA (Atlantic coast), Canada (Great Lakes)	Reviewed by Van der Putten and Van der Stoel, 1998; De Boer *et al.*, 1998a,b; Little and Maun, 1996; Seliskar and Huettel, 1993

	Shrub decline	*Hippophaë rhamnoides*	Pathogenic fungi and parasitic nematodes		The Netherlands	Oremus and Otten, 1981; Maas *et al.*, 1983; Zoon, 1995
	Clonal growth regulation	*Carex arenaria*	Most likely fungi		The Netherlands	D'Hertefeldt and Van der Putten, 1998
Old fields, grasslands, prairies and subtropical savannas	Co-existence and regulation plant species diversity	Grasses, legumes, and other forbs	Pathogenic fungi, one example of nematodes	One example of mycoparasitic fungus	USA, The Netherlands, Botswana	Bever, 1984; Mills and Bever, 1998; Holah and Alexander, 1999; Olff *et al.*, 2000; Blomqvist *et al.*, 2000; Van der Putten *et al.*, unpublished
	Annual plants: fitness reduction	*Vulpia ciliata*	*Fusarium oxysporum, Embellisia chlamydospora*	Arbuscular mycorrhizal fungi	UK	Carey *et al.*, 1992; Newsham *et al.*, 1994, 1995a
	Annual plants: induction seedling death	*Kummerowia stipulacea*	*Rhizoctonia solani, Pythium irregulare*		USA	Mihail *et al.*, 1998; Alexander and Mihail, 2000

Plant species succession in coastal foredunes

Vegetation succession in coastal dunes is driven by changes in the abiotic conditions (sand deposition, salt spray, leaching of calcium carbonate, accumulation of organic matter) (Sykes and Wilson, 1988, 1990; Olff *et al.*, 1993; Maun, 1998). In addition to these abiotic factors, herbivores, such as rabbits, may influence vegetation processes. Shrub encroachment may be the result of decline of the rabbit population by myxomatosis. Soil pathogens and their natural antagonists are also involved in vegetation processes, because of their role in local decline and replacement of dominant plant species.

In northwestern European coastal foredunes, soil pathogens contribute to the degeneration of dominant plant species, such as *Hippophaë rhamnoides* (sea buckthorn) and *Ammophila arenaria* (marram grass) (Oremus and Otten, 1981; Maas *et al.*, 1983; Van der Putten *et al.*, 1988; De Rooij-Van der Goes, 1995; Zoon *et al.*, 1993; Kowalchuk *et al.*, 1997). Pathogenic soil fungi and plant-parasitic nematodes are supposed to be involved in soil pathogen complexes of *A. arenaria* and *H. rhamnoides* (Van der Putten *et al.*, 1990; Zoon *et al.*, 1993). Specificity of the successive soil pathogen complexes contributes to succession (Van der Putten *et al.*, 1993) because it changes the competitive ability of host plants (Van der Putten and Peters, 1997).

Plant-parasitic nematodes have also been isolated from the root zone of the American beach grass *A. breviligulata* along the Atlantic coast of North America (Seliskar and Huettel, 1993) and from lacustrine sand dunes along the Great Lakes in Canada (Little and Maun, 1996). In the latter case, arbuscular mycorrhizal fungi may provide protection to the plants against the plant-parasitic nematodes *Pratylenchus* sp. and *Heterodera* sp.

Plant species diversity and spatio-temporal mosaics in old fields and grasslands

Bever (1984) cultured plant species from old fields with their own soil communities. *Kriga dandelion* showed significantly lower survival rate when exposed to its own soil community than to others. The growth of three grasses (*Danthonia spicata*, *Panicum sphaerocarpon* and *Anthoxanhum odoratum*) was also reduced when exposed to their own soil communities. Reciprocal transplant experiments showed some differential response to each others' soils, but effects were less strong than with their own soil. However, these differences did not result in reduced competitive ability (Bever, 1994). Mills and Bever (1998) isolated *Pythium* spp. from the roots of *D. spicata* and *Panicum sphaerocarpon* and inoculated these to all three plant species and *Plantago lanceolata*. The fungi reduced overall plant mass and root:shoot ratios, but *D. spicata* and *Panicum sphaerocarpum* were more susceptible than the other two plant species. Plants that were susceptible to *Pythium* were more likely to accumulate the fungus (Mills and Bever, 1998).

Olff *et al.* (2000) examined the relationship of spatio-temporal mosaics in grazed natural pasture on sandy river dunes which are alternatively dominated by sand sedge (*Carex arenaria*) and red fescue (*Festuca rubra*). Seedlings of *C. arenaria* planted in soil from sites where *C. arenaria* had decreased (and *F. rubra* increased) during the past decade were more reduced in growth than seedlings planted in soil where the abundance of *C. arenaria* was increasing (and where *F. rubra* decreased). Planted seedlings of *F. rubra* showed a similar pattern in relation to its own population dynamics. As growth of *C. arenaria* was reduced most at high plant-parasitic nematode densities and that of *F. rubra* at low densities, different soil pathogens seem to affect different plant species. Therefore, local variation in the composition (or abundance) of species in soil pathogen complexes may contribute to spatio-temporal dynamics of grassland vegetation (Olff *et al.*, 2000).

Blomqvist *et al.* (2000) carried out studies in the same grassland ecosystem as Olff *et al.* (2000) and studied effects of ant burrowing activity on re-colonization potential of *C. arenaria* and *F. rubra*. They showed that complex feedback mechanisms of plant traits, soil pathogens, and interactions with herbivores and soil animals may cause heterogeneity in natural vegetation. Since ants bring up fresh subsoil that may be relatively pathogen-free, plants could benefit from the ant activity. Differences in re-colonization ability of both species (*C. arenaria* has the ability of fast clonal spread) may contribute to subsequent mosaics as observed in the grassland (Blomqvist *et al.*, 2000).

Clonal expansion of *C. arenaria* is unidirectional when exposed to patches with soil pathogens (most likely soil fungi with a possible involvement of plant-parasitic nematodes) and rhizomes become more intensively branched when entering local pathogen-free patches of soil (D'Hertefeldt and Van der Putten, 1998). In Swiss mountain grassland, *Prunella* spp. responded to arbuscular mycorrhizal fungi by more intensive branching (Streitwolf-Engel *et al.*, 1997). Therefore, soil pathogens seem to have a contrasting effect on plant clonal growth patterns when compared to arbuscular mycorrhizal fungi.

Holah and Alexander (1999) studied plant–soil feedback of two native tall grass prairie species *Andropogon gerardii*, a perennial grass, and *Chamaecrista fasciculata*, an annual legume. Both plant species were cultured in their own soil and in soil from the other species. A second set of treatments involved the same soil types after partial soil sterilization by microwave treatment. Neither prairie grass species showed signs of negative feedback from their own soil community. However, *A. gerardii* plants were shorter and had fewer inflorescences in non-sterilized soil, especially in soil from *C. fasciculata*. Fungi that were collected from the roots of *A. gerardii* that were grown in *C. fasciculata* soil type, when re-inoculated to healthy *A. gerardii* plants, reduced tiller production and caused early growth reduction to *A. gerardii*. When potentially mycoparasitic fungi isolated from *A. gerardii* roots were added, negative effects of the pathogenic fungi were counteracted in part. Holah and Alexander (1999) concluded that the negative effect of fungi of the annual *C. fasciculata* on the perennial *A. gerardii* (more than on *C. fasciculata* itself) might contribute to the coexistence of these two tall grass prairie species.

Annual plant communities

Applying the fungicide benomyl to natural populations of the annual grass *Vulpia ciliata* ssp. *ambigua* affected the grass fecundity positively, however, this depended on the site where benomyl was applied (Carey *et al.*, 1992). As fecundity of the annual grass species was correlated to the presence of the pathogens *Fusarium oxysporum* and *Embellisia chlamydospora* (Newsham *et al.*, 1995a), it was supposed that there may be interaction of pathogens and arbuscular mycorrhizal fungi. *Vulpia* seedlings were infected with *F. oxysporum* and/or the arbuscular mycorrhizal fungus *Glomus* sp. and placed in the field. *Glomus* had a positive effect on plant shoot and root growth, but only in the pathogen-infected treatments. Therefore, the mycorrhizal fungus seemed to protect the plants against pathogens, rather than having a direct effect on the uptake of phosphorus (Newsham *et al.*, 1995a). Different sites with rather comparable abiotic environmental conditions showed different combinations of soil fungi in the root zone of *V. ciliata*, suggesting that differences between sites in plant performance were due to different combinations of soil pathogens (Newsham *et al.*, 1995b). Newsham *et al.* (1995c) reviewed effects of arbuscular mycorrhizal fungi on plants in relation to the morphology of the root system. Plants with poorly branched root systems seemed to benefit from mycorrhizal fungi by enhanced phosphorus uptake, whereas plants with highly branched roots benefited more from protection against soil pathogens.

Over a period of 4 years, a population of the annual legume *Kummerowia stipulacea* showed a large reduction in the number of plants within seasons (Mihail *et al.*, 1998). Seeds, as well as seedlings of this annual legume, were sensitive to *Rhizoctonia solani* and *Pythium irregulare* that had both been isolated from the field soil.

Thrall *et al.* (1997) made estimates of model parameters from data in the literature and showed that stable coexistence of host (natural plant populations) and pathogen (*Phytophthora* spp. and *Fusarium oxysporum*) populations is possible even when the pathogen has a positive intrinsic growth rate. For *Fusarium*, there were substantial ranges where the pathogen would coexist or be lost from the system. For *Phytophthora*, however, host persistence was most likely when disease transmission was described by a non-linear model. Thrall *et al.* (1997) concluded that long-term dynamics of annual plants interacting with soil pathogens may be difficult to predict, because small changes in parameter values lead to qualitatively different outcomes.

Neher *et al.* (1987) compared *Glycine max* (soja) and its progenitor *G. soja*, collected from China. The progenitor germinates more irregularly and the hypothesis was tested that this might reduce severity of damping-off by *Pythium* sp. Effects of damping-off in relation to age structure of the populations of both species (experimentally manipulated) were species-specific; however, the pathogen was not originally obtained from the native *G. soja* population, which does not allow this to be regarded as an example of a natural pathogen complex.

Invasive plants

The invasiveness of plants is difficult to predict and one of the possible causes of uncontrolled expansion of plants in new territories is the release from natural enemies (Williamson, 1996). Thus far, there have been no reported examples of plant invasiveness arising from a release from natural soil pathogen pressure. This would be an interesting area for future research. The possible consequence of releasing plants from the pressure of their natural soil pathogens is currently being examined for *Ammophila arenaria* (marram grass), which has been introduced for dune stabilization in South Africa, the west coast of North America, Australia and New Zealand (Lubke and Hertling, 1995; Hertling and Lubke, 1999). Since soil pathogens are involved in the ecology of *A. arenaria* (reviewed by Van der Putten and Van der Stoel, 1998), the introduction of *A. arenaria* as seeds may have resulted in escape from its natural soil pathogens.

Cenchrus biflorus is an invasive annual grass species near boreholes and in other disturbed sites of the Kalahari savanna in Botswana. It is not known if soil pathogens in its area of origin (most probably India) affect the ecology of this plant species. It is now being studied if soil pathogens are present in the Kalahari savanna and how *C. biflorus* may respond to these pathogens. Examination of the indigenous flora shows that at least one indigenous plant species from the Kalahari savanna (*Eragrostis lehmanniana*; Lehmann lovegrass) possess soil pathogens and that *C. biflorus* seems resistant to these (Van der Putten *et al.*, unpublished). Interestingly, *E. lehmanniana* is invasive in southern Arizona, USA, where it has transformed the structure and function of 145,000 ha since its introduction in 1932 (Anable *et al.*, 1992). Therefore, this species may also provide a model for studying the role of absence of soil pathogens in plant invasiveness.

Another model system that would enable examination of the contribution of escape from soil pathogens is the invasiveness of *Prunus serotina* in north-western Europe. This mid-successional tree species originates from North American forests and seedling establishment is affected by *Pythium* spp. that seem rather specific (Packer and Clay, 2000). If natural soil pathogens are lacking in European forests, this would allow a comparison of the ecology and invasiveness of the tree species with and without natural soil pathogens.

Invasive soil pathogens

Plant disease epidemics resulting from introductions of exotic fungal pathogens are a well-known phenomenon. Interspecific hybridization may result into new host specificities or completely new pathogen taxa. One example for soil pathogens is a new aggressive *Phytophthora* pathogen comprising a range of heteroploid-interspecific hybrids involving a *Phytophthora cambivora*-like species and an unknown taxon similar to *Phytophthora fragariae*, that has been found in alder species in Europe (Brasier *et al.*, 1999). Dramatic effects of soil pathogens on forest development have been reported for indigenous *Eucalyptus* forest in Australia, which are strongly affected by the introduction of *Phytophthora cinnamomi* (Weste, 1981). Although many tree and shrub species are affected by this pathogen, some species can be completely killed while others may survive (Peters and Weste, 1997). Another example from Australia concerns *Armillaria luteobubalina* that may kill off 125 (38%) of all species in the coastal vegetation (Shearer *et al.*, 1998). Enhanced global trade is a main driver of invasive pathogens because routine detection measures are not sufficient to detect all pathogens (Brasier *et al.*, 1999), so that we might expect more of these outbreaks in the near future.

Interactions of Soil Pathogens and Their Antagonists in Natural Ecosystems

Many of the examples of soil pathogens in natural ecosystems have focused on one or a few fungal species only (Table 15.1). Natural antagonists of the soil pathogens have not been studied with some exceptions (Newsham *et al.*, 1995a; De Boer *et al.*, 1998a,b; Holah and Alexander, 1999; reviews by Clay, 1991; Ingham and Molina, 1991; Roncadori, 1997). I will now take coastal foredunes as a model ecosystem where degeneration of dominant plant species is supposed to be due to complex soil pathogens consisting of plant-pathogenic fungi and plant-parasitic nematodes (Van der Putten *et al.*, 1990; Zoon, 1995). Possible interactions of pathogens with plant-parasitic nematodes as well as with natural antagonists will be discussed.

Coastal dunes are characterized by a sequence of dominant clonal plant species ranging from salt-resistant species along the beach to salt-tolerant plant species on the top and leeward side of the foredunes. Towards the inner dunes, salt deposition decreases, there is less deposition of windblown sand originating from the beach, organic matter accumulates in the soil profile and the pH decreases. *Ammophila arenaria* (marram grass) is one of the pioneers that occurs in the outer dunes. The plants are vigorous when buried regularly by windblown sand, whereas they degenerate in the absence of sand deposition. Sand burial enables *A. arenaria* to escape from soil pathogens (Van der Putten *et al.*, 1988; De Rooij-Van der Goes *et al.*, 1995a,b) which, in turn, are able to keep up with sand deposition (De Rooij-Van der Goes *et al.*, 1998; Fig. 15.1).

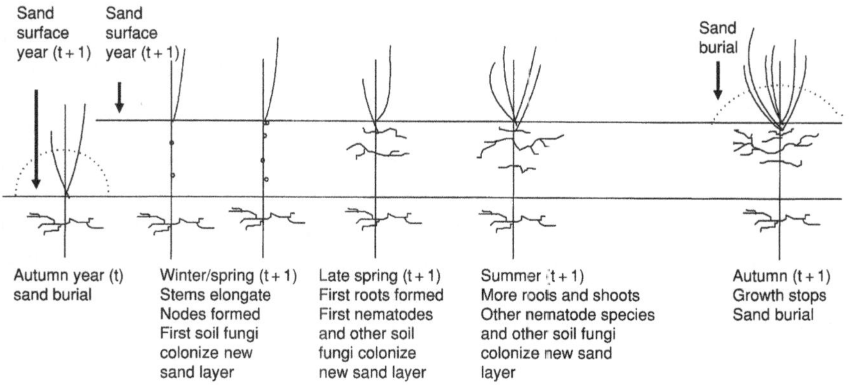

Fig. 15.1. Timing of sand burial, vertical development of *Ammophila arenaria* (marram grass), and colonization of the new soil layer by fungi and plant-parasitic nematodes. (Redrawn after De Rooij-Van der Goes, 1996.)

The majority of sand burial occurs in winter. In spring, plants emerge from the deposited sand layer by stem internode elongation. (Non-lethal sand burial may be up to 1.2 m; a review on the response of dune plants to sand burial is provided by Maun 1998). Some soil fungi are able to colonize the new sand layer before roots are formed, whereas plant-parasitic nematodes and other soil fungi colonize after root formation (Fig. 15.1). At the end of the growing season, the soil community may reduce the growth of *A. arenaria* seedlings when planted in a greenhouse in unsterilized soil and compared to sterilized soil (Van der Putten *et al.*, 1988). In the subsequent winter, sand deposition is required to enable plants to escape from soil pathogens. Continuous exposure to soil pathogens when sand deposition has stopped contributes to decline of plant vigour and, finally, to replacement by other plant species.

Hippophaë rhamnoides (sea buckthorn) dominates the vegetation in later successional stages. The shrub is resistant to pathogens of *A. arenaria*, but it sooner or later degenerates due to its own pathogens (Oremus and Otten, 1981). Both dominant dune-plant species (with the emphasis on *A. arenaria*) have been subject to detailed studies in order to unravel which soil organisms contribute to the soil pathogen complex.

The use of nematicides enhanced plant productivity in bioassays (Van der Putten *et al.*, 1990; Zoon, 1995). However, thus far addition of plant-parasitic nematodes to plants growing in sterilized soil (and comparison of growth reduction with that observed in non-sterilized field soil) has shown that plant-parasitic nematodes are not the only cause of growth reduction (Maas *et al.*, 1983; De Rooij-Van der Goes, 1995; De Rooij-Van der Goes *et al.*, 1997). In the case of *A. arenaria*, endoparasitic nematodes seemed to be involved in the specificity of the soil pathogen complex, however, they may not be regarded as major pathogens (Van der Stoel *et al.*, unpublished). Plant-pathogenic fungi could also be involved in the pathogen complexes of both *A. arenaria* (De

Rooij-Van der Goes, 1995) and *H. rhamnoides* (Zoon, 1995), but if and how they interact with plant-parasitic nematodes is unknown. Synergism could not be demonstrated (De Rooij-Van der Goes, 1995), however, the nematodes might act as vectors for plant-pathogenic fungi.

The soil pathogen complexes of successive plant species are affecting the later successional plant species less than their host plants or pre-successional plant species (Van der Putten *et al.*, 1993). Genetic variation within host plant populations has not yet been tested. It is not clear what may be the cause of the observed specificity. There are various possibilities for the ways in which soil organisms interact within the soil pathogen complexes: (i) all species are somewhat pathogenic and the complex as a whole is specific; (ii) some species are key species that cause specificity and activate all others; and (iii) some species are major pathogens and the rest is relatively unimportant (Van der Putten and Van der Stoel, 1998). Current studies attempt to further unravel the contribution of the different soil-pathogenic fungi and plant-parasitic nematodes to the successive pathogen complexes by analysing their occurrence, specificity, and effects on host plants, predecessors and successional plant species.

In the dune sand, many forms of nematode-suppressing biota have been observed, such as carnivorous nematodes (H. Duyts, personal observations), nematophagous fungi and parasitic *Pasteuria penetrans* (P.C.E.M. de Rooij-Van der Goes, personal observations), and arbuscular mycorrhizal fungi (Nicolson, 1960; Ernst *et al.*, 1984; Clapp, unpublished) that may reduce the effects of nematodes (Little and Maun, 1996). In addition, plant-pathogenic fungi isolated from the root zone of *A. arenaria* may also be suppressed by natural soil antagonism, as appeared when the fungi collected from dune sand were grown on sterilized and unsterilized dune soil (De Boer *et al.*, 1998b). Finally, plants may also escape from their soil pathogen and parasites by clonal growth (De Rooij-Van der Goes, 1995; D'Hertefeldt and Van der Putten, 1998). Escape of the host (bottom-up control) seems to play a major role in the maintenance of vigour of dominant foredune plant species. However, regarding the many potential forms of pathogen suppression in the dune soils, one of the questions to be answered is what may be the role of top-down control of soil pathogens in the ecology of foredune plant species (Fig. 15.2). The relative importance of the different control mechanisms of foredune soil pathogens needs to be examined in subsequent studies.

Conclusions

Natural ecosystems and plant traits that have been associated with soil pathogens

There are relatively few studies on the ecological role of soil pathogens in (semi) natural plant communities. Nevertheless, the available examples originate from a wide range of ecosystems, from tropical rainforests to

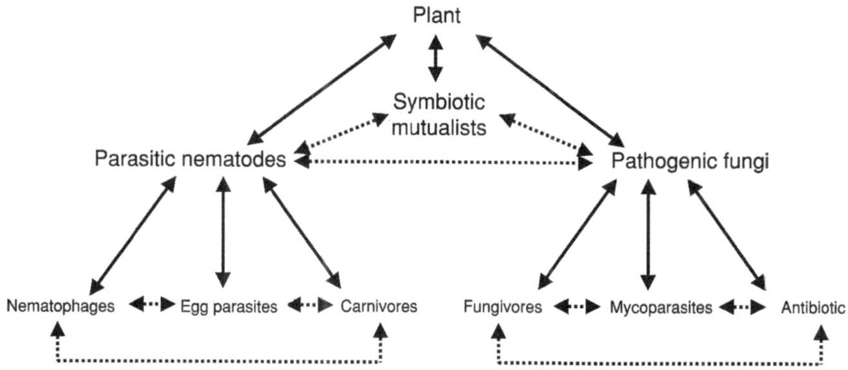

Fig. 15.2. Direct (closed arrows) and indirect (dotted arrows) interactions between plants, symbiotic mutualists (e.g. arbuscular mycorrhizal fungi or endophytes), plant-parasitic nematodes, plant-pathogenic fungi and non-plant-associated antagonists as might occur in the case of *Ammophila arenaria* (marram grass) in coastal foredunes. Determination of the relative strength of these interactions could elucidate the relative importance of different natural top-down control processes (versus bottom-up) in the soil pathogen complex.

temperate deciduous and evergreen forest, and from temperate species-rich grasslands, old fields and early succession stages of coastal foredunes to annual plant communities and southern African savannas. In some cases soil pathogens have contributed to mortality of seeds or seedlings, whereas in other cases soil pathogens are associated with reduced growth, decline or die-back of individual plants, reduced fitness, apparent or indirect competition, cyclical and directional succession, and coexistence or maintenance of plant species diversity. Soil pathogens have been detected in plant species representing a great variety of plant traits from annual to clonal plants, as well as trees.

Single pathogen species or disease complexes?

Most examples studied are concerned with one or a few fungal pathogen species, whereas few studies considered a possible role of plant-parasitic nematodes in the soil pathogen complexes. Biotic interactions between pathogenic fungal species or between pathogenic fungi and plant-parasitic nematodes have been barely examined with some exceptions. Few studies in natural ecosystems included bacterial pathogens and virtually none included soil-borne viruses.

Antagonists of soil pathogens and plant traits as natural defence mechanisms

In natural ecosystems, arbuscular mycorrhizal fungi may be more important for plant protection against soil pathogens than in agricultural ecosystems. There are only few studies on the role of arbuscular mycorrhizal fungi in protecting plants in natural ecosystems against pathogenic soil fungi or plant-parasitic nematodes. There are even fewer studies on natural soil suppressiveness against soil pathogens. Effects of nematophagous fungi have been examined only in relation to entomophagous nematodes (Koppenhofer *et al.*, 1997). Probably, the increasing number of studies on the role of soil pathogens in natural plant communities will encourage further research on natural pathogen suppression. Contrary to the case in agriculture, where natural monocultures may be artificially maintained (Weller *et al.*, 1995), natural soil suppressiveness may not easily develop in natural communities of annual plants. Loss of fitness will easily lead to outcompetition. Populations of many natural annual (and biennial) plant species seem short-lived. The unpredictable appearance of annual plant species may be a life history strategy that reduces the possible effectiveness of natural antagonists, as the plants have disappeared before the antagonists may become active. Probably, slow-growing perennial or clonal plant species are more likely to be associated with natural antagonists than annuals or biennials. However, the studies on the annual *Vulpia ciliata* (Carey *et al.*, 1992; Newsham *et al.*, 1994, 1995a) show that annuals may make use of arbuscular mycorrhizal fungi in protection against pathogenic fungi.

As well as protection of plants by mutualists or other antagonists, soil pathogen-sensitive plant species need dispersal as a means of escaping pathogen pressure, because of limited opportunities for selection of resistance. Specificity or different levels of susceptibility between plant species for soil pathogens may contribute to cyclical, or when the environment changes fast, directional succession. Therefore, in natural ecosystems soil pathogens seem to contribute to heterogeneity and possibly also to biodiversity of natural plant communities.

Relation to plant protection against soil pathogens in production ecosystems

Cyclic succession resembles crop rotation in agriculture or horticulture, showing that rotational cycles of ample length will be indeed a natural and efficient method to counteract soil pathogens in production systems. However, current agricultural intensification practices do not favour a wide range of crops to be grown in rotation. The practice of set-aside, in order to counteract overproduction of certain crops, might be useful in reducing the load of soil pathogens.

However, the short-term effects of set-aside could be that the density of soil pathogens is stimulated, rather than suppressed. In a study on set-aside arable land, the numbers of plant-parasitic nematodes were enhanced in response to non-tillage and free weed development, as well as sowing of plant mixtures of low or high species diversity (Korthals *et al.*, unpublished).

It is generally stated that many agricultural crops have been derived from relatively early successional, short-lived, fast-growing, r-selected species. Early successional species have different defence systems than later successional species (Price *et al.*, 1980) which could explain the sensitivity of crop plants for pathogens. Overviews on the domestication and the origin of crop species (Harlan, 1992; Smartt and Simonds, 1995), however, show that a wide variety of plant species have been used for cultivation, which provides some counterargument to the early successional species hypothesis. Additionally, as soil pathogens occur in a wide variety of natural ecosystems, the exclusiveness of soil pathogens to agricultural crops because they have been derived from early successional plant species does not seem to hold. An analysis of the evolution of resistance, however, may provide some insight, because for example some groups of plant-parasitic nematodes that are known as aggressive in agriculture do not seem to be that aggressive in natural soil (Van der Stoel and Van der Putten, unpublished).

Biological control of soil-borne diseases in agriculture has largely been investigated without much knowledge or recognition of the importance of these processes in natural soils. Comparison of the role of various forms of antagonism (symbiotic mutualists versus various sorts of free-living antagonists) in relation to the life history of natural plant species could be of interest for more efficient targeting of biological control programmes against soil pathogens in agriculture, horticulture or other production ecosystems.

Summary and Perspectives for Further Research

A review of known cases of soil pathogens in natural ecosystems demonstrates that soil pathogens are not limited to certain ecosystems or to certain types of plant life histories. Most work has focused on fungal soil pathogens and some on plant-parasitic nematodes. Bacterial diseases have been rarely studied. Soil-borne viruses in natural ecosystems seem to have been unexplored thus far.

There is strong need for a more comparative approach when studying soil pathogens in natural ecosystems. Ecosystem development (the stage of successional development) may be one form of stratification that can be used. Are soil pathogens as important in all kinds of successional stages, or do they drive vegetation succession in early stages and species diversity in later (climax) stages of vegetation succession?

Plant life histories may be another form of stratification: are annual plant species equally affected by soil pathogens as biennials, perennials or clonal plants? Are herbs more or less affected by soil pathogens than trees and are fast-growing plants less affected than slow-growing plants? Thus far, the number of examples studied is insufficient to address these questions.

There are only a few studies on potential antagonists of soil pathogens in natural ecosystems. There is also little known about the role of secondary plant compounds in protecting natural plants against soil pathogens. A case study on coastal foredunes is elaborated in order to demonstrate that in natural ecosystems plant species may be protected by their life strategy and/or by a variety of symbiotic mutualists and free-living antagonists. Identifying the conditions affecting the relative importance of one or other plant-influencing biotic agents may help to further understand the ecological role of soil pathogens. In addition, identifying the relative importance of natural antagonists of soil pathogens may be helpful in targeting research on the biological control of soil diseases in agriculture, horticulture or forestry. Soil pathogens drive plant dispersal rather than the development of resistance. Crop rotation, therefore, is the most natural approach for production ecosystems to counteract the accumulation of soil pathogens. If economic developments do not allow extensive crop rotation, set-aside might be used to counteract pathogen developments, however, there is a risk of maintaining or even enhancing inoculum densities of soil pathogens during set-aside.

Soil pathogens may be invasive when they are accidentally introduced in novel environments. There are examples of exotic soil pathogens having devastating effects on local native vegetation, probably because of poor plant resistance against these exotic pathogens. However, the absence of natural antagonists of the exotic pathogens in their new territories might also affect the uncontrolled nature of their activities. This aspect has not received much (if any) attention.

Invasiveness of plants could also be, to some extent, enhanced by the release from the pressure of the natural soil pathogens. There is no evidence yet that this has happened, however, potential model systems have now been identified: marram grass (*Ammophila arenaria*) and wild cherry (*Prunus serotina*). Probably, the number of potential cases of release from soil pathogens enhancing plant invasiveness may increase when plant–soil pathogen studies in natural ecosystems are also directed towards comparisons of invasive plant species in their native and new territories.

Research on soil pathogens, their interactions and interactions with their antagonists in natural ecosystems, may yield more basic information to be used for nature and biodiversity conservation, ecosystem restoration and the control of soil pathogens in production ecosystems. Thus far, ecologists have benefited largely from the knowledge developed by phytopathologists and agronomists. A re-investment of knowledge on biotic interactions involving soil pathogens in natural ecosystems may in return contribute to disease control in future production systems.

Acknowledgements

I thank Mike Jeger and the BSPP for the invitation to contribute to the 1999 presidential meeting at Oxford, which stimulated the writing of this chapter. Wietse de Boer and Jeff Harvey have critically commented on a previous version of the manuscript. Kees van Dijk provided additional literature on tree diseases.

References

Alexander, H.M. and Mihail, J.D. (2000) Seedling disease in an annual legume: consequences for seeding mortality, plant size, and population seed productions. *Oecologia* 122, 346–353.

Anable, M.E., McClaran, M.P. and Ruyle, G.B. (1992) The spread of introduced Lehmann lovegrass *Eragrostis lehmanniana* Nees in Southern Arizona, USA. *Biological Conservation* 61, 181–188.

Augspurger, C.K. (1983) Offspring recruitment around tropical trees: changes in cohort distance with time. *Oikos* 40, 189–196.

Augspurger, C.K. (1990) Spatial patterns of damping-off disease during seedling recruitment in tropical forests. In: Burdon, J.J. and Leather, S.R. (eds) *Pests, Pathogens and Plant Communities*. Blackwell Scientific Publications, Oxford, pp. 131–144.

Augspurger, C.K. and Kelly, C.K. (1984) Pathogen mortality of tropical tree seedlings: experimental studies of the effects of dispersal distance, seedling density, and light conditions. *Oecologia* 61, 211–217.

Bever, J.D. (1984) Feedback between plants and their soil communities in an old field community. *Ecology* 75, 1965–1977.

Bever, J.D., Westover, K.M. and Antonovics, J. (1997) Incorporation of the soil community into plant population dynamics: the utility of the feedback approach. *Journal of Ecology* 85, 561–573.

Blomqvist, M.M., Olff, H., Blaauw, M.B., Bongers, T. and Van der Putten, W.H. (2000) Interactions between above- and below-ground biota: contribution to small-scale vegetation mosaics in a grassland ecosystem. *Oikos* 90, 582–598.

Brasier, C.M., Cooke, D.E.L. and Duncan, J.M. (1999) Origin of a new *Phytophthora* pathogen through interspecific hybridization. *Proceedings of the National Academy of Sciences USA* 96, 5878–5883.

Burdon, J.J. (1987) *Diseases and Plant Population Biology*. Cambridge University Press, Cambridge.

Burdon, J.J. (1993) The structure of pathogen populations in natural plant communities. *Annual Review of Phytopathology* 31, 305–323.

Carey, P.D., Fitter, A.H. and Watkinson, A.R. (1992) A field study using the fungicide benomyl to investigate the effect of mycorrhizal fungi on plant fitness. *Oecologia* 90, 550–555.

Clay, K. (1991) Fungal endophytes, grasses, and herbivores. In: Barbosa, P., Krischik, V.A. and Jones, C.G. (eds) *Microbial Mediation of Plant-Herbivore Interactions*. John Wiley & Sons, New York, pp. 199–226.

Clay, K. and Kover, P. (1996) The Red Queen hypothesis and plant/pathogen interactions. *Annual Review of Phytopathology* 34, 29–50.

Connell, J.H. (1978) Diversity in tropical rain forests and coral reefs. *Science* 199, 1302–1310.

De Boer, W., Klein Gunnewiek, P.J.A., Lafeber, P., Janse, J.D., Spit, B.E. and Woldendorp, J.W. (1998a) Antifungal properties of chitinolytic dune soil bacteria. *Soil Biology and Biochemistry* 30, 193–203.

De Boer, W., Klein Gunnewiek, P.J.A. and Woldendorp, J.W. (1998b) Suppression of hyphal growth of soil-borne fungi by Dune soils from vigorous and declining stands of *Ammophila arenaria. New Phytologist* 138, 107–116.

De Rooij-Van der Goes, P.C.E.M. (1995) The role of plant-parasitic nematodes and soil-borne fungi in the decline of *Ammophila arenaria* L. Link. *New Phytologist* 129, 661–669.

De Rooij-Van der Goes, P.C.E.M. (1996) Helmduinen zijn gebaat bij zandverstuivingen. *Duin* 19, 16–18.

De Rooij-Van der Goes, P.C.E.M., Van der Putten, W.H. and Van Dijk, C. (1995a) Analysis of nematodes and soil-borne fungi from *Ammophila arenaria* (Marram grass) in Dutch coastal foredunes by multivariate techniques. *European Journal of Plant Pathology* 101, 149–162.

De Rooij-Van der Goes, P.C.E.M., Van der Putten, W.H. and Peters, B.A.M. (1995b) Effects of sand deposition on the interaction between *Ammophila arenaria*, plant parasitic nematodes and pathogenic fungi. *Canadian Journal of Botany* 73, 1141–1150.

De Rooij-Van der Goes, P.C.E.M., Van Dijk, C., Van der Putten, W.H. and Jungerius, P.D. (1997) The effects of sand movement by wind in coastal foredunes on nematodes and soil-borne fungi. *Journal of Coastal Conservation* 3, 133–142.

De Rooij-Van der Goes, P.C.E.M., Peters, B.A.M. and Van der Putten, W.H. (1998) Vertical migration of plant-parasitic nematodes and pathogenic fungi to new developing roots of *Ammophila arenaria* (L.) Link after sand accretion. *Applied Soil Ecology* 10, 1–10.

D'Hertefeldt, T. and Van der Putten, W.H. (1998) Physiological integration of the clonal plant *Carex arenaria* and its response to soil-borne pathogens. *Oikos* 81, 229–237.

Ernst, W.H.O., Van Duin, W.E. and Oolbekking, G.T. (1984) Vesicular-arbuscular mycorrhiza in dune vegetation. *Acta Botanica Neerlandica* 33, 151–160.

Gibbs, J.N., Lipscombe, M.A. and Peace, A.J. (1999) The impact of *Phytophthora* disease on riparian populations of common alder (*Alnus glutinosa*) in southern Britain. *European Journal of Forest Pathology* 29, 39–50.

Harlan, J.R. (1992) *Crops and Man*, 2nd edn. American Society of Agronomy, Crop Science Society of America, Madison, Wisconsin.

Hertling, U.M. and Lubke, R.A. (1999) Indigenous and *Ammophila arenaria* dominated dune vegetation on the South African Cape coast. *Applied Vegetation Science* 2, 157–168.

Holah, J.C. and Alexander, H.M. (1999) Soil pathogenic fungi have the potential to affect the co-existence of two tallgrass prairie species. *Journal of Ecology* 87, 598–608.

Holah, J.C., Wilson, M.V. and Hansen, E.M. (1993) Effects of a native forest pathogen, *Phellinus weirii*, on douglas-fir forest composition in western Oregon. *Canadian Journal of Forest Research* 23, 2473–2480.

Holah, J.C., Wilson, M.V. and Hansen, E.M. (1997) Impacts of a native root-rotting pathogen on successional development of old-growth Douglas fir forests. *Oecologia* 111, 429–433.

Ingham, E.R. and Molina, R. (1991) Interactions among mycorrhizal fungi, rhizosphere organisms, and plants. In: Barbosa, P., Krischik, V.A. and Jones, C.G. (eds) *Microbial Mediation of Plant–Herbivore Interactions.* John Wiley & Sons, New York, pp. 169–198.

Janzen, D.H. (1970) Herbivores and the number of tree species in tropical forests. *American Naturalist* 104, 501–528.

Jarosz, A.M. and Davelos, A.L. (1995) Effects of disease in wild plant populations and evolution of pathogen aggressiveness. *New Phytologist* 129, 371–378.

Jung, T., Blaschke, H. and Neumann, P. (1996) Isolation, identification and pathogenicity of phytophthora species from declining oak stands. *European Journal of Forest Pathology* 26, 253–272.

Koppenhofer, A.M., Jaffee, B.A., Muldoon, A.E. and Strong, D.R. (1997) Suppression of an entomopathogenic nematode by the nematode-trapping fungi *Geniculifera paucispora* and *Monacrosporium eudermatum* as affected by the fungus *Arthrobotrys oligospora. Mycologia* 89, 557.

Kowalchuk, G.A., Gerards, S. and Woldendorp, J.W. (1997) Detection and characterization of fungal infections of *Ammophila arenaria* (marram grass) roots by denaturing gradient gel electrophoresis of specifically amplified 18s rDNA. *Applied and Environmental Microbiology* 63, 3858–3865.

Little, L.R. and Maun, M.A. (1996) The "*Ammophila* problem" revisited: a role for mycorrhizal fungi. *Journal of Ecology*, 84, 1–7.

Lubke, R.A. and Hertling, U.M. (1995) Is *Ammophila arenaria* (marram grass) a threat to South African dunefields? *Journal of Coastal Conservation* 1, 103–108.

Maas, P.W.Th., Oremus, P.A.I. and Otten, H. (1983) Nematodes (*Longidorus* sp. and *Tylenchorhynchus microphasmis* Loof) in growth and nodulation of Sea buckthorn (*Hippophaë rhamnoides* L.). *Plant and Soil* 73, 141–147.

Maun, M.A. (1998) Adaptations of plants to burial in coastal sand dunes. *Canadian Journal of Botany* 76, 713–738.

Mihail, J.D., Alexander, H.M. and Taylor, S.J. (1998) Interactions between root-infecting fungi and plant density in an annual legume, *Kummerowia stipulacea. Journal of Ecology* 86, 739–748.

Mills, K.E. and Bever, J.D. (1998) Maintenance of diversity within plant communities: soil pathogens as agents of negative feedback. *Ecology* 79, 1595–1601.

Mortimer, S., Van der Putten, W.H. and Brown, V.K. (1999) Insect and nematode herbivory below-ground: interactions and role in vegetation development. In: Olff, H., Brown, V.K. and Drent, R.H. (eds) *Herbivores: Between Plants and Predators.* Blackwell Scientific Publications, Oxford, pp. 205–238.

Neher, D.A., Augspurger, C.K. and Wilkinson, H.T. (1987) Influence of age structure of plant populations on damping-off epidemics. *Oecologia* 74, 419–424.

Newsham, K.K., Fitter, A.H. and Watkinson, A.R. (1994) Root pathogenic and arbuscular mycorrhizal fungi determine fecundity of asymptomatic plants in the field. *Journal of Ecology* 82, 805–814.

Newsham, K.K., Fitter, A.H. and Watkinson, A.R. (1995a) Arbuscular mycorrhizae protect an annual grass from root pathogenic fungi in the field. *Journal of Ecology* 83, 991–1000.

Newsham, K.K., Watkinson, A.R. and Fitter, A.H. (1995b) Rhizosphere and root-infecting fungi and the design of ecological field experiments. *Oecologia* 102, 230–237.

Newsham, K.K., Fitter, A.H. and Watkinson, A.R. (1995c) Multi-functionality and biodiversity in arbuscular mycorrhizas. *Trends in Ecology and Evolution* 10, 407–411.

Nicolson, T.H. (1960) Mycorrhiza in the Gramineae. II. Development in different habitats, particularly sand dunes. *Transactions of the British Mycological Society* 43, 132–145.

Olff, H., Huisman, J. and Van Tooren, B.F. (1993) Primary succession in coastal sand dunes: species dynamics in relation to biomass and nutrient accumulation. *Journal of Ecology* 81, 693–706.

Olff, H., Hoorens, B., De Goede, R.G.M., Van der Putten, W.H. and Glidman, J.M. (2000) Consequences of soil pathogens for the species composition of a natural grassland community. *Oecologia* 125, 45–54.

Oremus, P.A.I. and Otten, H. (1981) Factors affecting growth and nodulation of *Hippophaë rhamnoides* L. ssp. *rhamnoides* in soils from two successional stages of dune formation. *Plant and Soil* 63, 317–331.

Packer, A. and Clay, K. (2000) Soil pathogens and spatial patterns of seedling mortality in a temperate tree. *Nature* 404, 278–281.

Peters, D. and Weste, G. (1997) The impact of *Phytophthora cinnamomi* on six rare native tree and shrub species in the Brisbane ranges, Victoria. *Australian Journal of Botany* 45, 975–995.

Price, P.W., Bouton, C.E., Gross, P., McPheron, B.A., Thompson, J.N. and Weis A.E. (1980) Interactions among three trophic levels: influence of plants on interactions between insect herbivores and natural enemies. *Annual Review of Ecology and Systematics* 11, 41–65.

Robin, C., Desprezloustau, M.L., Capron, G. and Delatour, C. (1998) First record of *Phytophthora cinnamomi* on cork and holm oaks in France and evidence of pathogencity. *Annals of Scientific Forestry* 55, 869–883.

Roncadori, R.W. (1997) Interactions arbuscular mycorrhizas and plant parasitic nematodes in agro-ecosystems. In: Gange, A.C. and Brown, V.K. (eds) *Multitrophic Interactions in Terrestrial Systems*. Blackwell Science, Oxford, pp. 101–114.

Seliskar, D.M. and Huettel, R.N. (1993) Nematode involvement in the dieout of *Ammophila breviligulata* (Poaceae) on the Mid-Atlantic coastal dunes of the United States. *Journal of Coastal Research* 9, 97–103.

Shearer, B.L., Crane, C.E., Fairman, R.G. and Grant, M.J. (1998) Susceptibility of plant species in coastal dune vegetation of southwestern Australia to killing by *Armillaria luteobubalina*. *Australian Journal of Botany* 46, 321–334.

Smartt, J. and Simmonds, N W. (1995) *Evolution of Crop Plants*, 2nd edn. Longman Scientific & Technical, Harlow.

Stanton, N.L. (1988) The underground in grasslands. *Annual Review of Ecology and Systematics* 19, 573–589.

Streitwolf-Engel, R., Boller, T., Wiemken, A. and Sanders, I.R. (1997) Clonal growth traits of two *Prunella* species are determined by co-occurring arbuscular mycorrhizal fungi from a calcareous grassland. *Journal of Ecology* 85, 181–191.

Sykes, M.T. and Wilson, J.B. (1988) An experimental investigation into the response of some New Zealand sand dune species to salt spray. *Annals of Botany* 62, 159–166.

Sykes, M.T. and Wilson, J.B. (1990) An experimental investigation into the response of New Zealand sand dune species to different depths of burial by sand. *Acta Botanica Neerlandica* 39, 171–181.

Szabo, K. (1999) Investigations into the specificity of specific replant diseases and possibilities to overcome it: are replant diseases of apple and rose caused by the same pathogen? *Zeitschrift für Pflanzenkrankheit und Pflanzenschutz* 106, 237–243.

Thrall, P.H., Bever, J.D., Mihail, J.D. and Alexander, H.M. (1997) The population dynamics of annual plants and soil-borne fungal pathogens. *Journal of Ecology* 85, 313–328.

Van der Putten, W.H. and Peters, B.A.M. (1997) How soil-borne pathogens may affect plant competition. *Ecology* 78, 1785–1795.

Van der Putten, W.H. and Van der Stoel, C.D. (1998) Effects of plant parasitic nematodes on spatio-temporal variation in natural vegetation. *Applied Soil Ecology* 10, 253–262.

Van der Putten, W.H., Van Dijk, C. and Troelstra, S.R. (1988) Biotic soil factors affecting the growth and development of *Ammophila arenaria*. *Oecologia* (Berlin) 76, 313–320.

Van der Putten, W.H., Maas, P.W.Th., Van Gulik, W.J.M. and Brinkman, H. (1990) Characterization of soil organisms involved in the degeneration of *Ammophila arenaria*. *Soil Biology and Biochemistry* 22, 845–852.

Van der Putten, W.H., Van Dijk, C. and Peters, B.A.M. (1993) Plant-specific soil-borne diseases contribute to succession in foredune vegetation. *Nature* 362, 53–56.

Weller, D.M., Thomashow, L.S. and Cook, R.J. (1995) Biological control of soil-borne pathogens of wheat: benefits, risks and current challenges. In: Hokkanen, H.M. and Lynch, J.M. (eds) *Biological Control: Benefits and Risks*. Plant and Microbial Biotechnology Research Series 4. Cambridge University Press, Cambridge, pp. 149–160.

Weste, G. (1981) Changes in the vegetation of sclerophyll shrubby woodland associated with invasion by *Phytophthora cinnamomi*. *Australian Journal of Botany* 29, 261–276.

Williamson, M. (1996) *Biological Invasions*. Chapman & Hall, London.

Zoon, F.C. (1995) Biotic and abiotic soil factors in the succession of Sea buckthorn, *Hippophaë rhamnoides* L. in coastal sand dunes. PhD thesis, Agricultural University Wageningen, The Netherlands.

Zoon, F.C., Troelstra, S.R. and Maas, P.W.Th. (1993) Ecology of the plant-feeding nematode fauna associated with Sea buckthorn (*Hippophaë rhamnoides* L. ssp. *rhamnoides*) in different stages of dune succession. *Fundamental and Applied Nematology* 16, 247–258.

Development of Methods and Models and Their Application to Disease Problems in the Perennial Citrus Crop System

16

G. Hughes,[1] T.R. Gottwald[2] and S.M. Garnsey[3]

[1]Institute of Ecology and Resource Management, University of Edinburgh, Edinburgh EH9 3JG, UK; [2]USDA-ARS, US Horticultural Research Laboratory, 2001 South Rock Road, Fort Pierce, FL 34945, USA; [3]USDA-ARS (retired), 2313 Sherbrooke Road, Winter Park, FL 32792, USA

Introduction

South-eastern Asia is the centre of origin of citrus. From there, it has now spread throughout the world wherever the climatic moisture and temperature conditions are appropriate. Present-day major centres of citriculture include Argentina, Australia, Brazil, China, Cuba, Egypt, India, Israel, Italy, Japan, Mexico, Morocco, South Africa, Spain, Turkey, the USA and Venezuela. There is continual movement of citrus among most of the citrus-growing regions of the world. As a consequence, there is also an enormous potential for transport of citrus pathogens and pests accompanying this plant material, resulting in new combinations of pathogens and their insect vectors. To lessen this potential threat, regulatory agencies must maintain a constant vigil against such introductions. For vector-borne citrus pathogens (Table 16.1), quarantine situations may arise: (i) when a vector is present but the pathogen is absent; (ii) when the pathogen is present but a vector is absent; (iii) when both a pathogen and its vector(s) are absent; or (iv) when severe isolates of a pathogen or efficient vectors are absent. Quarantine procedures such as confiscation of material at ports of entry, inspection of ships' cargoes at ports, and inspections at border crossings and airports, can reduce the rate of introduction of exotic species that are harmful to agriculture. However, international borders will always be 'leaky' as far as pathogens and pests are concerned, and efficiency of detection of introductions is often very low. Once an exotic species is introduced into an agro-ecosystem, the ability to detect it at very low incidence is fundamental to its containment. This involves the development of survey protocols that are often unique to the species being

Table 16.1. Vector-borne citrus pathogens.

Disease	Geographical distribution	Causal agent	Vector
Citrus chlorotic dwarf	Turkey	Uncharacterized virus	Whitefly
Citrus variegated chlorosis	South America	*Xylella fastidiosa*	Leafhoppers
Citrus yellow mosaic	India	Badnavirus	Mealybugs
Huanglongbing (greening)	Asia, Africa, Saudi Arabia	*Liberobacter asiaticum*, *L. africanum*	Psyllids
Leprosis	Brazil	Rhabdovirus?	Mite
Lime witches' broom	Oman, UAR, Iran, India?	Phytoplasma	Leafhoppers
Satsuma dwarf	China, Japan	Comoviridae?	Soil-borne
Stubborn	USA, North Africa, Middle East	*Spiroplasma citri*	leafhoppers
Tristeza	Worldwide	Closterovirus	Aphids
Vein enation	Widespread	Luteovirus	Aphids

surveyed, and requires knowledge of the biology and spatial pattern of the exotic species within the crop. Surveys also have an important part to play in disease management in endemic areas.

Citrus Tristeza

Citrus tristeza virus (CTV) is the most economically important virus of citrus. CTV poses a continuing problem for citrus production in a number of countries, and increasingly threatens other regions where damaging isolates are still either absent or rare. Development of rapid serological assays, such as enzyme-linked immunosorbent assay (ELISA), has allowed extensive surveys to determine spatial and temporal changes in CTV incidence. Data have been collected from Spain, Florida and California, where natural spread is mediated by the melon aphid, *Aphis gossypii*, a moderately effective vector, and from Costa Rica, the Dominican Republic, Puerto Rico and Taiwan, where the brown citrus aphid, *Toxoptera citricida*, a highly efficient vector, is present. *T. citricida* was discovered on the east coast of Florida in the autumn of 1995. Since then, it has spread throughout the citrus-producing areas of the state and acceleration of CTV spread has been noted recently in several areas.

Sampling in Citrus Groves

In the USA, an active CTV eradication programme is presently underway in the Central Valley of California. Until recently, the Central California Tristeza

Eradication Agency (CCTEA) used a systematic sampling scheme, in which every fifth tree in every fifth row was selected for sampling (Fig. 16.1), to conduct an extensive survey of the incidence of CTV in commercial citrus groves in the Central Valley. Laboratory-based ELISA of the material sampled in the field was used to classify individual trees as either CTV-positive or CTV-negative. Thus, CTV incidence (the proportion of CTV-positive trees) was estimated, as a basis for decision making. If incidence was above an adopted threshold level, each individual tree in a block was then tested, with subsequent removal of those found to be CTV-positive.

For an initial evaluation of the CCTEA sampling scheme (Hughes and Gottwald, 1998), field data from censuses made during 1992 and 1993 in California were available from CCTEA records. In the censused blocks, the location and CTV status of each tree had been recorded in the form of a map. CTV status of individual trees had been determined by ELISA, as either CTV-negative or CTV-positive. The range of CTV incidence recorded was 0.4–19%. The maps from 12 (from a total of 36) blocks, covering the whole observed range of incidence, were used as a basis for an evaluation of the CCTEA sampling scheme. In such systematic sampling schemes, an element of randomization is usually introduced by selecting at random the starting position for sampling. Thus, in this case, the first tree for sampling could be

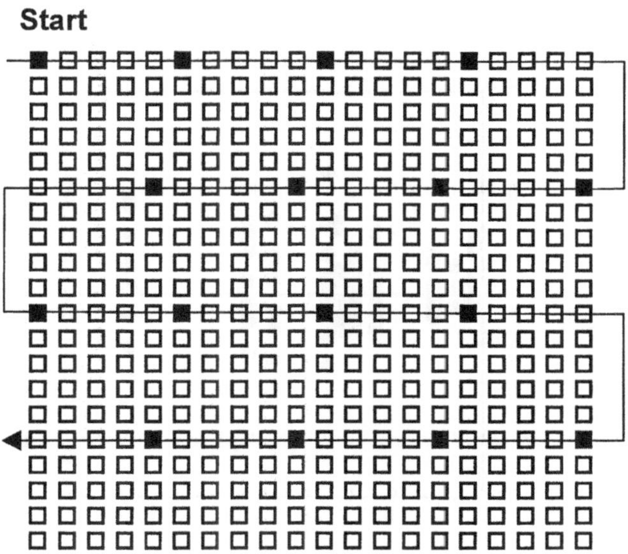

Fig. 16.1. Diagrammatic representation of the field implementation of the sequential sampling scheme used until recently by the CCTEA. Positions of individual trees are represented by squares. Black and white squares represent sampled and unsampled trees, respectively. The line represents the path traversed by the field sampling team. After the first tree, every fifth tree in every fifth row is sampled. Leaf petioles are collected from each sampled tree. From this, the CTV status of each individual tree sampled is determined by ELISA.

chosen at random from among those along and across the first five rows of a block. Subsequently, every fifth tree along every fifth row would be selected. For the purpose of evaluation, all 25 possible estimates of incidence were calculated for each of the 12 selected maps. This was achieved by systematically sampling every fifth tree in every fifth row, starting the sampling in turn from each of the 25 possible starting positions on each map.

The results of taking all 25 possible systematic samples of every fifth tree in every fifth row from each of 12 maps available from previous studies of CTV incidence are shown in Fig. 16.2. The range of estimates of incidence that may be made from a map is indicated by the vertical scatter of points at each actual CTV incidence. At around actual CTV incidences of 0.1, for example, estimated CTV incidences as low as zero and as high as 0.2 were obtained. It was concluded that the accuracy offered by the sampling scheme was insufficient for the purposes for which it was required (Hughes and Gottwald, 1998).

As an alternative, a sampling scheme was devised in which plant material is collected and assayed from groups of four trees, without distinguishing the individual trees. When such a 'group testing' approach is adopted, the procedure for estimating incidence at the scale of the individual can be summarized as follows. Calculations relating to the lower of the two scales (in this case, the individual tree scale) are denoted by the subscript '*low*', while those relating to the higher of the two scales (the group scale) are denoted by the subscript '*high*'. The mean proportion of infected trees per group (i.e. observed mean incidence measured at the tree scale) is denoted $\hat{p}_{low}$. When group size is constant:

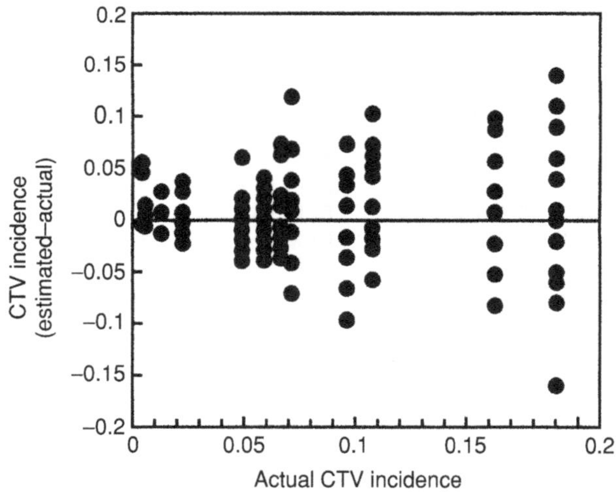

Fig. 16.2. For each of 12 maps of CTV incidence, the CCTEA sampling scheme (see Fig. 16.1) was used, beginning in turn in each of the 25 possible starting locations, to calculate all 25 possible estimates of incidence. The data points show differences between estimated and actual incidence (several data points may overlap at each actual CTV incidence). Copyright: American Phytopathological Society, reproduced by permission.

$$\hat{p}_{low} = \left(\sum_i \hat{p}_{low,i} \right) / N; i = 1, 2, ..., N \quad (16.1)$$

in which $\hat{p}_{low,i}$ is the proportion of trees infected in the ith group, and there are N groups. The proportion of groups with at least one infected tree (i.e. observed mean incidence measured at the higher of the two scales) is denoted $\hat{p}_{high}$. The proportions $\hat{p}_{low}$ and $\hat{p}_{high}$ are estimates of, respectively, p_{low} (the probability that an individual tree is infected) and p_{high} (the probability that a group has at least one infected tree).

In conventional group testing, material collected from individuals is assigned to groups at random, just prior to testing in the laboratory, and the disease status of one member of a group may therefore be assumed to be independent of the disease status of other members of the same group. In such cases, the binomial distribution may be assumed to provide an appropriate description of the frequency of infected trees per group. The zero term of the distribution (the probability that a group has no infected trees) is given by $(1 - p_{low})^n$ (n being the number of trees per group), so:

$$p_{high} = 1 - (1 - p_{low})^n. \quad (16.2)$$

By a straightforward rearrangement, $p_{low} = 1 - (1 - p_{high})^{1/n}$. Mean incidence at the lower scale may then be estimated from observations made at the higher scale:

$$\tilde{p}_{low} = 1 - (1 - \hat{p}_{high})^{1/n} \quad (16.3)$$

where the tilde is used to refer to an estimate that has been made from observations at a spatial scale other than the one denoted by its subscript.

However, in the alternative CTV sampling scheme outlined above, a group comprises a number of spatially adjacent trees. The CTV status (either positive or negative) of one member of a group cannot be assumed to be independent of the CTV status of other members of the same group. Because the CTV status of a tree may not be independent of the CTV status of its neighbours, use of equation 16.3 for estimation of $\tilde{p}_{low}$ requires more explicit justification than in cases where material collected in the field is randomly allocated to groups in the laboratory. Such justification would come in the form of an analysis of spatial pattern of field data for CTV incidence, showing that the binomial distribution provided an appropriate description of the frequency distribution of infected trees per group.

The data available from the censuses made during 1992 and 1993 were used for this purpose. Data from 36 rectangular blocks, varying in size between 472 and 3000 trees, had been recorded in the form of a map showing the location and CTV status of each tree. The map of each block was divided into N groups (in this case, N varies from block to block) of $n = 4$ trees, arranged two rows by two trees along rows. The frequency distribution of CTV-positive trees per group was then compiled for each block. The empirical (observed) variance of each of these distributions (v_{low}) was estimated from:

$$\hat{v}_{low} = \left(\sum_i [\hat{p}_{low,i}]^2 - \left[\sum_i \hat{p}_{low,i} \right]^2 / N \right) / (N-1) \tag{16.4}$$

and the corresponding theoretical (binomial) variance ($v_{low,bin}$) was estimated from:

$$\hat{v}_{low,bin} = \hat{p}_{low} \cdot (1 - \hat{p}_{low}) / n. \tag{16.5}$$

A linear relationship, on logarithmic axes, between $\hat{v}_{low}$ and $\hat{v}_{low,bin}$ can be interpreted as an indicator of the spatial pattern of incidence (Hughes and Madden, 1992; Madden and Hughes, 1995). For the 36 field plots for which data were available, the relationship between the observed variance ($\hat{v}_{low}$, calculated from equation 16.4) and the corresponding binomial variance ($\hat{v}_{low,bin}$, calculated from equation 16.5), was very close to the 'binomial line' line (i.e. observed variance = binomial variance) (Fig. 16.3). The slope and intercept of the least squares linear regression line fitted to the data (plotted on logarithmic axes) were, respectively, 1.03 (SE = 0.026) and 0.10 (SE = 0.054). There is thus no suggestion that the data can be distinguished from the binomial line, and this is taken as evidence that the binomial distribution provides an appropriate description of the frequency distribution of CTV-positive trees per group. That is to say, the pattern of CTV-positive trees may be taken as random at the within-group scale. A previous example of such an analysis for CTV incidence, using data from eastern Spain (Gottwald *et al.*, 1996a), also showed that the pattern of CTV-positive trees at the within-group scale was indistinguishable from random (Hughes *et al.*, 1997). In both the Spanish case and in that of

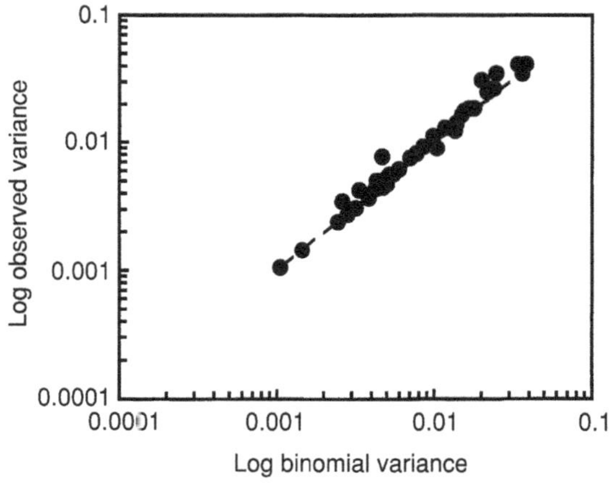

Fig. 16.3. The relationship between the observed and the theoretical binomial (random) variances of incidence of citrus tristeza virus at the tree scale. Each data point represents a CTV assessment in a block of citrus in the Central Valley of California. The dashed line (– – –) represents the binomial line (i.e. observed variance = binomial variance). Copyright: American Phytopathological Society, reproduced by permission.

the data set from California, the main vector of CTV was the melon aphid, *A. gossypii*. Thus, at least when the main vector is *A. gossypii*, estimates of CTV incidence at the scale of the individual tree can be made from data on incidence collected at the group scale, using a formula based on the zero term of the binomial distribution (equation 16.3).

However, when the brown citrus aphid, *T. citricida*, is the predominant vector, non-random patterns of CTV infection have been detected at several spatial scales (Gottwald *et al.*, 1996b). As the most efficient vector of CTV presently known, *T. citricida* represents a threat to production in many citrus-growing areas (Rocha-Peña *et al.*, 1995). A similar analysis of pattern to that described above was carried out for census data collected in Costa Rica between 1992 and 1996 (Gottwald *et al.*, 1998). The main vector of CTV in these plots was identified as *T. citricida*. The data represent repeated assessments on each of three blocks. In all, there were 17 assessments, each relating to a block of 400 trees in a 20×20 array. Data were recorded in the form of a map showing the location and CTV status of each tree. The map of each block was divided into N groups (here, $N = 100$ in each case) of $n = 4$ trees, arranged two rows by two trees along rows. The frequency distribution of CTV-positive trees per group was then compiled for each block, and the empirical (observed) variance of each of these distributions (v_{low}) and the corresponding theoretical (binomial) variance ($v_{low,bin}$) were estimated from equations 16.4 and 16.5, respectively (Hughes and Gottwald, 1999). As before, there was a linear relationship, on logarithmic axes, between $\hat{v}_{low}$ and $\hat{v}_{low,bin}$ (Fig. 16.4). In this case, however, the slope of 1.07 (SE = 0.027) and intercept of 0.23 (SE = 0.057)

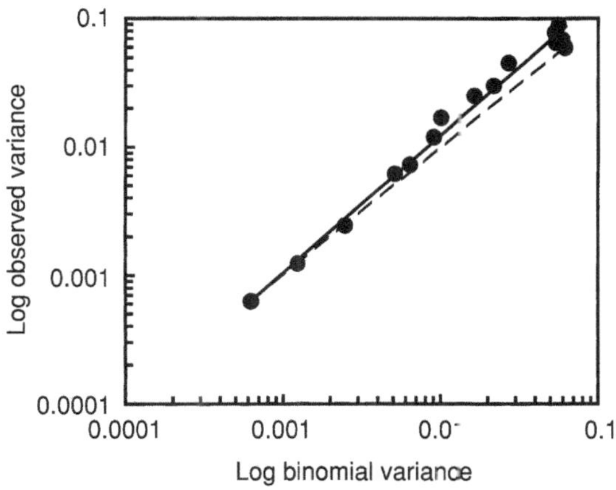

Fig. 16.4. The relationship between the observed and the theoretical binomial (random) variances of incidence of citrus tristeza virus disease at the tree scale. Each data point represents a CTV assessment in a block of citrus in Guanacaste, Costa Rica. The solid line represents the relationship: $\log(\hat{v}_{low}) = 0.23 + 1.07 \cdot \log(\hat{v}_{low,bin})$ fitted to the data by ordinary least squares regression. The dashed line (– – –) represents the binomial line (i.e. observed variance = binomial variance). Copyright: American Phytopathological Society, reproduced by permission.

placed the fitted regression line above the binomial line (Fig. 16.4), indicative of a weakly aggregated pattern of CTV incidence at the within-group scale (Hughes and Madden, 1992; Madden and Hughes, 1995).

The implications of the analyses outlined above, and illustrated in Figs 16.3 and 16.4, for a sampling scheme based on group testing can be seen by plotting the corresponding graphs of $\hat{p}_{high}$ against $\hat{p}_{low}$. When *A. gossypii* is the predominant vector of CTV, the observed data fall close to the line described by equation 16.2 with $n = 4$ (Fig. 16.5, see also figure 5 in Hughes *et al.*, 1997), allowing the use of equation 16.3 for estimation of $\tilde{p}_{low}$. However, when *T. citricida* is the predominant vector, the observed data tend to fall below the line described by equation 16.2 (Fig. 16.6). There is aggregation at the

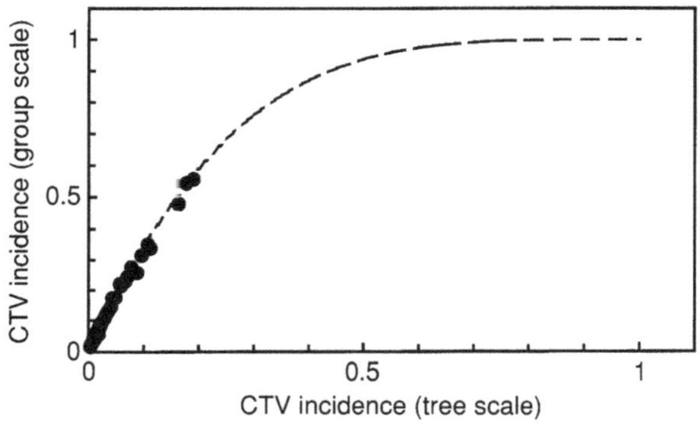

Fig. 16.5. The relationship between CTV incidence at the group scale and CTV incidence at the tree scale when *A. gossypii* is the main vector. Data derived from disease assessments in California are indicated by points (●). The curve indicated by the dashed line (– – –) is derived from the binomial distribution (equation 16.2 with $n = 4$).

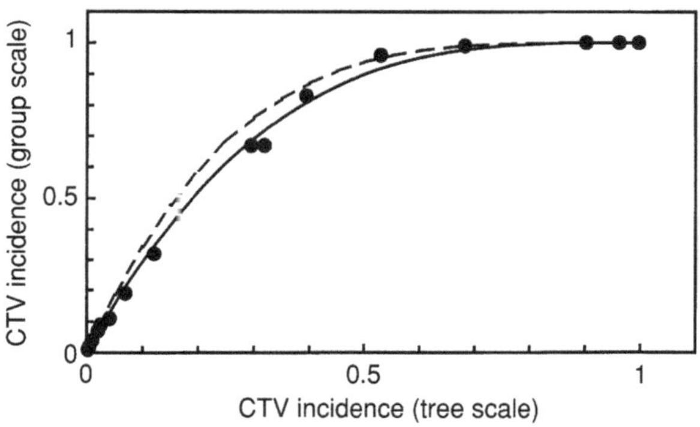

Fig. 16.6. The relationship between CTV incidence at the group scale and CTV incidence at the tree scale when *T. citricida* is the main vector. Data derived from disease assessments in Costa Rica are indicated by points (●). The curve indicated by the dashed line (– – –) is derived from the binomial distribution (equation 16.2 with $n = 4$). The curve indicated by the solid line is derived from equation 16.7 with $v = 3.3$).

within-group scale, so for any level of CTV incidence at the tree scale, there is lower incidence at the group scale than if the pattern of CTV incidence were random. The effect of using equation 16.3 for estimation of $\tilde{p}_{low}$ in such circumstances would be to consistently underestimate $\tilde{p}_{low}$.

Aggregated incidence data may be described by the beta-binomial distribution (Hughes and Madden, 1993; Madden and Hughes, 1995). If the beta-binomial is appropriate, the zero term of the distribution is given by $\prod_{i=0}^{n-1}\left(\frac{1-p_{low}+i\theta}{1+i\theta}\right)$, in which θ is an aggregation parameter ($\theta \geq 0$; $\theta = 0$ corresponds to the binomial; increasing θ indicates increasing aggregation). Thus, in this case:

$$p_{high} = 1-\prod_{i=0}^{n-1}\frac{1-p_{low}+i\theta}{1+i\theta}. \quad (16.6)$$

It is not difficult to calculate an estimate of the aggregation parameter from observed data (the maximum likelihood estimate for the data from Costa Rica is $\hat{\theta} = 0.116$ [SE = 0.021]), and so describe the relationship between $\hat{p}_{high}$ and $\hat{p}_{low}$ on the basis of equation 16.6. However, equation 16.6 is not generally helpful in the context of sampling, because its rearrangement to give p_{low} as a function of p_{high} is only possible for a limited range of group sizes.

Aggregation at the within-group scale is the tendency for trees in the same group to have the same CTV status. Because of this, less information about CTV incidence is obtained from within-group replicates than would be the case if the CTV status of a tree could be regarded as independent of the CTV status of others in the same group. Essentially, aggregation means that less than four trees-worth of information is obtained from a group of four trees. In such cases, an 'effective sample size', ν, can be calculated (Madden and Hughes, 1999). This provides a basis for approximating the form of equation 16.6 with a simple equation in the format of equation 16.2:

$$p_{high} = 1 - (1 - p_{low})^{\nu}. \quad (16.7)$$

Now, $p_{low} = 1 - (1 - p_{high})^{1/\nu}$, and mean incidence at the lower scale may be estimated from observations made at the higher scale:

$$\tilde{p}_{low} = 1-(1-\hat{p}_{high})^{1/\hat{\nu}}. \quad (16.8)$$

When *T. citricida* is the main vector of CTV, with $n = 4$ and $\hat{\theta} = 0.116$, equation 10 of Madden and Hughes (1999) gives $\hat{\nu} = 3.3$. Substituting this estimate into equation 16.7 then provides a good description of the observed data (Fig. 16.6), allowing the use of equation 16.8 for estimation of $\tilde{p}_{low}$.

For the implementation of the sampling scheme outlined above, a 20 × 20 tree block is considered to be 100 groups of four individual trees, each group arranged two by two. One out of the first four groups is selected at random, then every fourth group systematically after that. Four out of every 16 trees are sampled, but only CTV incidence at the group level (either no CTV-positive individuals or at least one CTV-positive individual) is recorded (Fig. 16.7). We refer to this as 'hierarchical sampling'. Although hierarchical sampling borrows its theoretical justification entirely from group testing, it differs from conventional group testing in that it involves a hierarchy of data with an explicit spatial component. Hierarchical sampling also differs from conventional group testing in the matter of choice of group size. In group testing, much attention has been devoted to the choice of an appropriate group size (Swallow, 1985, 1987; Hepworth, 1996). An experimenter has great flexibility in making this choice, since material is combined into groups in the laboratory after sampling has taken place in the field. In hierarchical sampling, the choice of sampling unit (i.e. the number of individuals per group) must take into account practical aspects of field sampling. A group size of four trees, arranged two rows by two trees along rows, was chosen mainly on the basis of practicability. However, smaller group sizes also tend to reduce statistical difficulties that can arise in group testing, such as bias of estimates (Swallow, 1985). The use of a larger group size would mean, of course, that fewer tests

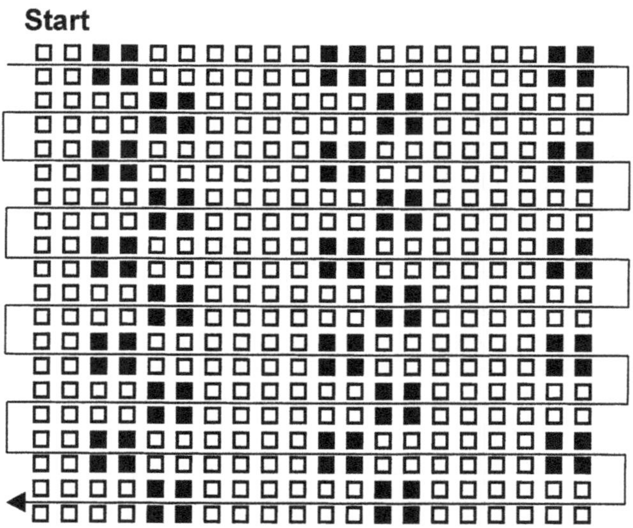

Fig. 16.7. Diagrammatic representation of the field implementation of the hierarchical sampling scheme proposed by Hughes and Gottwald (1998). Positions of individual trees are represented by squares. Black and white squares represent sampled and unsampled trees, respectively. The line represents the path traversed by the field sampling team. The scheme requires the random selection of one of the first four groups of four trees as a starting point. In this example, the second group of four trees was randomly selected as the starting point. Leaf petioles are collected from each tree within a group, and the collected material bulked. From this, the CTV status of each group sampled is determined by ELISA. The incidence of CTV-positive groups is related to the incidence of CTV-positive trees, as described in the text (see also Figs 16.5 and 16.6).

need to be carried out, so the budget for sampling could be reduced. Here, however, the main concern was to increase the effectiveness of the sampling effort rather than to reduce the budget.

The effectiveness of the sampling effort can be assessed by calculation of the 'operating characteristic (OC) curve'. OC curves are widely used in economic entomology to describe the performance of sampling schemes, particularly those based on sequential sampling (Nyrop and Binns, 1990; Binns and Nyrop, 1992; Jones, 1994). Although CTV sampling schemes are not sequential (because the virus-detection procedures are, of necessity, laboratory-based), the same concept is useful here. First, an appropriate decision threshold is defined. Here, for example, 10% CTV-positive trees was adopted as the threshold ($t_{low} = 0.1$) which, if exceeded, would trigger a programme of testing each individual tree in a block, and subsequent removal of those found to be CTV-positive. For sampling schemes based on the testing of individuals, such as the CCTEA scheme, T_{low} is then defined as the largest integer less than or equal to $t_{low}{\cdot}N$. T_{low} is, thus, the largest number of CTV-positive trees, out of a total of N trees, that can be observed without t_{low} being exceeded. The OC curve then shows the probability that the observed number of CTV-positive trees (X) is less than or equal to T_{low}:

$$\Pr(X \leq T_{low}) = \sum_{x=0}^{T_{low}} \Pr(X = x) \qquad (16.9)$$

where $\Pr(X = x)$ is based on the binomial distribution. This represents the probability of a decision (correct or otherwise) that the incidence of CTV-positive trees is less than or equal to the adopted threshold of 10%, for any actual value of incidence.

For a hierarchical sampling scheme, where the group (of $n = 4$ trees), rather than the individual tree, is the sampling unit, the definition of the threshold depends on the outcome of the analysis of pattern of virus incidence. If the pattern is indistinguishable from random, permitting use of equation 16.3 for the estimation of incidence at the individual scale from observations made at the group scale, the threshold is $t_{high} = 1-(1-t_{low})^n$. If the analysis is indicative of an aggregated pattern, the threshold is instead $t_{high} = 1-(1-t_{low})^{\hat{v}}$. Then, in either case, T_{high} is defined as the largest integer less than or equal to $t_{high}{\cdot}N$. T_{high} is, thus, the largest number of groups with at least one CTV-positive tree, out of a total of N groups, that can be observed without t_{high} being exceeded (and so, also, the largest number of groups with at least one CTV-positive tree that can be observed without t_{low} being exceeded). The OC curve then shows the probability that the observed number of groups with at least one CTV-positive tree (X) is less than or equal to T_{high}:

$$\Pr(X \leq T_{high}) = \sum_{x=0}^{T_{high}} \Pr(X = x) \qquad (16.10)$$

where, as in equation 16.9, $\Pr(X = x)$ is based on the binomial distribution. As with equation 16.9, this represents the probability of a decision (correct or otherwise) that the incidence of CTV-positive trees is less than or equal to the adopted threshold (at the individual tree scale) of 10%, for any actual value of incidence at that scale.

The OC curve is a graph of either $\Pr(X \leq T_{low})$ or $\Pr(X \leq T_{high})$, as appropriate, against actual CTV incidence. For values of the actual incidence less than or equal to the adopted decision threshold, the OC curve represents values of the true negative proportion (TNP); i.e. the proportion of decisions, based on sampling, that the incidence was less than or equal to the threshold when the actual incidence was less than or equal to the threshold. For values of the actual incidence above the adopted decision threshold, the OC curve represents the false negative proportion (FNP); i.e. the proportion of decisions, based on sampling, that the incidence was less than or equal to the threshold when the actual incidence was above the threshold.

Typically, values on an OC curve are near 1 when the actual incidence is much less than the value of incidence defined as the decision threshold, near 0.5 when the actual incidence is near the decision threshold, and near 0 when the actual incidence is much larger than the decision threshold. Figure 16.8 shows these characteristics for the sampling schemes discussed here, calculated from either equation 16.9 (the CCTEA sampling scheme) or equation 16.10 (the hierarchical sampling scheme). The hierarchical sampling scheme is clearly superior to the CCTEA sampling scheme, providing both higher TNPs and lower FNPs. There is little difference between the OC curves for the hierarchical sampling scheme described, whether *T. citricida* or *A. gossypii* is the predominant vector of CTV.

The hierarchical sampling scheme may successfully be employed in estimating CTV incidence at the scale of the individual tree when either *A. gossypii* or *T. citricida* is the main vector. When *A. gossypii* is the main vector, estimates of CTV incidence at the scale of the individual tree may be made from observations made at the group scale using equation 16.3, because the pattern of CTV incidence at the within-group scale is indistinguishable from random. When *T. citricida* is the main vector, the pattern of CTV incidence at the within-group scale is sufficiently aggregated for the use of equation 16.3 to result, on average, in an underestimate of CTV incidence at the scale of the individual tree. This problem is overcome by the empirical device of using an effective sample size instead of the nominal group size. Thus, when *T. citricida* is the main vector, estimates of CTV incidence at the scale of the individual tree may be made from observations made at the group scale using equation 16.8. In the field, the operation of the sampling scheme is the same whichever of the vectors is predominant (Fig. 16.7).

Sampling in Citrus Nurseries

Monitoring of plant health also takes place in citrus nurseries. Preventing the distribution of infected plants from nurseries is a basic measure that can be

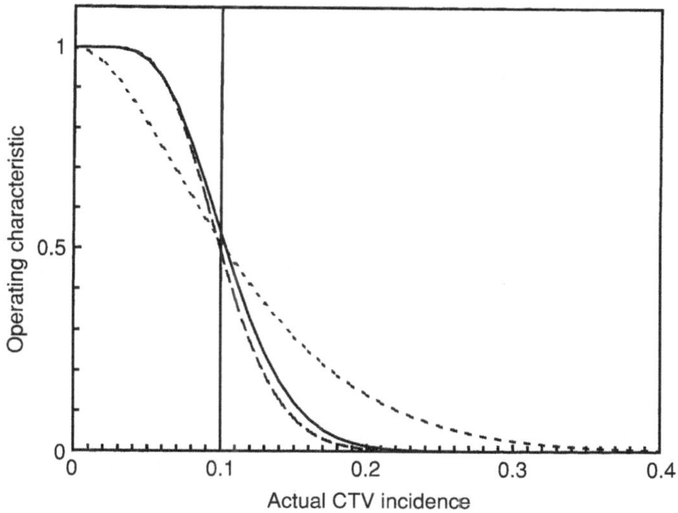

Fig. 16.8. Operating characteristic curves. The curve indicated by the dotted line (---) is based on equation 16.9, and represents the OC curve for the previously used CCTEA sampling scheme (see Fig. 16.1). The curve indicated by the dashed line (– – –) is based on equation 16.10 and represents the OC curve for the hierarchical sampling scheme proposed by Hughes and Gottwald (1998) (see Fig. 16.7), with T_{high} calculated using the nominal group size, $n = 4$. This is appropriate when *A. gossypii* is the main vector. The curve indicated by the solid line is also based on equation 16.10, and represents the OC curve for the hierarchical sampling scheme proposed by Hughes and Gottwald (1998) (see Fig. 16.7), with T_{high} calculated using the effective sample size, $\nu = 3.3$. This is appropriate when *T. citricida* is the main vector. The vertical line at actual CTV incidence = 0.1 represents the adopted decision threshold.

employed for the control of citrus virus diseases, both exotic and endemic. Growers regularly remove dead, declining or non-productive trees in mature plantings and replant with young trees. If infected nursery trees are used to replace trees removed from mature plantings, an opportunity is created for the introduction of a pathogen that may not have been present in the original planting, but can now move into the existing trees from the replants.

Commercial propagation of citrus usually involves the budding (grafting) of a desirable scion variety on a rootstock selected for its horticultural and disease-resistance properties. Rootstocks are typically produced from seed and budded when pencil-sized. The entire propagation process generally takes 18–24 months. Budwood is obtained from selected mother trees or from budwood-increase nurseries. Several thousand buds can be cut from a single large budwood mother tree, while individual nursery-increase plants may yield 50–100 buds over a season. In either case, the logistics of producing the millions of nursery trees required in countries with large citrus industries involves use of multiple budwood sources. Various strategies are employed to help ensure that the nursery trees produced and sold are free of graft-transmissible pathogens. While testing of individual trees just prior to sale would be the most accurate approach, it is generally not practical because of time and cost. The common approach is to bud young seedlings (which are assumed to have remained disease-free) using budwood from mother trees that have undergone testing at intervals designed to provide reasonable assurance that they were not infected at the time budwood was cut, or from designated budwood-increase nurseries. Budwood-increase nurseries are propagated from a specific tested source of budwood and generally can be used for only a limited period of time during which it is expected that levels of pathogen ingress will be low. Budwood-increase nurseries may be tested at some predetermined intervals for ingress of natural infection or propagation of an unrecognized infection in the mother tree.

Even though field-grown budwood source trees may be isolated from known sources of infection and/or grown in large screen houses, chances remain for infections that can be substantially increased in subsequent propagation operations. Recent infections in large trees may be irregularly distributed and escape detection. Accordingly, the need has increased for sampling protocols that can be used to detect low levels of pathogen infection in budwood mother sources, increase-nurseries and seedling populations. These protocols must address both conventional field nursery production systems, where tree position is fixed from the time of budding until sale, and indoor container production systems, where trees are grown in individual containers which may be moved repeatedly during their time in the nursery.

From a sampling perspective, we will consider the problem of virus infection in daughter plants that might arise if budwood has been taken from a mother tree that is infected, but not yet systemically so. For the purpose of illustration, we assume that the budwood takes the form of 'budsticks', that all such sticks provide ten buds, and that infection is systemic at the level of the stick, so that all the buds on any one stick are either infected or healthy. In a field nursery system, where the rootstock seedlings are budded sequentially along rows, all ten buds from a stick will naturally be grouped together. In this case, sampling systematically from every tenth daughter plant along a row is, in effect, conducting a census of the population at the level of the budstick.

In an indoor container system, the situation is different. Plants are grown in individual containers which may be moved around the benches on which they are kept, both during the budding process, and in subsequent agronomic operations. It is therefore not possible to identify a group of daughter plants as the progeny of a particular budstick, simply by their location on a bench. As an illustration, we consider a simple random sample (SRS) comprising material from $m = 5$ daughter plants taken from a population of $n \cdot N = 50$ daughter plants which are known to be the progeny of $N = 5$ budsticks, each of which had $n = 10$ buds. The sampling rate is 10%. Inevitably, there is likely to be a degree of redundancy in such samples. That is to say, although the sampling unit of interest is the budstick, the progeny of a stick may be represented more than once in a sample. It is the corollary – that another stick may therefore not be represented at all – that is of concern.

The total possible ways of selecting $m = 5$ plants at random out of a population of $n \cdot N = 50$ is $\binom{50}{5} = \frac{50!}{5! \cdot (50-5)!} = 2{,}118{,}760$. To calculate the proportion of samples in which, on average, the progeny of at least one budstick are not represented (in such a way as may be generalized to be applicable to other % sampling rates), we use the 'inclusion–exclusion (I–E) principle' (Marcus, 1998). We refer to the five budsticks that provided the buds as *A*, *B*, *C*, *D* and *E*, although of course the progeny of these sticks cannot be distinguished from one another at the time of sampling (or there would not be a problem). Let *V* be the set of all samples with the progeny of at least one stick not represented ($V = V_A \cup V_B \cup V_C \cup V_D \cup V_E$). We then calculate the quantity $\#(V)$, the number of elements in set *V*, using the I–E principle as follows:

$$\begin{aligned}
\#(V) &= \#(V_A \cup V_B \cup V_C \cup V_D \cup V_E) \\
&= [\#(V_A) + \#(V_B) + \#(V_C) + \#(V_D) + \#(V_E)] \\
&- [\#(V_A \cap V_B) + \#(V_A \cap V_C) + \#(V_A \cap V_D) + \#(V_A \cap V_E) + \#(V_B \cap V_C) + \\
&\quad \#(V_B \cap V_D) + \#(V_B \cap V_E) + \#(V_C \cap V_D) + \#(V_C \cap V_E) + \#(V_D \cap V_E)] \\
&+ [\#(V_A \cap V_B \cap V_C) + \#(V_A \cap V_B \cap V_D) + \#(V_A \cap V_B \cap V_E) + \#(V_A \cap V_C \cap V_D) + \\
&\quad \#(V_A \cap V_C \cap V_E) + \#(V_A \cap V_D \cap V_E) + \#(V_B \cap V_C \cap V_D) + \#(V_B \cap V_C \cap V_E) + \\
&\quad \#(V_B \cap V_D \cap V_E) + \#(V_C \cap V_D \cap V_E)] \\
&- [\#(V_A \cap V_B \cap V_C \cap V_D) + \#(V_A \cap V_B \cap V_C \cap V_E) + \#(V_A \cap V_B \cap V_D \cap V_E) + \\
&\quad \#(V_A \cap V_C \cap V_D \cap V_E) + \#(V_B \cap V_C \cap V_D \cap V_E)].
\end{aligned}$$

We can write:

$$\begin{aligned}
\#(V_1) &= [\#(V_A) + \ldots + \#(V_E)] \\
\#(V_2) &= [\#(V_A \cap V_B) + \ldots + \#(V_D \cap V_E)] \\
\#(V_3) &= [\#(V_A \cap V_B \cap V_C) + \ldots + \#(V_C \cap V_D \cap V_E)] \\
\#(V_4) &= [\#(V_A \cap V_B \cap V_C \cap V_D) + \ldots + \#(V_B \cap V_C \cap V_D \cap V_E)].
\end{aligned}$$

In this simplified notation, the number of samples with the progeny of at least one stick not represented is:

$$\#(V) = \#(V_1) - \#(V_2) + \#(V_3) - \#(V_4). \qquad (16.11)$$

Numerically, then, the number of samples with the progeny of at least one stick not represented in a SRS of five plants sampled from a population of 50 is (from equation 16.11):

$$\#(V) = \left[\binom{40}{5} \cdot \binom{5}{1}\right] - \left[\binom{30}{5} \cdot \binom{5}{2}\right] + \left[\binom{20}{5} \cdot \binom{5}{3}\right] - \left[\binom{10}{5} \cdot \binom{5}{4}\right] = 2{,}018{,}760.$$

The proportion of samples with the progeny of at least one stick not represented is 2,018,760/2,118,760 = 0.953. That is to say, in about 95% of samples of five plants out of 50 (comprising the progeny of five budsticks), the progeny of at least one budstick will be missing from the sample. It is immediately clear that sampling in indoor container nurseries is a very different proposition to sampling in field nurseries.

We can compare the above result for a 10% SRS of a population of 50 with that for a 20% SRS of a population of 50. The number of samples with the progeny of at least one budstick not represented in a SRS of ten pots sampled from a population of 50 is (from equation 16.11):

$$\#(V)=\left[\binom{40}{10}\cdot\binom{5}{1}\right]-\left[\binom{30}{10}\cdot\binom{5}{2}\right]+\left[\binom{20}{10}\cdot\binom{5}{3}\right]-\left[\binom{10}{10}\cdot\binom{5}{4}\right]=3{,}939{,}700{,}045.$$

Now, the total number of possible samples is $\binom{50}{10} = 10{,}272{,}278{,}170$ and so the proportion of samples with the progeny of at least one stick not represented is 3,939,700,045/10,272,278,170 = 0.384. In this case, in rather less than 40% of samples of ten plants out of 50 (comprising the progeny of five budsticks), the progeny of at least one budstick will be missing from the sample.

The above calculations are not concerned with whether or not any of the budsticks are infected. Now consider the problem when there may be $d = 0, 1, \ldots, N$ (here, $N = 5$) infected budsticks among those from which the 50 daughter plants are produced. As previously, for the purpose of illustration, we deal with infection that is systemic at the level of the stick (but not at the level of the mother tree), and assume that each stick provides $n = 10$ buds. The total number of daughter plants is $n{\cdot}N$ (= 50), the number of infected daughter plants is $n{\cdot}d$ (= 0, 10, 20, 30, 40 or 50), and the sample size (number of daughter plants sampled) is m (= 5 for a 10% SRS, = 10 for a 20% SRS). The (hypergeometric) probability distribution of X, the number of infected daughter plants in the sample, is given by:

$$\Pr(X=x)=\frac{\binom{n\cdot d}{x}\cdot\binom{n\cdot N-n\cdot d}{m-x}}{\binom{n\cdot N}{m}} \qquad x = 0, 1, \ldots, m. \tag{16.12}$$

Of particular interest is the case when $X = 0$ and $d > 0$, because this provides the operating characteristic, showing the probability of reaching the conclusion, based on sampling, that the population of 50 daughter plants contains no infection when in fact the progeny of at least one stick are infected. This simple illustrative example shows that the probability of concluding that the population of 50 plants contains no infection when one of the budsticks (out of five) was infected is about 0.3 for a sample of five plants (10% of the population) and about 0.1 for a sample of ten plants (20% of the population) (Fig. 16.9). Note that this result is not dependent on how the virus testing is carried out after the plant material has been obtained by sampling. Group testing methodology may be used to reduce the number of assays required to test larger samples of plant material.

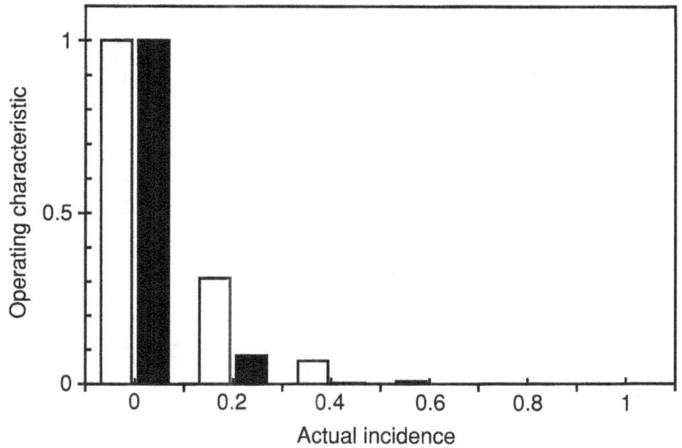

Fig. 16.9. Sampling in citrus nurseries using the indoor container system. The operating characteristic shows the probability of reaching the conclusion, based on sampling, that a population of 50 daughter plants contains no infection when in fact the progeny of at least one of the five budwood source sticks, each of which provided ten buds, are infected (see equation 16.12). The white bars denote a simple random sample of five plants, the black bars denote a simple random sample of ten plants.

Urban Citrus

In many countries where citrus is cultivated commercially, there are, near the commercial groves, urban areas in which citrus is frequently grown in private gardens. In Florida, for example, there are large numbers of various types of citrus trees in metropolitan Miami and the surrounding areas, grown as ornamentals and/or for home fruit production. The importance of this for the areas of commercial citrus production to the north is twofold. First, the urban citrus represents a potential reservoir of pathogens and vectors which may threaten commercial groves in the short term. Second, the existence of so much urban citrus alongside other types of plants grown for similar purposes represents an opportunity for virus ingress into citrus, providing a longer-term threat. Perennial growth means that both urban and commercial citrus populations are exposed to the possibility of virus ingress over an extended period of time. The irregular pattern of urban citrus (Fig. 16.10) means that the conventional methods used by plant pathologists in agricultural crops for analysing patterns of disease, and for sampling, are no longer appropriate. It will be a major challenge to develop methods for survey of urban citrus that will provide increased protection for neighbouring commercial citrus groves.

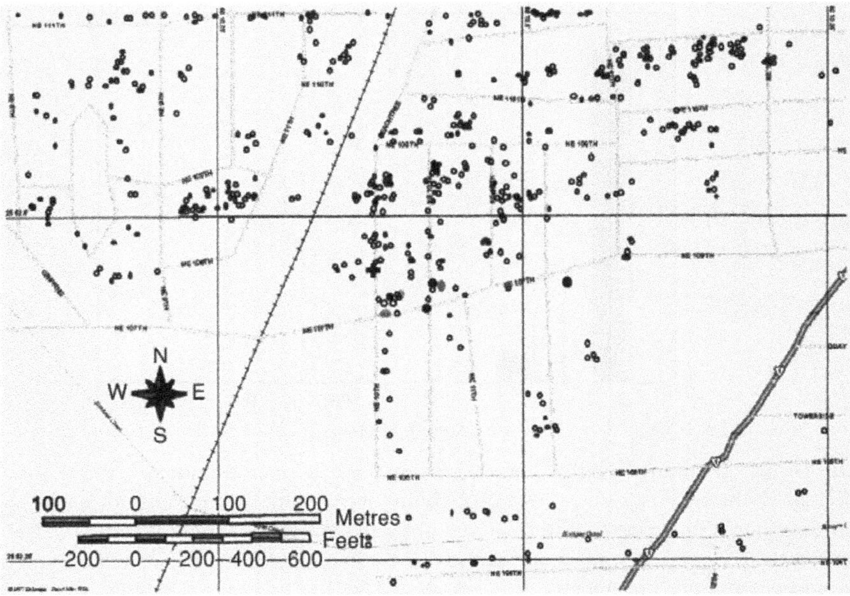

Fig. 16.10. The irregular pattern of urban citrus. Circles indicate the positions of individual citrus trees located in a typical neighbourhood in eastern Miami, Florida. Trees were located by means of a differential global positioning system (GPS) to ± 7 m.

Pathogen Ingress into Citrus

Consider first the situation when primary infections occur within the citrus crop system. The major concern has been with the introduction of pathogen-infected plants into a new area, where the pathogen was not previously present. If vectors are present, there is an immediate chance of spread of the pathogen within citrus. The temporal and spatial characteristics of this spread will reflect the behavioural ecology of the vector. Secondary spread via propagation may also occur, and this may have an impact on both local and long-distance spread. The threats of severe stem-pitting isolates of CTV and of huanglongbing (citrus greening) are examples of problems facing the Florida citrus industry, following the recent introductions of *T. citricida* and the Asian citrus psyllid, *Daiphorina citri*, the latter a vector of the pathogen responsible for huanglongbing (Table 16.1).

It is also possible that problems of pathogen ingress may arise if a new cultivar with an unrecognized disease susceptibility is introduced into an existing citrus-growing area where a pathogen is present, but does not cause damage on existing tolerant cultivars. In this case, the new cultivar is the receptor and the temporal and spatial characteristics of pathogen spread will reflect both continuing primary infections from existing plantings and secondary infections within the new cultivar.

In citrus pathology, most attention has been focused on diseases established on citrus, and on the movement of pathogens and vectors within the citrus crop system. It is increasingly apparent that for some diseases, this is only part of the picture. In this situation, the concern is that primary infections can be introduced from a non-citrus crop. For example, infections may arise from chance visits by a vector that does not normally inhabit citrus. The level of primary infection may vary, but the chance for secondary vector-mediated spread of the pathogen is low. If the resulting symptoms are severe, diseased plants will be self-eliminating and the chance of secondary pathogen spread via propagation is low. However, if the pathogen does not cause severe symptoms, and primary infections continue to accumulate over time, the chance of secondary spread via propagation is increased. If the symptoms of pathogen infection are recognizable, and appropriate detection methods are available, implementation of a certification programme can prevent secondary spread.

Another such example is infection of citrus via movement of a pathogen by a vector that can inhabit both the donor crop and citrus. In this case, there is opportunity for vector-mediated secondary spread within citrus, in addition to the primary infections. Temporal and spatial spread of the pathogen will reflect both types of infection. The level of inoculum in the non-citrus donor crop will influence the incidence of primary infections, and aspects of the behavioural ecology of the vector will influence incidence of both primary and secondary infections. Secondary spread via propagation may also occur. Citrus chlorotic dwarf and citrus variegated chlorosis are probably both examples of this type of pathogen ingress, although important information on the identity of the non-citrus donor crop is still missing in these cases.

Development and Deployment of Detection Methodology

Access to rapid, sensitive, accurate and inexpensive assays is important to most disease management strategies. Methods for detecting citrus viruses have continued to evolve as new information on the viruses and new detection technology have become available. Inoculating test plants of citrus cultivars especially reactive to certain viruses, and characterization of their reactions, was the first step to rapid detection. This remains the only reliable method for detection of some viruses, and for distinguishing strains of others. Herbaceous hosts have been used in the detection of citrus viruses that are easily mechanically transmitted. Electron microscopy was the first detection tool based on a specific virus property. As citrus viruses have been purified, serological detection methods have been developed, and ELISA is now widely used for detection of CTV and other citrus viruses. Methods based on molecular probes, such as the polymerase chain reaction (PCR) and nucleic acid hybridization assays, are now being developed as information on the sequences of citrus viruses becomes available.

Extending new detection technologies into practical use for disease management requires extensive testing to determine how to apply the technology accurately, and extensive comparison with biological testing. Development of appropriate controls for each type of assay also remains a challenging problem. Also, in the context of disease management decision making, it is not only the characteristics of the assay used for detection that is important, but also the characteristics of the sampling scheme in which the assay is deployed.

For example, Mathews *et al.* (1995, 1997) discussed the use of PCR-based methods for the detection of CTV. It was shown that the PCR-based methods provided a more sensitive assay for CTV than the standard ELISA procedure. Now, consider the implications of these findings if PCR-based methods were to be deployed in sampling. Mathews *et al.* (1995) noted that in the Central Valley of California, 86% of groves are believed to be CTV-free. A 5% sample of a typical 2000 tree grove provides material from 100 trees for testing. This material could be combined into a single group and tested using PCR-based CTV detection methods. Rapid screening out of CTV-negative groups would enable more resources to be devoted to sampling in those groves that provided a CTV-positive sample.

Of interest here is the extent to which such a procedure would provide false negatives – indications based on sampling that the grove was CTV-free when, in fact, there actually was CTV infection. Assuming that the 100-tree samples could reasonably be regarded as random samples of the groves, we need to calculate the proportion of decisions, based on sampling, that a grove is CTV-free when actual incidence is greater than zero. This is shown by plotting $(1 - p_{low})^{100}$ against p_{low} over an appropriate range of values of p_{low}. The proportion of false negative decisions that would arise if groves were classified as CTV-free on the basis of an assay of material from 100 trees combined into a single group is shown in Fig. 16.11. This shows, for example, that if the actual incidence of CTV infection is 2% (the threshold mentioned by Mathews *et al.* (1997), above which every tree in a grove would be tested) the probability of reaching a decision that the grove is CTV-free is around 13%.

This example indicates that the deployment of PCR-based detection methods in screening of groves for presence or absence of CTV infection would require careful consideration of the acceptable rate of false negative decisions in advance of the implementation of such a scheme. This would be particularly important if *T. citricida* were the main vector, since rates of increase from low to high incidence of CTV appear to be much quicker with *T. citricida* than with *A. gossypii* (Gottwald *et al.*, 1996b).

Conclusion

Virus and virus-like diseases of citrus are a dynamic system. Dispersal of pathogens and vectors to new areas, changes in the properties of viruses and introduction of new citrus cultivars are all factors. The proximity of

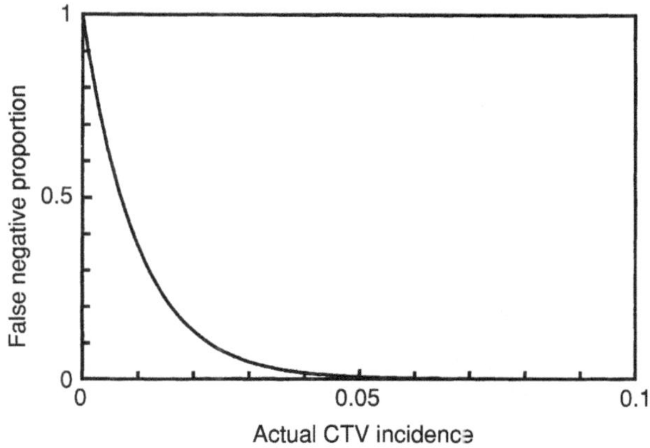

Fig. 16.11. The relationship between the proportion of false negative decisions and actual CTV incidence, if groves were classified as CTV-free on the basis of an assay of material from 100 trees combined into a single group. Copyright: American Phytopathological Society, reproduced by permission.

commercial citrus groves and nurseries to rapidly expanding urban population centres where citrus is a popular ornamental may also present problems. Citrus is a long-lived perennial into which viruses from outside sources can be introduced and accumulated. The impact of these ingress events is determined by the potential for secondary spread, and the pathogen and vector reservoirs in other crops. The complexity of these interactions poses many challenging questions. The development of detection methodology, and the deployment of this methodology in sampling protocols founded on epidemiological models, will play an important part in the management of new citrus disease problems as they continue to arise.

Acknowledgement

The authors are grateful to the American Phytopathological Society for permission to reproduce Figs 16.2 and 16.3 from Hughes and Gottwald (1998), and Figs 16.4 and 16.11 from Hughes and Gottwald (1999).

References

Binns, M.R. and Nyrop, J.P. (1992) Sampling insect populations for the purpose of IPM decision making. *Annual Review of Entomology* 37, 427–453.

Gottwald, T.R., Cambra, M., Moreno, P., Camarasa, E. and Piquer, J. (1996a) Spatial and temporal analyses of citrus tristeza virus in eastern Spain. *Phytopathology* 86, 45–55.

Gottwald, T.R., Garnsey, S.M., Cambra, M., Moreno, P., Irey, M. and Borbón, J. (1996b) Differential effects of *Toxoptera citricida* vs. *Aphis gossypii* on temporal increase and spatial patterns of spread of citrus tristeza. In: De Graça, J.V., Moreno P. and Yokomi, R.K. (eds) *Proceedings of the 13th Conference, International Organization of Citrus Virologists*. IOCV Riverside, pp. 120–129.

Gottwald, T.R., Garnsey, S.M. and Borbón, J. (1998) Temporal increase and spatial patterns of spread of citrus tristeza virus infections in Costa Rica and the Dominican Republic in the presence of *Toxoptera citricida*. *Phytopathology* 88, 621–636.

Hepworth, G. (1996) Exact confidence intervals for proportions estimated by group testing. *Biometrics* 52, 1134–1146.

Hughes, G. and Gottwald, T.R. (1998) Survey methods for assessment of citrus tristeza virus incidence. *Phytopathology* 88, 715–725.

Hughes, G. and Gottwald, T.R. (1999) Survey methods for assessment of citrus tristeza virus incidence when *Toxoptera citricida* is the predominant vector. *Phytopathology* 89, 487–494.

Hughes, G. and Madden, L.V. (1992) Aggregation and incidence of disease. *Plant Pathology* 41, 657–660.

Hughes, G. and Madden, L.V. (1993) Using the beta-binomial distribution to describe aggregated patterns of disease incidence. *Phytopathology* 83, 759–763.

Hughes, G., McRoberts, N., Madden, L.V. and Gottwald, T.R. (1997) Relationships between disease incidence at two levels in a spatial hierarchy. *Phytopathology* 87, 542–550.

Jones, V.P. (1994) Sequential estimation and classification procedures for binomial counts. In: Pedigo, L.P. and Buntin, G.D. (eds) *CRC Handbook of Sampling Methods for Arthropods in Agriculture*. CRC Press, Boca Raton, Florida, pp. 175–205.

Madden, L.V. and Hughes, G. (1995) Plant disease incidence: distributions, heterogeneity, and temporal analysis. *Annual Review of Phytopathology* 33, 529–564.

Madden, L.V. and Hughes, G. (1999) An effective sample size for predicting plant disease incidence in a spatial hierarchy. *Phytopathology* 89, 770–781.

Marcus, D.A. (1998) *Combinatorics: a Problem Oriented Approach*. The Mathematical Association of America Washington, DC.

Mathews, D.M., Riley, K. and Dodds, J.A. (1995) Comparison of ELISA and PCR for the sensitive detection of citrus tristeza virus (CTV) in pooled leaf samples from sweet orange groves with a low incidence of infection. In: De Graça, J.V., Moreno, P. and Yokomi R.K. (eds) *Proceedings of the 13th Conference, International Organization of Citrus Virologists*. IOCV, Riverside, pp. 12–16.

Mathews, D.M., Riley, K. and Dodds, J.A. (1997) Comparison of detection methods for citrus tristeza virus in field trees during months of nonoptimal titer. *Plant Disease* 81, 525–529.

Nyrop, J.P. and Binns, M.B. (1990) Quantitative methods for designing and analyzing sampling programs for use in pest management. In: Pimentel, D. and Hanson, A.A. (eds) *CRC Handbook of Pest Management in Agriculture*, 2nd edn, Vol. II. CRC Press, Boca Raton, Florida, pp 67–132.

Rocha-Peña, M.A., Lee, R.F, Lastra, R., Niblett., C.L., Ochoa-Corona, F.M., Garnsey, S.M. and Yokomi, R.K. (1995) Citrus tristeza virus and its aphid vector *Toxoptera citricida*: threats to citrus production in the Caribbean and Central and North America. *Plant Disease* 79, 437–444.

Swallow, W.H. (1985) Group testing for estimating infection rates and probabilities of disease transmission. *Phytopathology* 75, 882–889.

Swallow, W.H. (1987) Relative mean squared error and cost considerations in choosing group size for group testing to estimate infection rates and probabilities of disease transmission. *Phytopathology* 77, 882–889.

17

Observation and Theory of Whitefly-borne Virus Disease Epidemics

John Holt and John Colvin

Natural Resources Institute, University of Greenwich, Central Avenue, Chatham Maritime, Chatham, Kent ME4 4TB, UK

Introduction

Movement of agricultural and horticultural plant material across international boundaries has often resulted in the inadvertent introduction to new regions of vectors, such as the whitefly, *Bemisia tabaci* (Genn.), and the viruses they transmit (Bos, 1992; Brown, 1994; Polston and Anderson, 1997; Banks *et al.*, 2000). The increased movement of plant material has occurred in conjunction with a trend towards increasing agricultural intensification that has altered ecosystems to favour the rapid spread of whitefly-borne virus disease epidemics. In the current cassava mosaic virus disease pandemic in East Africa (Otim-Nape *et al.*, 2000), for example, large areas of one or two cassava varieties were grown in continuous cultivation. These proved particularly susceptible to both single infection by the new, recombinant Uganda variant geminivirus (UgV) and even more so to a double infection with both African cassava mosaic geminivirus (ACMV) and UgV (Harrison *et al.*, 1998). Large populations of emigrant viruliferous *B. tabaci*, originating from infected cassava fields, were able to colonize new cassava fields, situated ahead of the epidemic front (Colvin *et al.*, unpublished), which resulted in the rapid spread of this epidemic (Legg and Ogwal, 1998; Colvin *et al.*, 1999; Holt *et al.*, 1999b; Zhang *et al.*, 2000). The mechanism that drives this epidemic is, as yet, incompletely understood, although an interaction mediated through a fecundity boost to whiteflies feeding on CMD-affected plants may play a crucial role.

The Interaction Mechanism

Increases in the size of *B. tabaci* populations and virus disease outbreaks often coincide, and a one-way, cause and effect relationship is often assumed, i.e. that an increase in vector number causes an increase in disease spread. Factors cited as responsible for increased vector number include altered agricultural practices, climate, insecticide resistance and population resurgence, and the invasion of a new vector biotype. An additional possibility and a more general explanation for the mechanism driving some epidemics is that an interaction occurs between the vector, virus, host plants and the environment – the net effect of which is increased virus disease spread.

Phloem sap-feeding insects are very sensitive to host plant quality and their survival and oviposition rates vary significantly on different healthy and virus-infected host plant species (Kennedy, 1951; Fereres *et al.*, 1989; Colvin *et al.*, 1999). In a study on the effects of viral host plant infection on *B. tabaci* fecundity and survival, adults that had been reared on a succession of pumpkin plants for more than 5 years were exposed in clip cages to six plant species infected with one of four whitefly-transmitted plant viruses. Survival to adulthood was significantly higher on the diseased than on the healthy pumpkin, as was the concentration of total free amino acids in the diseased plants. For other plant species and virus combinations, infection either had no apparent effect on oviposition and survival or affected these life history traits negatively. Compared to healthy lettuce, pumpkin, tomato, zucchini, cotton and cantaloupe, total free amino acid concentrations were significantly higher in all the virus-infected plant species, although no simple relationship could be detected between total free amino acid levels and the oviposition or survival rates of *B. tabaci* (Costa *et al.*, 1991).

These results indicate that for any given host plant–virus–*B. tabaci* biotype combination, there is not necessarily an obvious mutually beneficial relationship between the virus and the whitefly, as preference, rate of oviposition and survival vary with the particular virus–host plant combination (Costa *et al.*, 1991).

It is probable that much of the variability in the influence of virus-infected plants on vectors is due to differences in the nitrogenous compounds mobilized in these plants (Power, 1992), as these form an important constituent of the food of phloem sap-feeding insects (Auclair, 1963; Kunkel, 1977). The amino acid content in the phloem sap of diseased plants typically increases in response to virus infection and it was suggested that each virus alters the amino acid concentrations of the plant host in a unique way (Selman *et al.*, 1961). Harpaz and Applebaum (1961) listed 11 virus diseases, including maize rough dwarf virus, which caused an accumulation of asparagine and suggested that this might be involved in the biosynthesis of the virus material itself. Asparagine was also found to occur in higher concentrations in UgV-infected cassava than in healthy plants (Colvin *et al.*, unpublished). In the same experiment, *B. tabaci* adults were used to transmit the UgV virus to healthy plants. Colonies of *B. tabaci* were then maintained on these plants and their population growth monitored. Where the Uganda variant was transmitted successfully, disease symptoms appeared in plants 3–7 weeks after inoculation and numbers of nymphs increased more rapidly than on virus-free plants, an effect that was apparent even before all of the plants had developed symptoms (Colvin *et al.*, unpublished). *Bemisia tabaci* and the UgV virus, therefore, apparently interact in a mutually beneficial manner that could promote the spread of this virus.

The Biotype Mechanism

In the mid-1980s in the New World, an introduced *B. tabaci* population, referred to as the B-biotype, caused enormous agricultural yield losses as a direct pest (Perring, 1996). Its arrival also altered the epidemiologies of many known plant virus diseases and led to the introduction of previously unknown geminiviruses from weeds into agricultural crops (Costa and Brown, 1991; Brown *et al.*, 1996; Poulston and Anderson, 1997). This *B. tabaci* biotype was considered sufficiently different from the indigenous north American population, termed the A-biotype, to be regarded as a separate species, *Bemisia argentifolii* Bellows and Perring, by some researchers (Perring *et al.*, 1993; Bellows *et al.*, 1994), although this was considered premature by others (Bedford *et al.*, 1994).

In north America, the population of the B-biotype rapidly displaced that of the indigenous A-biotype because of its impressive ability to colonize an extremely broad range of plant species, including those comprising the A biotype's niche (Perring, 1996). Its arrival into a new region, however, was not always followed by an increase in plant virus disease incidence. The severity of the lettuce infectious yellows virus (LIYV) problem in Arizona and California, for instance, actually decreased, probably because the B-biotype is a relatively inefficient vector of this virus (Cohen *et al.*, 1992; Falk and Klaassen, 1996).

The B-biotype has continued to spread around the world and is now present in South America, Australia and most recently in south India (Banks *et al.*, 2000). In the latter case, the arrival of the B-biotype was associated with a severe outbreak of disease that occurred throughout the tomato crop of Kolar district, Karnataka State. Disease symptoms were observed in 100% of plants within 30 days after planting, resulting in almost no fruit set and complete crop failure. Prior to the arrival of the B-biotype, two to four adults per tomato plant was the seasonal norm (Ramappa *et al.*, 1998), which increased by up to 1000-fold in the disease outbreak region. The arrival of the B-biotype in India is of great concern, as it probably has a broader host range than the indigenous populations and could potentially acquire and transmit other viruses into crops from those present in infected weeds. As a result, *B. tabaci* and tomato leaf curl virus disease management strategies in south India (Holt *et al.*, 1999) may need to be re-evaluated.

The Model

We develop an epidemiological model that allows the effects of the vector fecundity changes associated with the two mechanisms to be compared. The model specifies the dynamics of the populations of the vector and the infected host. It is more parsimonious than the models of Holt *et al.* (1997) and Jeger *et al.* (1998) but uses the same idea of linking equations for the host and vector. Consider a field of the host plant, which over the period of crop growth we assume contains a constant number of plants, K. We model the change in the number of infected hosts, Y, and the change in the number of vectors, V. The number of healthy hosts is simply the total host number minus the number infected $(K - Y)$. It is assumed that, for the vectors feeding in the crop, the proportion that is infective is related to the proportion of hosts that are infected (Y/K). Thus the number of infective vectors is given by $q(Y/K)V$ where q is a constant relating the proportions of vectors infective to hosts infected. We now define a contact rate between healthy hosts and infective vectors. Assuming no host losses, the rate of change of infected hosts is given by this contact rate as

$$\frac{\mathrm{d}Y}{\mathrm{d}t} = k_1 \frac{Y}{K} V(K - Y) \tag{17.1}$$

where k_1 is the virus transmission rate (per vector per day) and which subsumes the constant q. Equation 17.1 can be rearranged as the product of vector number, V, and a logistic function of Y:

$$\frac{\mathrm{d}Y}{\mathrm{d}t} = k_1 Y\left(1 - \frac{Y}{K}\right)V. \tag{17.2}$$

Two important processes occur in the interaction between the whitefly vector and the CMV-infected host. The effects are most evident in, but not restricted to, the new UgV strain of CMV (Colvin *et al.*, 1999). The fecundity of the vectors is affected by host infection probably because this alters nutrient availability in the host. Fecundity is generally greater on CMV-infected host plants, but this also depends on feeding site and plant age (Zhang *et al.*, 2000). For simplicity it is assumed that the proportion of the vector population in the field which experience enhanced fecundity is equal to the fraction of hosts which are infected, Y/K. The fecundity of the vectors is partitioned between those exposed to low nutrient, $r_1(1 - Y/K)$, and high nutrient, r_2Y/K, conditions, where r_1 and r_2 are the vector birth rates (per day) under the respective conditions. Thus,

$$\text{fecundity} \propto r_1\left(1 - \frac{Y}{K}\right) + r_2 \frac{Y}{K}. \tag{17.3}$$

It is likely that the stimuli for whitefly vector emigration from the crop are crowding and competition for suitable feeding sites (Holt *et al.*, 1999b; Zhang *et al.*, 2000). Because of the effects of infection on the host and therefore on the availability of suitable feeding sites, it can be assumed that crowding is proportional to infection and therefore

$$\text{emigration} \propto m\frac{Y}{K} \tag{17.4}$$

where m (per day) is the emigration rate. By combining equations 17.3 and 17.4 we can now specify the rate of change in vector numbers as

$$\frac{dV}{dt} = \left[r_1 \left(1 - \frac{Y}{K} \right) + (r_2 - m)\frac{Y}{K} \right] V \tag{17.5}$$

The rate of advance of the epidemic by the spread of the disease to new cassava fields depends upon the rate of production of infective emigrant vectors. From equation 17.5 we have the rate of production of emigrants as $m(Y/K)V$. As discussed above we can assume that the proportion which is infective is qY/K and, therefore, the rate of production of infective emigrants is:

$$\frac{dW}{dt} \propto m\left(\frac{Y}{K}\right)^2 V \tag{17.6}$$

where W is initially zero and, at the end of the crop period, provides a relative measure of the total production of infective emigrants; m subsumes the constant q. In an analysis of the model, equations 17.2, 17.5 and 17.6, we examine the conditions under which W is maximized. In determining the parameter values that maximize W, we consider whether the changes in *B. tabaci* biology and behaviour, brought about by CMV infection of the cassava host, do indeed enhance virus spread to new hosts. If so, it could provide the driving mechanism for the devastating epidemic associated with the new UgV strain of CMV in East Africa.

Results

A typical time course of the dynamics of the model system from the time of planting of the cassava crop (Fig. 17.1) shows a logistic increase in disease incidence to reach an asymptote at *c.* 130 days. Parameter values were based on the work with earlier models (Holt *et al.*, 1997; Zhang *et al.*, 2000). Vector numbers rise from the start of the crop to reach a peak at *c.* 100 days then decline because rising host infection causes emigration to increase. Infective emigrants are produced in significant numbers from *c.* 100–120 days during the period when the vector population in the crop is high and incidence is increasing rapidly. The emigration of infective vectors declines despite the high level of infection in the crop because the crop is no longer able to support a high vector population. The timing of events depends on the choice of parameter values but Fig. 17.1 illustrates a general pattern: that an increase in infection and the departure of vectors are concurrent processes. A build-up of infection is required in order to produce infective emigrants but at the same time, the whiteflies are forced to depart. The parameter values that maximized the production of infective emigrants were instructive.

The effect of varying vector emigration rate, *m*, was examined over a wide range of values (Fig. 17.2). The result was, at first sight, unexpected: the higher the emigration rate, the *fewer* the number of infective emigrants. This occurred for the following reason: the longer whiteflies can be retained on the host, the more their numbers can increase and the more opportunity exists for virus acquisition before the insects are finally forced to leave due to declining host quality. The action of selection on the virus is expected to maximize disease spread and should therefore act to delay for as long as possible the

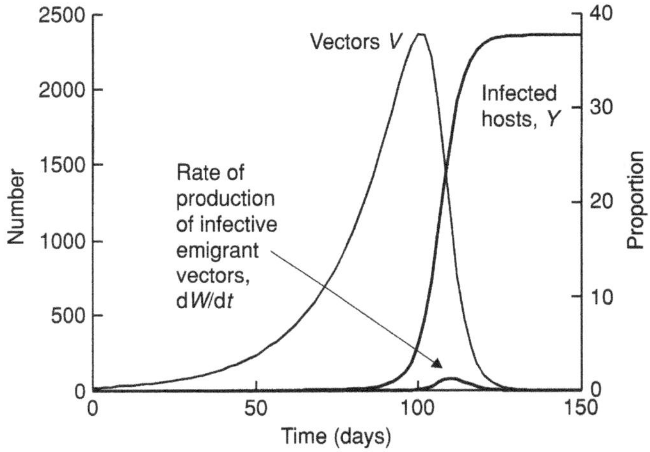

Fig. 17.1. Typical time course of numerical results from the system of equations 17.2 and 17.5. Parameter values were, $r_1 = 0.05$, $r_2 = 0.2$, $K = 100$, $m = 1$, $k_1 = 0.0001$. Initial values of the variables were, $V_0 = 20$, $Y_0 = 0.02$.

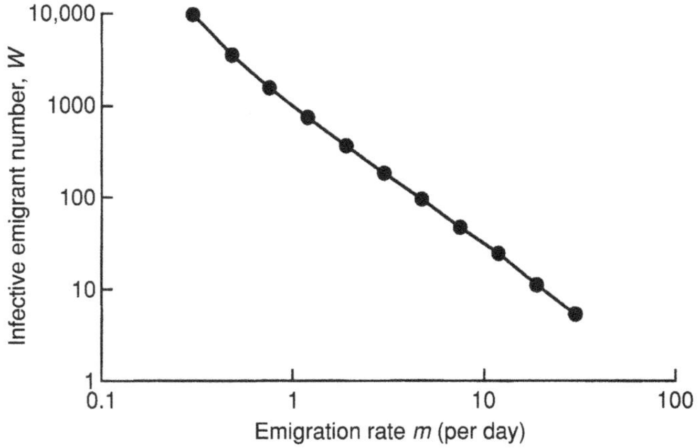

Fig. 17.2. The effect of changes in vector emigration rate, *m*, on the total number of infective emigrants, *W*, produced by the cassava crop in the 200 days following planting. Other parameters and initial states as Fig. 17.1.

deleterious effects of the virus on its host. Cassava plants take 3–7 weeks to develop symptoms after inoculation, which may be an adaptation of the virus to increase the build-up and retention of vectors on diseased plants.

Variation in the transmission rate k_1 also led to a surprising result. There was an optimum transmission rate that maximized the total number of infective emigrants, W (Fig. 17.3a). Intuition might suggest that higher infection rates would lead to faster disease spread but this was not the case. Again the reason lies in the advantage of providing the conditions for the whitefly population to increase. If the infection rate is too high, then disease progress in the host population quickly reaches an advanced stage, so driving off the whiteflies before their population has a chance to build up. In contrast, if the

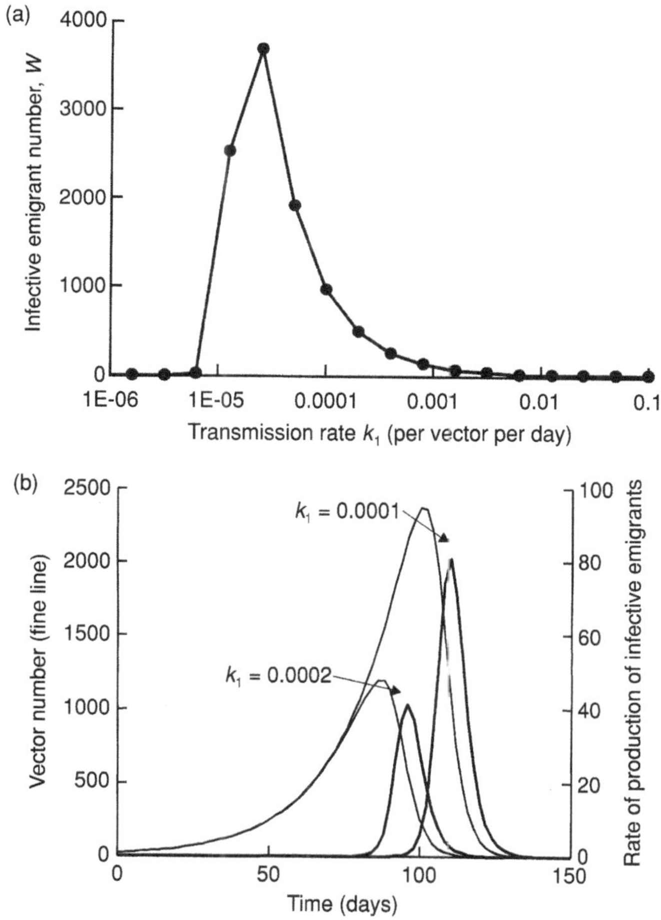

Fig. 17.3. (a) The effect of changes in infection rate k_1 on the total number of infective emigrants, W, and (b) a comparison of the rate of production of infective emigrants, dW/dt (number per day), with two different infection rates. Other parameters and initial states as Fig. 17.1.

infection rate is too low then incidence simply stays too low to produce a high proportion of infective vectors. To maximize spread, therefore, it is to the advantage of a virulent virus to have an infection rate that is neither too high nor too low. The effect of a change in transmission rate on the time course of the model system is illustrated in Fig. 17.3b. With a higher transmission rate, infective emigrants occur earlier in the crop period but also decline earlier than with a lower transmission rate.

As might be expected, the total number of infective emigrants, W, was positively correlated with vector fecundity (Fig. 17.4) but it is interesting to compare the influence of the two components of fecundity, r_1 and r_2. The first component r_1 is that associated with non-infected hosts, whereas the second component r_2 is that associated with infected hosts. For $r_2 > r_1$ there is a boost to fecundity associated with host infection. As can be seen from Fig. 17.4, changes in r_1 have a much greater effect on W than do changes in r_2. Thus, a high fecundity independent of infection generates more infective vectors than does one dependent on infection.

Discussion

An apparently paradoxical conclusion was reached with regard to vector emigration rate. To maximize virus fitness, there should be very strong selection for the virus to make the host as favourable as possible for the vectors. In this way, vectors are retained on the infectious host for as long a period as possible so allowing numbers to increase. With a virulent virus which causes the host to become stunted and necrotic there should be strong selection to

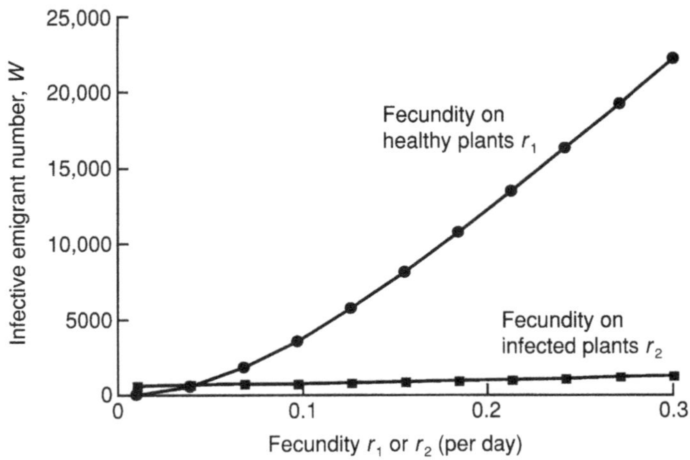

Fig. 17.4. The effect of changes in vector fecundities, r_1, on healthy hosts and r_2 on infected hosts, on the total number of infective emigrants, W. Other parameters and initial states as Fig. 17.1.

make the vectors as tolerant as possible to crowding and so prevent dispersal for as long as possible. Tolerance to crowding of the vector on infected plants would be expected to be a feature of successful virulent viruses and this hypothesis could be examined experimentally. Further, some guidance is provided for experimental work to test the supposition that emigration is enhanced from infected plants. It would be necessary to follow the plants over the later stages of necrosis in order to detect any increase in emigration compared to populations of whiteflies on healthy plants.

A surprising conclusion was also reached with regard to the infection rate of the virus. We predict that an intermediate infection rate is of advantage to a virulent virus. If virus titre and virulence are unavoidably linked, this raises the interesting issue that evolution may favour non-extreme values of either parameter, even for viruses that rely solely on vectors for their transmission. In a theory of optimum virulence, it has been suggested that a parasite should balance its reproduction ($\propto$ infection rate) and virulence in order to maximize its lifetime transmission success (Anderson and May, 1982; Ebert, 1994). Thus, ideas put forward in the context of animal pathogens may also hold for plant viruses.

In the case of the UgV virus, a selective advantage is provided by the boost to vector fecundity in infected plants. Any increase in vector fecundity whether or not associated with host infection acts to increase virus spread provided, of course, that virus infection is present. The results suggest, however, that a stimulus to fecundity that is independent of virus infection in the host has far more impact on virus disease spread than does one associated with infection. The reason is that the stimulus to fecundity occurs (in the model) at the same time as the stimulus for increased emigration. In fact, this is a simplification and there may be some time lag between the effect on fecundity and the effect on emigration, but this has yet to be examined in detail (Zhang *et al.*, 2000). Nevertheless, it is clear that a high rate of vector population increase prior to infection is more effective in spreading inoculum than one dependent on infection. As a general conclusion we therefore suggest that the 'biotype mechanism' is more likely to be associated with virus disease epidemics than is the 'interaction mechanism'.

An increase in fecundity and reduction in emigration rate would each be expected to follow an improvement in host plant quality. We have shown here that each effect will increase the spread of a virulent virus epidemic. With the UgV, an increase in host suitability for the whiteflies occurs, and we show here that this is likely to be a major cause of the observed epidemic. Further, we show that a moderate infection rate is an advantage and some researchers have reported cassava mosaic viruses to be relatively difficult to transmit under laboratory conditions (reviewed in Thresh *et al.*, 1998). The model offers an explanation why this might be so.

The B-biotype is relatively mobile (Byrne and Blackmer, 1996), transmits a wide range of viruses and in some cases is a more efficient vector than the A-biotype (Markham *et al.*, 1996). It is interesting to speculate that even for the B-biotype, selection may apply to both virus and vector for a moderate rather than a high transmission rate.

As well as their epidemiological implications, these findings highlight the potential dangers of moving diseased plant material to new locations. An important implication of the existence of a virus–host–vector interaction in the cassava mosaic disease pathosystem is that the majority of vectors are generated on infected plants, which raises the possibility that this mutually beneficial relationship could be exploited for disease control. Prompt removal of host plants directly following symptom expression, on an area-wide basis, probably has the potential to generate a 'virtuous cycle' of both vector and disease reduction. Sustainable solutions are needed for a disease that currently causes estimated annual losses of US$1.2–2.3 billion in Africa alone (Thresh *et al.*, 1997).

The two epidemic mechanisms discussed here are not mutually exclusive and the 'biotype hypothesis' is just one possibility of a more general 'interaction hypothesis'. It is possible that the B-biotype, as well as having an inherently higher fecundity, could also benefit from feeding on particular combinations of virus infected host plant material, thus leading to an even greater boost to vector fecundity and virus spread.

Conclusions

Two mechanisms associated with the occurrence of whitefly-borne virus disease epidemics are examined in this chapter. In the first, a non-indigenous *Bemisia tabaci* (Genn.) biotype, with a greater fecundity than the indigenous biotype, enters a new geographical region and causes an increase in virus spread. In the second, a whitefly-transmitted virus causes changes to the biochemical composition of the host plant's phloem sap that, in turn, increases vector fecundity and thus virus spread. Evidence is presented from the literature that both these mechanisms can occur and a mathematical model is developed to compare the potential rate of disease spread under the two mechanisms. It is concluded that a vector biotype with a higher fecundity is the more effective mechanism for stimulating an epidemic of a virulent viral disease. This is because an increase in vector fecundity that is independent of virus infection has more impact on the vector population size than one dependent on virus infection. Other seemingly paradoxical findings emerged which were applicable to either mechanism. The spread of a virulent virus is increased if the propensity for vectors to emigrate is reduced; this is because the longer vectors are retained on the host, the more their numbers can increase. A very high transmission rate is found to be a disadvantage to a virulent virus and instead, an optimum transmission rate exists which maximizes virus spread; this is because at very high transmission rates the host is destroyed before vector numbers have time to increase. The two mechanisms are not mutually exclusive and it is suggested that virus–host–vector interactions play a larger role in the epidemiology of plant virus diseases than has been recognized previously.

Acknowledgements

We are grateful to Prof. J.M. Thresh and Prof. R.J. Cooter for constructive criticism of this chapter. This publication is an output from an activity part-funded by the United Kingdom Department for International Development (DFID) for the benefit of developing countries. The views expressed are not necessarily those of DFID (Crop Protection Programme).

References

Anderson, R.M. and May, R.M. (1982) Coevolution of hosts and parasites. *Parasitology* 85, 411–426.

Auclair, J.L. (1963) Aphid feeding and nutrition. *Annual Review of Entomology* 8, 439–490.

Banks, G.K., Colvin, J., Chowda Reddy, R.V., Maruthi, M.N., Mumiyappa, V., Vantatesh, H.M., Kiran Kumar, M., Padmaja, A.S., Beitia, F.J. and Seal, S.E. (2000) First report of the *Bemisia tabaci* B biotype in India and an associated tomato leaf curl virus disease epidemic. *Plant Disease*, in press.

Bedford, I.D., Markham, P.G., Brown, J.K. and Rosell, R.C. (1994) Geminivirus transmission and biological characterisation of whitefly (*Bemisia tabaci*) biotypes from different world regions. *Annals of Applied Biology* 125, 311–325.

Bellows, T.S., Perring, T.M., Gill, R.J. and Headrick, D.H. (1994) Description of a species of *Bemisia* (Homoptera: Aleyrodidae). *Annals of the Entomological Society of America* 87, 195–206.

Bos, L. (1992) New plant virus problems in developing countries: a corollary of agricultural modernisation. *Advances in Virus Research* 38, 349–407.

Brown, J.K. (1994) Current status of *Bemisia tabaci* as a pest and virus vector in world agroecosystems. *FAO Plant Protection Bulletin* 39, 5–23.

Brown, J.K., Bird, J., Frohlich, D.R., Rosell, R.C., Bedford, I.D. and Markham, P.G. (1996) The relevance of variability within the *Bemisia tabaci* species complex to epidemics caused by subgroup III geminiviruses. In: Gerling, G. and Mayer, R.T. (eds) *Bemisia: 1995 Taxonomy, Biology, Damage, Control and Management.* Intercept, Andover, pp. 77–89.

Byrne, D.N. and Blackmer, J.L. (1996) Examination of short range migration by *Bemisia*. In: Gerling, G. and Mayer, R.T. (eds) *Bemisia: 1995 Taxonomy, Biology, Damage, Control and Management*. Intercept, Andover, pp. 17–28.

Cohen, S., Duffus, J.E. and Liu, H.T. (1992) A new *Bemisia tabaci* in the southwestern United States and its role in silverleaf of squash and transmission of lettuce infectious yellows virus. *Phytopathology* 82, 86–90.

Colvin, J., Otim-Nape, G.W., Holt, J., Omongo, C., Seal, S., Stevenson, P., Gibson, G., Cooter, R.J. and Thresh, J.M. (1999) Abstract: Factors driving the current epidemic of severe cassava mosaic disease in East Africa. In: Fereres, A. (ed.) *VIIth*

International Symposium of Plant Virus Epidemiology, April 11–16, 1999. International Society of Plant Pathology, Almeria, pp. 76–77.

Costa, H.S. and Brown, J.K. (1991) Variation in biological characteristics and esterase patterns among populations of *Bemisia tabaci*, and association of one population with silverleaf induction. *Entomologia Experimentalis et Applicata* 61, 211–219.

Costa, H.S., Brown, J.K. and Byrne, D.N. (1991) Life history traits of the whitefly, *Bemisia tabaci* (Homoptera: Aleyrodidae), on six virus-infected or healthy plant species. *Environmental Entomology* 20, 1102–1107.

Ebert, D. (1994) Virulence and local adaption of a horizontally transmitted parasite. *Science* 265, 1084–1086.

Falk, B.W. and Klaassen, V. (1996) Lettuce infectious yellows virus: a bi-partite closterovirus transmitted by *Bemisia*, and representative of a new genus of plant viruses. In: Gerling, G. and Mayer, R.T. (eds) *Bemisia: 1995 Taxonomy, Biology, Damage, Control and Management.* Intercept, Andover, pp. 265–275.

Fereres, A., Lister, R.M., Araya, J.E. and Foster, J.E. (1989) Development and reproduction of the English grain aphid (Homoptera: Aleyrodidae) on wheat cultivars infected with barley yellow dwarf virus. *Environmental Entomology* 18, 388–393.

Harpaz, I. and Applebaum, S.W. (1961) Accumulation of asparagine in maize plants infected with maize rough dwarf virus and its significance in plant virology. *Nature* 192, 780–781.

Harrison, B.D., Zhou, X., Otim-Nape, G.W., Liu, Y. and Robinson, D.J. (1998) Role of a novel type of double infection in the geminivirus-induced epidemic of severe cassava mosaic in Uganda. *Annals of Applied Biology* 131, 437–448.

Holt, J., Jeger, M.J., Thresh J.M. and Otim-Nape, G.W. (1997) An epidemiological model incorporating vector population dynamics applied to African cassava mosaic virus disease. *Journal of Applied Ecology* 34, 793–806.

Holt, J., Colvin, J. and Muniyappa, V. (1999a) Identifying control strategies for tomato leaf curl virus disease using an epidemiological model. *Journal of Applied Ecology* 36, 1–10.

Holt, J., Zhang, X.-S. and Colvin, J. (1999b) Abstract: A new general model of plant disease infection which incorporates vector aggregation. In: Fereres, A. (ed.) *VIIth International Symposium of Plant Virus Epidemiology, April 11–16, 1999.* International Society of Plant Pathology, Almeria, pp. 93–94.

Jeger, M.J., van den Bosch, F., Madden, L.V. and Holt, J. (1998) A model for analysing plant-virus transmission characteristics and epidemic development. *IMA Journal of Mathematics Applied in Medicine and Biology* 15, 1–18.

Kennedy, J.S. (1951) Benefits to aphids from feeding on galled and virus-infected leaves. *Nature* 168, 825.

Kunkel, H. (1977) Membrane feeding systems in aphid research. In: Harris, K.F. and Maramorosch, K. (eds) *Aphids as Virus Vectors.* Academic Press, New York, pp. 311–338.

Legg, J.P. and Ogwal, S. (1998) Changes in the incidence of African cassava mosaic virus disease and the abundance of its whitefly vector along south-north transects in Uganda. *Journal of Applied Entomology* 122, 169–178.

Markham, P.G., Bedford, I.D., Liu, S., Frolich, D.R., Rosell, R. and Brown, J.K. (1996) The transmission of geminiviruses by biotypes of *Bemisia tabaci* (Gennadius). In: Gerling, G. and Mayer, R.T. (eds) *Bemisia: 1995 Taxonomy, Biology, Damage, Control and Management.* Intercept, Andover, pp. 69–75.

Otim-Nape, G.W., Bua, A., Thresh, J.M., Baguma, Y., Ogwal, S., Ssemakula, G.N., Acola, G., Byabakama, B., Colvin, J., Cooter, R.J. and Martin, A. (2000) The current pandemic of cassava mosaic virus disease in East Africa and its control. *Natural Resources Institute Catalogue Services No. PSTC28*. University of Greenwich, Chatham.

Perring, T.M. (1996) Biological differences of two species of *Bemisia* that contribute to adaptive advantage. In: Gerling, G. and Mayer, R.T. (eds) *Bemisia: 1995 Taxonomy, Biology, Damage, Control and Management*. Intercept, Andover, pp. 3–16.

Perring, T.M., Cooper, A.D., Rodriguez, R.J., Farrar, C.A. and Bellows, T.S. (1993) Identification of a whitefly species by genomic and behavioural studies. *Science* 259, 74–77.

Polston, J.E. and Anderson, P.K. (1997) The emergence of whitefly-transmitted geminiviruses in tomato in the western hemisphere. *Plant Disease* 81, 1358–1369.

Power, A.G. (1992) Patterns of virulence and benevolence in insect-borne pathogens of plants. *Critical Reviews in Plant Sciences* 11, 351–372.

Ramappa, H.K., Muniyappa, V. and Colvin, J. (1998) The contribution of tomato and alternative host plants to tomato leaf curl virus inoculum pressure in different areas of south India. *Annals of Applied Biology* 133, 187–198.

Selman, I.W., Brierley, M.R., Pegg, G.F. and Hill, T.A. (1961) Changes in the free amino acids and amides in tomato plants inoculated with tomato spotted wilt virus. *Annals of Applied Biology* 49, 601–615.

Thresh, J.M., Otim-Nape, G.W., Legg, J.P. and Fargette, D. (1997) African cassava mosaic disease: the magnitude of the problem. *African Journal of Root and Tuber Crops* 2, 13–19.

Thresh, J.M., Otim-Nape, G.W., Thankappan, M. and Muniyappa, V. (1998) The mosaic diseases of cassava in Africa and India caused by whitefly-borne geminiviruses. *Review of Plant Pathology* 77, 935–945.

Zhang, X.-S., Holt, J. and Colvin, J. (2000) A general model of plant-virus disease infection to incorporate vector aggregation. *Plant Pathology* 49, 435–444.

Index